Practical methods in electron microscopy

Volume 15

Practical methods in electron microscopy

Volume 15

Series editor:

Audrey M. Glauert

Clare Hall
University of Cambridge
England

PORTLAND PRESS
London and Chapel Hill

Vacuum methods in electron microscopy

Wilbur C. Bigelow

Department of Materials Science and Engineering
University of Michigan
Ann Arbor
Michigan, USA

PORTLAND PRESS
London and Chapel Hill

Published by Portland Press, 59 Portland Place,
London W1N 3AJ, U.K.

ISSN 1353-6494
ISBN 1 85578 052 6(pbk) ISBN 1 85578 053 4(hbk)

British Library Cataloguing in Publication Data
A catalogue record for this book is available from
the British Library

Printed in Great Britain by the University Press, Cambridge

Editor's preface to the series

Electron microscopy is a fundamental technique with wide applications in all branches of science and technology. Every year large numbers of students, technicians and research workers start to use the electron microscope and need an introduction to the instrument, and to the many and varied methods for the preparation of specimens.

Many books have been published which describe the techniques of electron microscopy in general terms, but when this series was originally conceived, there was an urgent need for a set of practical laboratory handbooks in which these techniques were described in sufficient detail to enable the isolated worker to carry them out successfully. Now, over 20 years later, *Practical Methods in Electron Microscopy* has gained an international reputation as the unique source of practical information for all electron microscopists and is still fulfilling its original purpose. This achievement largely results from the fact that all the authors are recognized experts in their particular fields and are thus able to describe the techniques of electron microscopy in sufficient detail for them to be understood and applied accurately by beginners and experienced microscopists alike. In contrast, certain single author texts cease to be of any practical value once the authors move outside the field of their own practical experience. Perhaps the best compliment that *Practical Methods in Electron Microscopy* has received is a statement in a review of one book that "The book leaves the reader with the impression of not having read a textbook, but having talked to a practising microscopist".

Each book in the series starts from first principles, assuming no specialist knowledge, and is complete in itself. The books are designed to be guides through the often bewildering choice of methods that are available, and every attempt is made to simplify this choice. Consequently, in general, only well-established methods which have been used successfully outside their laboratory of origin are described in detail. We do not provide descriptions of the latest exciting 'break through', although each book does indicate the most promising of developments for the future.

The present volume is the first in the series to be published by Portland Press, the publishing subsidiary of the Biochemical Society, and the high standards attained by the previous volumes (published by Elsevier, Amsterdam) have been maintained. In addition, the opportunity has been taken to improve and modernize the appearance of the books and I look forward to a future in which authors, editor and publisher continue to contribute to the 'Science and Art of Electron Microscopy'.

Audrey M. Glauert, Sc.D.
Series Editor, 1994
Clare Hall
University of Cambridge

Author's preface and acknowledgements

Like so many other books, this one grew out of a set of notes on vacuum methods that was originally prepared for use in a course on electron microscopy that I taught for more than 30 years. In the beginning it was necessary to deal only with rotary-vane rough vacuum pumps and oil diffusion high vacuum pumps, and to discuss the operation of the standard vacuum systems for vacuum evaporators and electron microscopes based on these pumps. In the 1970s, however, it became necessary to introduce sputter-ion pumps and turbomolecular pumps. Then cryogenic and bulk-getter pumps had to be included, and the vacuum systems became increasingly complex. By 1989 the set of notes had grown to over 50 pages in length, and it became necessary to take the formal discussion of vacuum methods out of the course and to devote several special evening seminars to the subject. About this time I contacted Audrey Glauert about the possibility of expanding the notes into a book in her series *Practical Methods in Electron Microscopy*. She graciously agreed, and the project was underway. Like so many other authors, if I had known just how much work would be involved before the book was finished I would never have undertaken writing it.

Having been involved for more than a quarter of a century in the operation and management of electron microscopy and microprobe analysis laboratories that were used by faculty and graduate students throughout the College of Engineering and in most of the biological and physical science departments of the University of Michigan, it has been abundantly evident that most of these scientists and engineers are not very comfortable about operating the vacuum systems on the instruments they use. The purpose of this book is to provide the information that hopefully will help alleviate this situation. In the first two chapters I have described the characteristics of a vacuum and of the evacuation process in a way that I hope will be interesting and useful to all readers, whatever their previous experience with vacuum systems may be. These chapters are of particular importance, because they develop concepts and principles which are needed for understanding the operating procedures described in later chapters. The next six chapters describe the way the various vacuum gauges and pumps function, and summarize the operating capabilities and characteristics of these devices. In these chapters I have included such relatively new types of pumps as molecular drag pumps, bulk getter pumps, and various combination and hybrid pumps, because of their potential for use in oil-free vacuum systems on instruments of the future. Procedures for operating several simple vacuum systems are given in these chapters, while Chapter nine contains a discussion of six vacuum systems of the type that are on recent models of electron microscopes, some of which are surprisingly complex. The final chapter discusses procedures for detecting and repairing leaks in vacuum systems.

This book is perhaps somewhat different from most other books on vacuum methods, because it contains descriptions of a number of typical vacuum problems, and discusses methods for dealing with and preventing them. Although the emphasis in all of these discussions is oriented around the interests and needs of electron microscopists, I think the book will be equally useful to scientists and engineers working in most other fields, because the vacuum systems on electron microscopes have a great deal in common with those on many other types of scientific instruments, and the characteristics of vacuum pumps and gauges are the same whatever instrument they are used on.

I am indebted to a large number of people and organizations for assistance in this project. Audrey Glauert, the editor of the series *Practical Methods in Electron Microscopy*, has been extremely helpful in guiding the preparation of the book at every stage of its development. Lawrence Allard and Andrè Amy, who I am proud to claim as previous graduate students, have read the initial draft of every chapter and made many helpful comments and suggestions. John Mardinly and David Blake provided me with information and illustrations of their work in modifying the vacuum systems on electron microscopes. Alan Agar has reviewed a number of sections of the book, and contributed to Chapter 1. George Brooks, Ed Birko, Phil Danielson, and John Fahey have contributed several examples of practical applications and problems. Numerous companies have furnished catalogues, specifications, and operating instructions, which have been the source of most of the technical data cited in characterizing various pumps, gauges, and vacuum apparatus. Several of these companies also furnished the photographs and drawings that are used in a number of figures, while United Media provided the cartoon that follows this preface. These sources are cited and acknowledged throughout the book. I deeply appreciate the contributions of all these individuals and organizations, and sincerely thank them for their willingness to be of such great assistance to me.

Wilbur C. Bigelow, Ph.D.
Professor Emeritus, 1994
Materials Science and Engineering
The University of Michigan

Frank & Ernest reprinted by permission of NEA, Inc.

Contents

1 Vacuum and its characteristics

Our word 'vacuum' is derived from the Latin word *vacuus*, which means empty. The question of whether or not an empty space (i.e. a vacuum) could exist, and the consequences should one exist, were topics that concerned natural philosophers for centuries. The often-quoted statement that "Nature abhors a vacuum" is generally considered to have been made by Aristotle over 20 centuries ago while explaining the reason it is necessary to have a vent hole in the top of a wine barrel in order for wine to flow freely out through the tap hole in the bottom of the barrel. More contemporary points of view are contained in the statement made a few years ago by one of my graduate students that "Vacuum sucks", and in a recent comment by one of my neighbours to the effect that she couldn't understand why I would expend so much time and energy "writing about nothing".

Aristotle may well have had the most lasting influence on the field of vacuum technology of any person to date, because his conclusion that a vacuum was illogical and therefore impossible was generally accepted and supported by leading natural philosophers, including the famous and highly influential Descartes, until the early part of the 17th century. Then, however, observations by such experimentalists as Galileo, Pascal, Kepler, and Torricelli accumulated a body of evidence showing that a vacuum could indeed be produced. The most convincing and dramatic results of this kind were those of Otto von Guericke who developed the first effective vacuum pump and who, with his famous Magdeburg experiment in 1654, dramatically demonstrated the reality of a vacuum by showing that two teams of eight horses could barely separate two evacuated copper hemispheres. It will remain for the readers of this book to determine how the two more contemporary viewpoints mentioned above fare.

The references cited in this chapter are listed in Appendix 1, while manufacturers and suppliers of equipment are listed in Appendices 2 and 3.

1.1 Vacuum and electron microscopes

The involvement of electron microscopists with vacuum and vacuum systems began the moment the first instruments were built in the 1930s. Electron microscopes can be considered to consist of three major operating systems: the electron optical system consisting of the electron gun, which produces the electron beam, and the electron lenses, which produce the electron image; the electrical system, which consists of the high voltage and lens power supplies that energize the components of the electron optical system; and, the vacuum system, which removes air molecules from the instrument so that the electron beam can traverse the length of the electron optical column unimpeded to produce sharp, clear images. These three systems support and interact with one another, and each is critically important to the functioning of the instrument. Electron microscopes simply would not work without a vacuum system.

The first electron microscopes were built in the laboratories of individual scientists during the 1930s. The vacuum systems on these instruments were quite primitive by modern standards. Mercury diffusion pumps and good rotary-vane pumps had been in use for a number of years, and oil diffusion pumps had been introduced about 5 years earlier, and so it was possible to produce dynamically pumped systems with moderate pumping speeds of the general type needed for electron microscopes. However, O-rings were not yet available, and so it was necessary to cut gaskets from rubber sheet material, which usually was not of uniform thickness and which certainly was not formulated for vacuum applications. In fact, the demountable vacuum joints on some of the earliest instruments did not use rubber gaskets at all, but were produced by using grease to make a seal between matching flat or cone-shaped surfaces, which had been lapped together to obtain a close fit. Vacuum grease was often made by dissolving rubber bands in heated petroleum jelly, and a popular vacuum sealing wax was made by melting together equal amounts of beeswax and rosin. The camera on most early instruments held only one photographic plate, and so the entire instrument had to be let up to atmospheric pressure each time the specimen or the photographic plate was changed. Pumpdown was very slow, and the vacuum ultimately attained was probably not much below 10^{-3} Torr.

The first commercial electron microscopes were introduced in Europe by the Siemens company in about 1938, and in the United States by the Radio Corporation of America in about 1940. Details of these early developments are reviewed elsewhere (Hall, Section 1.2). By this time it was recognized that if a useful amount of work was going to be done it would be necessary to provide airlock systems (Section 9.1.2, p. 377) to facilitate the exchange of specimens and photographic plates, and so these first commercial models had more sophisticated vacuum systems that included these and some other convenience features. However, some of these early instruments did not have such luxuries as aperture centring controls. Usually there was only a single aperture in each lens, held in place in the pole piece by a threaded washer. This arrangement made it necessary to open the column and remove the pole pieces each time the apertures were cleaned and whenever an aperture needed centring. Centring was accomplished by a trial and error process, which made alignment of an instrument quite tedious. I vividly remember what a relief it was when someone discovered that if the apertures were not clamped tightly by the threaded retaining washers they could be moved around enough to achieve centring by tapping gently on the outside of the column with the handle of a screwdriver. The vacuum systems of this period were tricky to operate, and vacuum problems were frequently encountered and difficult to solve. A typical example is one problem caused by the oil diffusion pumps available at that time. Large bubbles of oil vapour formed on the hot surface of the boilers of these pumps and then suddenly broke free with an audible cracking sound. This 'bumping' of the diffusion pumps produced enough mechanical vibration to cause serious image instability. One cure for this problem involved putting glass beads in the boilers of the pumps to prevent the large bubbles from forming. Following World War II, standardized O-rings became available for making demountable and moveable vacuum seals and the performance of oil diffusion pumps was slowly improved, and so the vacuum systems on electron microscopes gradually became more reliable and trouble-free. However, there was little direct concern about the quality of vacuum in these early instruments. In fact, there was probably little that could have been done in this regard because of the relatively primitive high vacuum equipment that was available then.

During the first 2 decades after the first commercial electron microscopes were introduced, primary emphasis was placed on refining the electrical and electron optical systems and improving the thermal and mechanical stability of the specimen stage to achieve improved resolution. By the middle of the 1950s specialists in several laboratories demonstrated a resolution better than one nanometre (10 Å) using instruments that were 'tuned up' for the purpose. Micrographs showing a resolution better than 2 nm (20 Å) could be obtained more or less routinely under average laboratory conditions on most instruments produced by the early 1960s. Those who had the privilege and good fortune to experience these exciting times will also recall that the development of improved specimen preparation techniques, such as carbon replica and thin film methods for metallurgical and ceramic specimens and ultramicrotomy for biological specimens, also made critically important contributions to this progress.

While the vacuum system provided the fundamental benefit of making it possible to produce the unscattered beams of electrons that are necessary for image formation, it also brought with it one of the most nagging problems that has had to be dealt with by manufacturers and users of electron microscopes: the problem of specimen contamination. As we all now know, this infamous phenomenon occurs because the hydrocarbon molecules from the oils used in the vacuum pumps, from the greases and gaskets used in the vacuum seals, and from a variety of miscellaneous sources, collect on the specimen and under bombardment by the electron beam form a carbonaceous lacquer that can rapidly become thicker than the resolvable features in the specimen. In reviewing the early work on specimen contamination, Hall (p. 272) tells of encountering this problem in an electron microscope he built prior to 1940, and notes that it had also been observed in electron diffraction instruments even prior to that. In his Nobel Lecture, which is reprinted in the *Bulletin of the Electron Microscopy Society of America* (Vol. 18:2, Nov. 1988), Ernst Ruska, 'the father of the electron microscope', recounts the disappointment this problem caused when the new Elmiskop high-resolution instrument was first tested in 1954 — the fine detail gained by the improved resolution of the instrument was obscured so quickly it was difficult to observe and almost impossible to photograph.

Thus, serious concern about the quality of the vacuum in electron microscopes arose in the middle of the 1950s when it became evident that, from an operational point of view, the problem of specimen contamination was likely to cancel the advances that were being made in electron optics.

Ironically, as noted by Ruska, it was a significant improvement in the electron optics of electron microscopes, namely the introduction of the double condenser lens system, that finally caused microscopists to realize the full importance of this problem. Although a high concentration of hydrocarbon vapours had existed inside all electron microscopes produced up to this time, it could be tolerated because these instruments had only a single condenser lens, and this produced such a high level of electron irradiation over such a large area of the specimen that the specimen was heated enough to keep the rate of specimen contamination low enough to be acceptable at the level of resolution of these instruments. With the development of higher resolution instruments having double condenser lens systems, such as Ruska's Elmiskop mentioned above, however, the problem of specimen contamination became truly critical. The double condenser lens was introduced to reduce overall specimen damage from electron bombardment by making it possible to adjust the area of the specimen that was irradiated by the electron beam to about the size of the field of view. This reduced both the total number of electrons that struck the specimen per second and the total area of the specimen over which they were distributed, and so total specimen damage was indeed diminished. However, specimen heating was also cut down to the point where carbonaceous contamination built up so rapidly in the small area that was irradiated that it was difficult to work fast enough to obtain good micrographs.

Within a very short time, a measure of control was achieved by surrounding the specimen with metal fixtures that were cooled with liquid nitrogen. The relatively high molecular mass hydrocarbon molecules that were responsible for the contamination readily condensed on these cold anticontamination devices, leaving the specimen clean enough so that acceptable images could again be obtained. These devices saved the day until the late 1960s when a new generation of electron microscopes with further improvements in the electrical and electron optical systems were produced. Not only were these instruments capable of achieving a resolution well below half a nanometer (5 Å) and operating at magnifications above 200 000, but they also combined a refined double condenser lens system with a single-field condenser-objective lens to produce a high intensity focused beam less than 10 nm (100 Å) in diameter for operation in a variety of scanning-transmission and microbeam modes. Again, specimen contamination became a seriously limiting factor that nearly cancelled out the benefits gained from the advances in optics that had been achieved. When operated in one of the stationary microbeam modes, for example, it was not uncommon to have a cone-shaped deposit of contamination products several micrometres high develop at the site where the beam struck the specimen.

Although the vacuum systems on microscopes produced in the 1960s were more fully automated and did have somewhat improved oil diffusion pumps, they were not fundamentally much different from those on the commercial instruments of the late 1940s. And while it was widely accepted that changes in the vacuum systems were needed, there were certain problems that stood in the way. The situation from the manufacturers' point of view is related in the following private communication from Alan Agar, who was actively involved at this time in designing the AEI EM6B and EM7 instruments:

> "The vacuum systems of electron microscopes seemed to develop more slowly than other aspects of the instruments for good reason. The electron gun was a relatively easy vacuum problem, because there was plenty of room to accommodate a large bore pumping tube directly into the gun chamber. This was needed at an early date, both to ensure stable operation of the high voltage source, and to minimize the ionization of residual vapour molecules which could give rise to positive ions which would destroy the hot cathode."
>
> "The viewing chamber was likewise easy, but greatest problems arose in the electron optical column itself. The objective lens and specimen area had by definition to have small dimensions to achieve a reasonable resolution, and consequently, there was only space for a very narrow bore pumping tube, resulting in a very slow pumping speed at the specimen. The small bore pole pieces also resulted in metal surfaces which could degas very close to the specimen. No baking was feasible. For a long time it was necessary to allow for mechanical adjustments between each image forming lens, and hence a lining tube was not then feasible. (AEI included a screening tube in the double condenser lens from 1956, but all other lenses had to be able to move independently.) It was

> only much later when lens design had improved considerably, and misalignments had been reduced to a low level, that electronic alignment became feasible, and Philips were able to introduce a screening tube through the image forming system in the EM 400".
>
> "The improvement in pumping at the specimen only became feasible after the development of the symmetrical objective lens of Riecke and Ruska (*Proceedings of the International Congress on Electron Microscopy*, Kyoto, 1966, pp. 19—20)."
>
> "Thus, it really only became possible to make major improvements in the vacuum system after significant improvements had been made in electron optics, and I believe that any microscope designer would say that the vacuum problems were all too apparent, but that the technology to do something effective was not initially available."

In the late 1960s several electron microscopists grew impatient for a solution to the contamination problem, and undertook to improve the quality of the vacuum in their instruments. I particularly remember the pioneering work of Richard and Roberta Hartman which began prior to 1965, and which was summarized in a paper presented at the Annual Meeting of the Electron Microscopy Society of America (EMSA) in 1969 (because I was president of the society that year and because the Hartmans were good friends of mine) in which they described a prototype instrument that had been especially designed and constructed for them by the Hitachi company with the goal of obtaining a major improvement in the cleanliness of the vacuum system. At that same meeting J. C. Milikel presented a paper in which he discussed the advantages of using ion pumps on electron microscopes as a means of combatting the contamination problem, an approach that had already been adopted by the Hartmans. At the EMSA meeting in 1968 D. N. Braski of the Oak Ridge National Laboratory had also described extensive modifications to the vacuum system of a Hitachi HU-11B microscope for the same purpose. I also vividly recall a visit to the laboratory of Alton R. Taylor at the Parke Davis Company in Detroit in 1969 when he demonstrated the achievement of a remarkable decrease in contamination rate, and a noticeable improvement in image quality, by installing turbomolecular pumps on RCA EMU3 and EMU4 microscopes.

Improvements in instrument design and the increasing need to overcome the contamination problem finally resulted in the introduction of a new generation of electron microscopes in the 1970s with vacuum systems that were radically different from those found on earlier instruments. Among the first of these was the Model EM-400 microscope that was introduced by the Philips company in 1975. In this instrument the use of rubber O-rings was minimized, the internal surface area and volume of the column were greatly reduced, and the cleanliness of the column was greatly improved by using stainless steel column liners, by using indium metal seals instead of O-rings for all static seals in the column, and by using mechanisms sealed with metal bellows for centring apertures. The electron optical column and the electron gun were evacuated by a sputter-ion pump, thereby eliminating the oil diffusion pump as a direct source of oil vapours in the specimen chamber. While a high-speed oil diffusion pump was still used to evacuate the viewing and photographic chambers, this part of the vacuum system was effectively isolated from the electron optical column by a small aperture (200 μm) that was located at the bottom of the projector lens system. This very important advance was achieved by innovatively designing the lens system so that the electron beam was always focused to a crossover at the level of this aperture, not only while operating over the full range of magnification provided in the standard transmission mode, but also during operation in the scanning transmission mode and in several hybrid modes. Interestingly, this may have been the first instance in which vacuum requirements had a significant influence on the basic design of the electron optical system of an electron microscope. A very efficient specimen anticontamination device was also provided to effect rapid removal of water vapours and other volatile materials that might be introduced with the specimen. In these instruments the partial pressures of hydrocarbons and water vapour were below 3×10^{-10} Torr and 6×10^{-8} Torr, respectively. With this clean ultra-high vacuum environment it became possible to observe specimens under all modes of operation for a period of at least 15–20 min without the appearance of significant carbonaceous contamination, and it was possible to observe specimens cooled to near liquid nitrogen temperature for similar periods without having ice form on them.

Most models of transmission electron microscopes manufactured since 1980 have vacuum systems of this general type. Whereas the Philips EM-400 used a sputter-ion pump to obtain oil-free evacuation of the specimen chamber and electron optical column, other models employ cryogenic or turbomolecular pumps for this purpose. In addition, different types of oil-free roughing pumps are being tested as substitutes for the common oil-sealed rotary-vane pumps that have been standard components of most vacuum systems for so many years. One of the latest trends is the introduction of field-emission electron guns, which require bakeable, ultra-high vacuum systems capable of maintaining pressures in the 10^{-9} Torr range. Whereas in the past the vacuum system could be relegated to a secondary role by users of electron microscopes, in the same way that it had been by the designers and manufacturers, users must now give the vacuum system on their instruments the same care and attention they give to the optical and electronic systems, and to specimen preparation. A single, momentary, careless act can contaminate a modern instrument so badly that it is effectively ruined. This is the reason electron microscopists now need to be well acquainted with the principles of vacuum science, and with the characteristics of the several new types of pumps and gauges that are being used on electron microscopes, and on the other kinds of vacuum instruments that are now often associated with electron microscopy laboratories. Before proceeding to discuss these devices, however, it will be useful to describe some of the more interesting and important characteristics of the medium with which we are dealing, a vacuum.

1.2 Units of pressure

Practically, and operationally, here on Earth we consider we have produced a vacuum when we have managed to reduce the number of molecules per unit volume inside a vacuum vessel to a level that is measurably less than exists in the surrounding atmosphere. Although the molecules are then fewer in number, those remaining inside the vessel still strike its walls and exert a *pressure* on them, and the concepts and terminology that have developed are such that it is more pleasing to describe the level of the vacuum existing in the vessel in terms of this reduced pressure, using units of pressure, rather than directly as a vacuum, using some kind of reciprocal pressure (i.e. vacuum) units. We are presently in a period when a transition is being made from the use of a traditional 'practical' unit for this reduced pressure, the Torr, to the use of 'rational' units which are consistent with the Système International d'Unités (SI units). The fundamental SI unit for pressure is the *pascal* (Pa), which is named in honour of Blaise Pascal, a French mathematician, philosopher, and physicist of the 1600s who first demonstrated the variation in atmospheric pressure with altitude. Since the two sets of units are not simply related, and since scientists and manufacturers around the world are not following a uniform practice, this may lead to some confusion and discomfort. It is hoped that the following discussion of these units, their characteristics, and their relationships will help alleviate this situation somewhat.

For many years gas pressures in vacuum work were expressed in *millimetres of mercury*, 1 mmHg being the pressure needed to support a column of mercury 1 mm high. The standard atmosphere is 760 mmHg. This was a very practical unit that arose because of the common use of mercury as the working fluid in the manometers and McLeod gauges that were used to measure gas pressures in early vacuum work. Because it was ultimately considered aesthetically unsatisfactory to have a unit of pressure whose name suggested that it had the dimensions of length, the millimetre of mercury was renamed the *Torr* in honour of Evangelista Torricelli, an Italian physicist who developed the mercury manometer and first measured the pressure of the atmosphere in the 1640s. One Torr equals 133.32 pascals.

Two other units derived from the mmHg were once widely used in the United States. These are the micron (μ, also called the millitorr) which is equal to 10^{-3} mmHg, and the millimicron (mμ) which is equal to 10^{-3} μ or 10^{-6} mmHg. The practical significance of the micron is that pressures down to this level are relatively easy to obtain by very ordinary methods using rotary mechanical pumps, which are commonly called rough vacuum pumps. A pressure of 1 μ was thus considered to be the lower end of the rough or *low vacuum range*. In a similar vein, a vacuum of 1 mμ was about the best that could be obtained with oil diffusion pumps in early laboratory systems, and was commonly considered to be the lower end of the *high vacuum range*. Presently, most authors extend the high vacuum range down to 10^{-7} Torr (10^{-5} Pa), and consider all pressures below that to be in the *ultra-high vacuum range*.

Atmospheric pressure, which is about 760 Torr at sea level, decreases by a factor of about two with each 5 km (3 miles) increase in altitude up to about 100 km (60 miles). At this altitude it reaches about 10^{-3} Torr (about 0.1 Pa), the bottom of the low vacuum range. It then decreases more slowly at higher altitudes, dropping to 10^{-6} Torr (10^{-4} Pa), the bottom of the high vacuum range, at about 200 km and 10^{-9} Torr (10^{-7} Pa) at about 900 km. At altitudes near 30 000 ft (9 km or, 6 miles), where most commercial airliners fly, atmospheric pressure is about 200 Torr (3×10^{4} Pa). Most older electron microscopes currently in use probably operate in the 10^{-5} Torr (10^{-3} Pa) range. Newer instruments designed for analytical or ultra-high resolution applications usually have better two-stage vacuum systems that reach the 10^{-7} Torr (10^{-5} Pa) range in the gun and specimen chamber and the 10^{-5} Torr (10^{-3} Pa) range in the viewing and photographic chambers. Instruments with field-emission guns usually have multi-stage vacuum systems that allow 10^{-10} Torr (10^{-8} Pa) to be reached in the electron gun.

The Torr has fallen into disfavour because it is not a 'coherent' or 'rational' unit in the sense that it is not fundamentally derived from the basic physical units of mass, length, and time. However, the pascal fulfills these requirements, as do all other preferred SI units. One pascal is defined as a pressure of one newton per square metre, which reduces to 1 N/m^2 = 1 $kg/(m{\cdot}s^2)$ in the basic SI units. One pascal is equal to 7.50×10^{-3} Torr. Although the pascal is the fundamental and preferred SI unit, there is another that is rated as being acceptable, probably because it is universally used in recording meteorological data. This unit is the bar, which originated in an earlier attempt to bring about the common use of coherent pressure units consistent with the cgs (centimetre-gram-second) system. The bar is defined as 10^{6} dynes per square centimetre, which is equal to 10^{6} $g/(cm{\cdot}s^2)$ in the basic cgs units. In SI units this is equivalent to 10^{-1} MPa or 10^{5} $kg/(m{\cdot}s^2)$. The *millibar* (mbar) is the related unit that is now widely used in vacuum work. One millibar is equal to 100 Pa and 0.75 Torr.

It is somewhat unfortunate that there are two acceptable SI units for pressure because this tends to make the situation more confusing and uncomfortable, and somewhat defeats the underlying goal of uniformity. Fortunately, however, these two SI units differ by only a factor of 100, so conversion between them involves only the relatively simple operation of moving the decimal point two figures to the right or left. It appears that the millibar is now being used quite uniformly in the European vacuum community, while the pascal appears to be preferred in Japan. The Torr is still the most widely used unit in the United States. Table 1.1 gives factors for converting among these units. Berman (p. 10) and Roth (p. 42) give more extensive tables which include several additional older units, such as pounds per square inch and centimetres of water.

Table 1.1 Conversion factors for units of pressure

		Pa	Torr	mbar
1 Pa	=	1	0.0075	0.010
1 Torr	=	133	1	1.33
1 mbar	=	100	0.750	1
1 micron	=	0.133	0.001	0.00133
1 atm	=	101 324	760	1013.24

Since the pascal is the fundamental, preferred, and internationally recommended SI unit, it is the one that will be used in this book. However, equivalent pressures in Torr will be given parenthetically because many existing vacuum systems still have gauges that are calibrated in Torr and many people presently working with electron microscopes still think and work in terms of Torr. For these people, the usual problem involves converting from the less familiar units of millibar and pascal to the old familiar unit, the Torr. Actually, this is not too difficult if it is recognized that most vacuum measurements are not highly accurate and usually do not merit more than one significant figure. Usually pressures in vacuum work are expressed in scientific notation; that is, as a mantissa (a number from 1 to 9) multiplied by the number 10 raised to an exponent (e.g. 7×10^{-5}). To make the conversion, first decrease the mantissa to about three-quarters of the given value, and then subtract 2 from the exponent to convert pascal to Torr, or leave the exponent unchanged to convert millibar to Torr. Thus, 7×10^{-5} mbar converts to 5×10^{-5} Torr, 9×10^{-7} Pa converts to 7×10^{-9} Torr, and 1×10^{-5} Pa = 10×10^{-6} Pa converts to 7.5×10^{-8} Torr, with sufficient accuracy for most practical purposes.

When orders of magnitude are sufficient, the values for Torr and millibar can be taken as being the same, and a factor of 10^{-2} can be used for conversions involving pascal. Thus the 10^{-4} mbar range is about the same as the 10^{-4} Torr range, and the 10^{-1} Pa range is about the same as the

10^{-3} Torr range, for most practical vacuum operations. These simple relationships will be sufficient for most of the conversions required throughout the remainder of this book. In this general sense the term *order of magnitude* is often used to mean 'a factor of 10'. Thus, saying that the pressure increased by 'an order of magnitude' means that it increased by a factor of 10, while saying that it decreased by 'three orders of magnitude' means that it decreased by a factor of 10×10×10, or 1000, or 10^3.

The preferred SI unit for volume is the cubic metre (m^3); however, this is an inconveniently large and awkward unit, and is not yet uniformly used in manufacturers' specifications of vacuum apparatus. Therefore, the litre ($1\ l = 10^{-3}\ m^3 = 1\ dm^3 = 10^3\ cm^3 = 10^6\ mm^3$), which is the unit of volume most widely encountered in vacuum work, will be used here.

In closing this section I simply cannot resist recounting a story which was recently printed in *Chemical & Engineering News* (4 Jan, 1993, p. 51) about a student in a laboratory class in a French university who was asked the meaning of the letters 'psi' that were printed on the pressure gauge of an apparatus he was using. The student didn't know the answer, but decided to put forth his best effort anyway, and so after a bit of deliberation he replied that it probably meant 'pressure in SI units'. In relating this incident, the professor remarked that this answer undoubtedly caused Pascal to turn over in his grave, but noted that the joke is a particularly interesting one inasmuch as it can be understood in both the French and English languages, because the French word for pressure is 'pression'.

In case you don't know either, psi is the common abbreviation for 'pounds per square inch', a pressure unit generally used for pressures above one atmosphere, and therefore seldom encountered in vacuum work (1 psi = 51.7 Torr = 6876 Pa, 1 atm = 14.7 psi). And, since the subject of atmospheric pressure has come up, I can't help mentioning that many people have little appreciation of the magnitude of the force that the atmosphere exerts on the components of vacuum systems. For example, a 5×8 inch (125×200 mm) viewing window on an electron microscope has an area of 40 $inch^2$ (25 000 mm^2), and so it is subjected to a force of 40×14.7 = 588 pounds (lb) when the specimen chamber is evacuated. This is equivalent to having three sturdy athletes, each weighing about 195 lb (89 kg) standing on the window. It will be evident why it is not necessary to provide clamps to hold the window in place, nor to hold the doors on most scanning electron microscopes firmly shut.

1.3 The ideal gas equation

Our present understanding of the characteristics of gases stems from two parallel lines of work that began in the late 1600s and extended into the early 1900s. The first line of work, which consisted of the experimental measurement and empirical description of the macroscopic properties of gases, and which yielded the basic mathematical equations that were so necessary for making effective use of gases in all kinds of practical applications, was followed by such famous scientists as Amagat, Boyle, Cavendish, Charles, Dalton, Gay-Lussac, Graham, and Torricelli. The results of this work are now combined into the *ideal gas equation*:

$$P = \frac{nRT}{V} \tag{1.1}$$

which adequately describes the macroscopic properties of gases at the low pressures of concern in vacuum work where the molecules are widely separated and exert little influence on one another. This equation gives the pressure P that will be exerted by n moles of gas if confined in a container whose volume is V at a temperature of T kelvins (K = °C+273). R is a constant of proportionality known as the *ideal gas constant*. The number of molecules in a mole of gas is known as Avogadro's number ($N_A = 6.023\times10^{23}$ molecules/mol). The molar mass of a gas M is the mass of 1 mole or 6.023×10^{23} of its molecules. Some readers may recall from studying introductory chemistry and physics that 1 mole of any gas occupies a volume of 22.415 l at a pressure of one standard atmosphere (760 mmHg = 101 324 Pa) and a temperature of 273.16 K (0°C).

For those readers who have not practised their physical chemistry recently, it may be helpful to go through some simple calculations using the ideal gas equation. First, let us use the above data to calculate the numerical value of the ideal gas constant R. To do this we rearrange eqn. 1.1 to solve for R, and then use the pressure, volume, and temperature data just given to obtain:

$$R = \frac{PV}{nT} = \frac{101324 \times 22.415}{1 \times 273.16} = 8314.4 \quad \text{Pa} \cdot \text{l} / \text{mol} \cdot \text{K}$$

If volume is expressed in the preferred SI unit of cubic metres, $R = 8.3144$ Pa·m³/mol·K, since 1 litre = 10^{-3} m³ and the standard molar volume is 22.415×10^{-3} m³. Or, if you prefer to express pressure in units of Torr while using litres for volume:

$$R = \frac{PV}{nT} = \frac{760 \times 22.415}{1 \times 273.16} = 62.36 \quad \text{Torr} \cdot \text{l} / \text{mol} \cdot \text{K}$$

It is interesting to note that if we analyse the quantity 'Pa·l', which appears in the numerator of the first of the above equations, in terms of the basic SI units involved in defining its components we find that:

$$1 \text{ Pa-l} = (1 \text{ N/m}^2) \cdot (1 \times 10^{-3} \text{m}^3) = 10^{-3} \text{ N·m} = 10^{-3} \text{ J}$$

since 1 Pa equals 1 newton (N) per square metre, 1 litre (l) = 10^{-3} metre³, and 1 joule (J) equals 1 newton·metre. On this basis it is possible to write values of R in units of work or energy. That is, $R = 8.314$ J/mol·K = 1.987 cal/mol·K (since 1 calorie = 4.185 J). These values for R are widely used in thermodynamic calculations.

Now, let us calculate the number of gas molecules in 1 litre of gas at a pressure of 1 Pa and a temperature of 20°C (293 K). To do this we first calculate the number of moles of gas per litre n by rearranging the ideal gas equation to solve for n, and using the appropriate numerical values for P, V, R, and T, as follows:

$$n = \frac{PV}{RT} = \frac{1 \times 1}{8314.4 \times 293} = 4.10 \times 10^{-7} \text{ mol}$$

Since there are Avogadro's number of molecules in each mole of a gas, the number of molecules N in the litre of gas is obtained by multiplying the above number by N_A:

$$N = nN_A = (4.10 \times 10^{-7}) \times (6.023 \times 10^{23}) = 2.47 \times 10^{17} \text{ molecules.}$$

Athough simple, this is actually a fairly important calculation, for it illustrates *a basic principle*; namely, giving a value to the product of pressure and volume is equivalent to specifying the number of molecules of a gas, provided the temperature can be assumed to be known. As we have just shown, one *pascal-litre* (1 Pa-l) of gas at room temperature consists of about 4.10×10^{-7} moles or 2.47×10^{17} molecules. The above equation shows that n varies linearly with pressure; therefore, 1 litre of gas at a pressure of 10^{-7} Pa equals 10^{-7} Pa-l and consists of about 4×10^{-14} moles or 2.5×10^{10} molecules.

As we will see throughout the remainder of this book, the term '*pascal-litre*' (Pa-l) is commonly used in this way in vacuum work (as also are the terms 'Torr-litre' and 'mbar-litre'), under the assumption that the gas is near room temperature. There are two other terms that are sometimes used in this same context, particularly in describing leak rates and the capacities of cryogenic pumps. These are the standard litre and the standard cubic centimetre. These terms arose because in physical chemistry it is customary to refer to 0°C and 760 Torr as standard temperature and pressure (STP). On this basis, one standard litre of gas is the amount of gas contained in a volume of 1 litre at standard temperature and pressure, 1 l (STP). As noted above, 1 mole of gas occupies a volume of 22.415 l at STP, and so 1 l will contain $n = 1/22.415 = 4.46\times10^{-2}$ moles or

Table 1.2 The definitions of frequently-used pressure–volume quantities and the factors for converting among them.

		n	N	Pa-l	Torr-l	l (STP)
1 Pa-l	=	4.10×10^{-7}	2.47×10^{17}	1	7.50×10^{-3}	9.18×10^{-6}
1 mbar-l	=	4.10×10^{-5}	2.47×10^{19}	1.00×10^{2}	7.50×10^{-1}	9.18×10^{-4}
1 Torr-l	=	5.47×10^{-5}	3.29×10^{19}	1.33×10^{2}	1	1.23×10^{-3}
1 SCC	=	4.46×10^{-5}	2.69×10^{19}	1.09×10^{2}	8.15×10^{-1}	1.00×10^{-3}
1 l (STP)	=	4.46×10^{-2}	2.69×10^{22}	1.09×10^{5}	8.15×10^{2}	1

n = the number of moles of gas (1 mol = 6.023×10^{23} molecules); N = the number of gas molecules; l = litre; Pa = pascal; 1 Pa-l = one litre of gas at a pressure of one Pa (assuming a temperature near 20°C); 1 Torr-l = one litre of gas at a pressure of one Torr (assuming a temperature near 20°C); 1 l (STP) = one litre of gas at conditions of standard temperature and pressure (i.e. at STP, or 1 atmosphere pressure and 0°C); 1 SCC = one standard cubic centimetre (i.e. one cubic centimetre of gas at standard temperature and pressure).

$N = nN_A = 2.69\times10^{22}$ molecules. Thus, 1 l (STP) = 2.69×10^{22} molecules of gas, while a standard cubic centimetre is one-thousandth of this number (i.e. 1 SCC = 2.69×10^{19} molecules). Although these terms are not consistent with SI conventions they still appear in much advertising material and industrial literature on vacuum topics. These pressure-volume terms are used so frequently in vacuum work that their characteristics, and the factors for converting among them, are summarized in Table 1.2.

The gas of concern in most vacuum work is the ordinary atmospheric mixture. Dry air consists of about 21 percent (by volume) oxygen (O_2, $M = 32$), 78 percent nitrogen (N_2, $M = 28$), and about 1 percent of other gases, such as carbon dioxide (CO_2, $M = 44$), hydrogen (H_2, $M = 2$), helium (He, $M = 4$), neon (Ne, $M = 20$), and argon (Ar, $M = 40$). Berman (Table 1.1, p. 3) gives more detailed compositional information. It is a matter of very considerable importance, particularly for our physical comfort, that ordinary air also contains one to three percent water (H_2O, $M = 18$). For most purposes an averaged value of $M = 29$ g/mol can be used as the molar mass for air. Using the results of the above calculation, we find that one pascal-litre of air has a mass of $m = nM = (4.10\times10^{-7})\times29 = 1.19\times10^{-5}$ g.

1.4 The kinetic molecular model

The second line of work that contributed to our understanding of the properties of gases involved the theoretical development of the kinetic molecular model of gases. This work was carried out by such notables as Avogadro, Bernouilli, Boltzmann, Clausius, Jeans, Maxwell, and Van der Waal. The revolutionary contribution of the kinetic molecular model was the concept that gases are not continuous fluids, but rather consist of a large number of tiny particles of matter, the gas molecules, that are in constant random motion. The macroscopic property known as pressure is then explained as a manifestation of the force exerted by these molecules as they collide with the walls of their container. The quantity we commonly refer to as temperature is proportional to the average kinetic energy of the molecules, and the collisions are considered to be mechanically elastic in the sense that the molecules do not experience a net change in kinetic energy in these collisions when they are at the same temperature as the container. When applied to the measured properties of gases, this theory showed that at ordinary and low pressures the average distance between the molecules in a gas is much greater than the diameter of the molecules themselves. We tend to take these concepts more or less for granted today; however, the kinetic molecular model of gases was one of the most outstanding achievements of early theoretical physicists and chemists, because it provided a mechanistic basis for understanding the behaviour and properties of gases, and at the same time strongly supported the atomic–molecular model for the structure of matter that was then in the early stages of development.

Combining the ideal gas equation with the kinetic molecular model yields a number of mathematical relationships which elegantly describe the characteristics of the medium with which we are dealing. These are derived and discussed in some detail by Berman (Chapter 3), Dushman (Chapter 1), Roth (Chapter 2), and Van Atta (Chapter 1). Several of these relationships that will be particularly useful in developing and understanding topics covered later in this book are presented below.

1.5 The concentration of gas molecules

While we all are probably aware that there is a tremendous number of gas molecules in each unit volume of the atmosphere around us, there may be some tendency to assume that there are very few gas molecules in a vacuum. This is far from being true, however. As shown earlier, we can calculate β, the number of gas molecules per litre, using the ideal gas equation (eqn. 1.1). The following equations summarize the overall result of the calculation:

$$\beta = \frac{nN_A}{V} = \frac{N_A P}{RT} = \frac{6.023 \times 10^{23} P}{8314.4T}$$
$$= 7.24 \times 10^{19}\left(\frac{P}{T}\right) \text{ molecules / l} \qquad (1.2a)$$

where n is the number of moles of gas contained in a volume of V litres, P is the pressure (Pa) and T is the temperature (K) of the gas, and N_A is Avogadro's number or the number of molecules in a mole of gas. This equation shows that if a gas is not constrained by a rigid container the

Table 1.3 Values of physical parameters for air at low pressures

P	β	λ	γ
10^5	2.5×10^{22}	6.6×10^{-5}	2.8×10^{21}
10^2	2.5×10^{19}	6.6×10^{-2}	2.8×10^{18}
10^1	2.5×10^{18}	6.6×10^{-1}	2.8×10^{17}
10^{-1}	2.5×10^{16}	6.6×10^{1}	2.8×10^{15}
10^{-5}	2.5×10^{12}	6.6×10^{5}	2.8×10^{11}
10^{-7}	2.5×10^{10}	6.6×10^{7}	2.8×10^{9}
10^{-9}	2.5×10^{8}	6.6×10^{9}	2.8×10^{7}

β = the number of gas molecules per litre, λ = the average distance (mm) gas molecules travel between collisions, and γ = the number of collisions per second per square millimetre gas molecules make with solid surfaces, at various pressures P (Pa) and at 20°C (293 K).

number of molecules per litre varies with both temperature and pressure, but does not depend on M, the molar mass of the gas. This equation becomes:

$$\beta = 2.47 \times 10^{17}\ P \text{ molecules / l} \qquad (1.2b)$$

at ambient temperature (20°C or 293 K). As shown in Table 1.3, at atmospheric pressure ($\approx 10^5$ Pa) there are about $\beta = (2.47\times10^{17})\times101\,324 = 2.5\times10^{22}$ molecules in each litre of air around us. Each time we breathe we inhale about half a litre of air consisting of some 10^{22} molecules. There are more than 10^{12} molecules per litre at 10^{-5} Pa (10^{-7} Torr), the bottom of the high vacuum range, and even at 10^{-9} Pa (10^{-11} Torr), which is well into the ultra-high vacuum range, there are more than 10^8 gas molecules per litre. While the concentration of gas molecules is many orders of magnitude less in a vacuum than at atmospheric pressure, even the best vacuum that is likely to be encountered in most electron microscopes will have many millions of gas molecules in each cubic millimetre of volume.

1.6 The speed of gas molecules

Gas molecules are constantly in motion, and they move with surprisingly high speeds. The average speed σ of gas molecules is independent of pressure, but varies with temperature T (K) and molar mass M (g/mol) as follows:

$$\sigma = 145.5\sqrt{\frac{T}{M}} \text{ m/s} \tag{1.3}$$

At room temperature the oxygen and nitrogen molecules in the air around us are moving at an average speed of about 450 m/s (1000 miles/h) while the helium atoms, which are much lighter, have speeds in excess of 1200 m/s (2500 mph)!

Note that this equation gives the numerical average molecular speed. Actually, the distribution of molecular speeds is not strictly symmetrical, but is described by the asymmetric Maxwell–Boltzmann distribution function, and so two other 'average' values can be used to describe the speed of gas molecules. These are the most probable speed $\sigma_p = 0.89\sigma$ and the root mean square speed $\sigma_r = 1.09\sigma$ (see Roth, p. 35). While these speeds are often cited in discussions of gas kinetics, they will not be required for our purposes.

1.7 The mean free path

The average distance gas molecules travel between collisions with each other is a very important parameter since it strongly influences the way they behave when they flow through the components of vacuum systems, which, in turn, becomes an essential factor in determining operating procedures for most vacuum systems. Fundamentally, this *mean free path* λ is inversely proportional to β, the number of gas molecules per litre (eqn. 1.2), and according to the kinetic molecular model is given by:

$$\lambda = \frac{1}{\beta\delta^2\pi\sqrt{2}} = \frac{V}{nN_A\delta^2\pi\sqrt{2}} \tag{1.4a}$$

which can be converted to the following form by substituting $V/n = RT/P$ from the ideal gas equation:

$$\lambda = \frac{R}{N_A\delta^2\pi\sqrt{2}}\left(\frac{T}{P}\right) \tag{1.4b}$$

In these equations n is the number of moles of gas contained in a volume of V litres at a pressure P (Pa) and temperature T (K), R is the ideal gas constant (8.314 Pa·m^3/mol·K), N_A is Avogadro's number (6.023×10^{23} molecules/mol), and δ is the effective diameter of the gas molecules in collision processes. This collision diameter has a value of about 0.37×10^{-9} m (i.e. 0.37 nm or 3.7 Å) for nitrogen and oxygen molecules, the most abundant molecules in air, and so this equation becomes:

$$\lambda = \frac{6.6 \times 10^{-3}}{P} \text{ m} \approx \frac{6.6}{P} \text{ mm} \quad (1.4c)$$

for air at 293 K (20°C) when pressure is given in pascals (if P is given in Torr, $\lambda \approx 0.049/P$ mm). Berman (p. 50) and Roth (p. 40) show that in mixtures of gases, molecules of smaller size and smaller mass have longer mean free paths. This effect, and the fact that helium molecules (with $\delta \approx 0.22$ nm and $M = 4$ g/mol) move at a higher speed than air molecules, is used to advantage in some leak detection methods by spraying a jet of helium around the outside of an apparatus (see Section 10.5, p. 434). When the jet passes over a leak the helium diffuses through the leak, and into and throughout the vacuum system very rapidly, and so it will reach a helium detector, and thereby indicate the site of the leak, quickly and with high sensitivity.

At atmospheric pressure and ambient temperatures air molecules travel, on the average, only $\lambda \approx 6.6/101\,324 \approx 6.5 \times 10^{-5}$ mm $\approx$ 65 nm, or less than 200 atomic diameters, between collisions. As shown in Table 1.3, however, λ increases as the pressure decreases, becoming about 70 mm, which is of the order of the internal dimensions of typical components of vacuum systems, at 10^{-1} Pa (10^{-3} Torr). Above this pressure gas molecules collide more frequently with one another than with the walls of the vacuum system, and so the movement of any one molecule is strongly influenced by the movement of neighbouring molecules; that is, one molecule can be swept along by the other molecules around it. The flow of gases under these conditions is commonly called *viscous flow*. At lower pressures λ becomes larger than the internal dimensions of the pumping lines of most vacuum systems (e.g. $\lambda \approx 700$ mm at 10^{-2} Pa) and the molecules move independently of one another. Flow at these pressures involves only the random thermally-activated motion of the individual molecules and is called *molecular flow*.

The numerical quantity commonly used to delineate the conditions under which gases show these different flow characteristics is the *Knudsen number*, which is the ratio of the mean free path of the gas molecules λ to the internal diameter D of the tube through which they are flowing (i.e. $K_n = \lambda/D$). Specifically, gases exhibit:

viscous flow	if	$K_n<0.01$	(i.e. $D>100\lambda$)
molecular flow	if	$K_n>1.0$	(i.e. $\lambda>D$)
transitional flow	if	$0.01>K_n<1.0$.	

These criteria will prove to be useful for a number of purposes in subsequent sections of this book, because the character of the gas flow process is such an important factor in determining operating procedures and in understanding how various devices function. Note that K_n is a dimensionless quantity, and so when it is being calculated λ and D must both be expressed in the same units of length.

A word of caution may be in order before leaving this section. When using eqn. 1.4c it is necessary to keep in mind the fact that it was derived under the assumption that the gas temperature was near 20°C. If this precaution is not observed incorrect conclusions can be reached under some circumstances. For example, on the basis of this equation it might be concluded that λ would decrease if the gas in a closed container were heated, because obviously increasing its temperature would cause the pressure of the gas to increase. This would be incorrect, however, because the number of gas molecules per unit volume, which is the fundamental parameter in determining λ, would not change appreciably. Both eqn. 1.4a and eqn. 1.4b, which do not contain an inherent assumption about the temperature of the gas, would give a more correct result. Actually, as noted by Berman (p. 49) and Roth (p. 39), increasing the temperature in this way does cause a small change in the mean free path because of the change in the speed of the gas molecules it produces; however, the effect is small and of little consequence for our purposes.

1.8 Collisions with surfaces

According to the kinetic molecular model (see Weston, p. 277) the number of gas molecules in the gas phase that move through a unit area in unit time in any arbitrary direction is $\gamma = \beta\sigma/4$. If we consider motion toward the surface of the vessel containing the gas, and use Eqns. 1.2 and 1.3 given above for β and σ, the rate at which gas molecules strike the surface becomes:

$$\gamma = 2.64 \times 10^{18}\left(\frac{P}{\sqrt{MT}}\right) \text{ collisions/mm}^2\cdot\text{s} \qquad (1.5a)$$

where P is the pressure of the gas (Pa), T is its temperature (K), and M its molar mass (g/mol). For air at room temperature $M = 29$ g/mol and T = 293 K, and so this simplifies to:

$$\gamma = 2.86 \times 10^{16} P \quad \text{collisions/mm}^2\cdot\text{s} \qquad (1.5b)$$

This number is so large it is difficult to appreciate. At atmospheric pressure each square millimetre of our bodies is struck by more than 10^{21} gas molecules each second. At 10^{-5} Pa (10^{-7} Torr), the bottom of the high vacuum range, all surfaces in a vacuum system experience 10^{11} collisions/s·mm^2, and even at 10^{-9} Pa (10^{-11} Torr), deep in the ultra-high vacuum range, γ has a value of 10^7, as shown in Table 1.3. This quantity γ is very important to the flow of gases in the high and ultra-high vacuum ranges because there it determines the rate at which gas molecules enter tubes and apertures, as will be shown in Section 2.3 (p. 32).

When using this equation it is important to keep in mind that the area involved in it is the projection onto the container of the area through which the gas molecules move as they approach its surface. This is the macroscopic surface area, or the area which would be used in making a pressure measurement or calculation. In particular, it is not the microscopic surface area, which may be several times greater than the macroscopic area because of surface roughness.

1.9 The time for monolayer formation

Vacuum systems are widely used to provide an environment in which the physical and chemical effects of the atmospheric gases are greatly reduced. One measure of the extent to which this is achieved is the time τ required for a freshly-formed surface to become covered with a monolayer of gas molecules. An idealized value for τ can be obtained by assuming that the surface is atomically smooth, so that the macroscopic area equals the microscopic area, and that every gas molecule that strikes it sticks to it. As noted above, gas molecules have collision diameters of about $\delta \approx 3.7\times10^{-7}$ mm, and so each molecule would occupy an average area of about $A = \pi\delta^2/4 \approx 1.08\times10^{-13}$ mm^2. On this basis it would require about $\omega = 1/A \approx 9.3\times10^{12}$ (roughly 10^{13}) molecules/mm^2 to form a monolayer. Dividing this number by the value for γ given above yields:

$$\tau \approx \omega/\gamma \approx 3.3\times10^{-4}/P \quad \text{s} \qquad (1.6a)$$

as the time for monolayer formation, where P is the gas pressure (Pa), and assuming the temperature is near 20°C. This equation suggests that if a piece of mica or a rock salt crystal is cleaved in a vacuum, the new surface formed will become covered with adsorbed gas molecules in:

a few seconds	at	10^{-4} Pa (10^{-6} Torr)
a few minutes	at	10^{-6} Pa (10^{-8} Torr)
and about an hour	at	10^{-7} Pa (10^{-9} Torr).

This makes it quite clear why sensitive surface analysis instruments, which are capable of analysing the outermost two or three layers of atoms on the surface of a specimen, must operate at pressures well into the ultra-high vacuum range.

As noted above, this equation is highly idealized. Most surfaces are not atomically smooth, but have areas on the atomic scale that are many times greater than their apparent, macroscopic areas. This *surface roughness factor* g can be as great as 100 for some surfaces. In addition, only a fraction f of the gas molecules that strike a surface usually stick to it, and this fraction may decrease as the degree of surface coverage increases. Incorporating this *sticking coefficient* and the surface roughness factor into the above equation gives:

$$\tau \approx \frac{g\omega}{f\gamma} \approx \frac{3.3\times10^{-4}g}{Pf} \quad \text{s} \qquad (1.6b)$$

as a more practical expression. As noted above, g can have values ranging from 10 to 100 for machined and ground surfaces. The sticking coefficient for nitrogen is about 0.3 on most clean, fresh metal surfaces, but decreases by a factor of about 10 after 10 percent of the surface becomes covered with gas molecules. For oxygen f remains nearly constant at about 0.8, even for nearly complete surface coverage. These values suggest that the times needed for the adsorption of a monolayer of gas molecules on most surfaces found in practical vacuum systems can be somewhat longer than the idealized values deduced above.

1.10 The density of gases

The density ρ of a gas with molar mass M (g/mol), at a pressure P (Pa) and temperature T (K), is given by:

$$\rho = 1.2 \times 10^{-7}\left(\frac{M\,P}{T}\right)\ \mathrm{Mg/m^3} \tag{1.7a}$$

(note that 1 $\mathrm{Mg/m^3}$ = 1 $\mathrm{g/cm^3}$ = 1 $\mathrm{mg/mm^3}$). For air at room temperature this simplifies to:

$$\rho = 1.1 \times 10^{-8}\,P \quad \mathrm{g/cm^3} \tag{1.7b}$$

At atmospheric pressure the density of air is only 1×10^{-3} $\mathrm{g/cm^3}$, 1000 times less than the density of water.

1.11 Gas viscosity

Viscosity is a measure of the resistance of a fluid to flow. It results from interactions between the molecules themselves and between the molecules and the walls of the tube through which they are flowing. The viscosity η of a gas depends on the temperature T, and the size of the gas molecules as represented by their collision diameter δ and their molar mass M as follows:

$$\eta \approx 500\left(\frac{\rho\,\sigma\,\lambda}{\delta^{2}}\right) \approx 2.7 \times 10^{-26}\left(\frac{\sqrt{MT}}{\delta^{2}}\right)\ \mathrm{Pa\cdot s} \tag{1.8a}$$

Using $M = 29$ g/mol, $T = 293$ K, and $\delta = 0.37 \times 10^{-9}$ m for air at room temperature, this simplifies to:

$$\eta \approx 1.82 \times 10^{-5} \text{ Pa·s} \approx 1.82 \times 10^{-4} \text{ poise} \tag{1.8b}$$

It is important to note that the viscosity of a gas does not vary with pressure. As we will see in the next chapter, this has an important influence on the characteristics of gas flow in the viscous flow range where intermolecular interactions, which can be described by the viscosity of the gas, play a significant role.

1.12 Concluding remarks

There are many interesting principles involved in describing the properties of gases at low pressures, and it is surprising the extent to which a familiarity with these principles can contribute to understanding the design and proper operation of vacuum systems. The speed with which gas molecules move and the number of molecules present per unit volume determine the frequency with which they collide with other molecules and with surfaces inside the vacuum vessel. We are now in a position to see how these factors also influence the characteristics gases exhibit when they flow through the components of vacuum systems, which is the topic of the next chapter.

2 The evacuation process

Electron microscopists carry out an evacuation process each time they perform such common operations as changing specimens, exchanging photographic film, replacing a filament in an electron microscope, or when pumping down a vacuum evaporator, a sputter coater, or a freeze-etching unit. An old Chinese proverb supposedly says something to the effect that, "Even a journey of a thousand miles begins with but a single step". Somewhat similarly, even the evacuation process that produces the best ultra-high vacuum attainable begins at atmospheric pressure. Thus, to produce a high vacuum, a *rough vacuum pump* must first be used to remove the free atmospheric gas molecules from the vacuum vessel, reducing the pressure to about 10 Pa (0.1 Torr). Then, a *high vacuum pump* is used to further lower the pressure to the desired operating level. There are a number of interesting and operationally important phenomena involved in each stage of this rather complex process, and since we carry out evacuation operations so frequently in electron microscopy laboratories a knowledge of these phenomena provides a better understanding of the proper way to operate the instruments involved. It is therefore useful to examine this evacuation process in some detail in a general way before proceeding to discuss the types of pumps that are available for use in carrying it out, and the types of gauges that are used to monitor its progress.

The references cited in this chapter are listed in Appendix 1. The manufacturers and suppliers of vacuum equipment referred to here are listed in Appendices 2 and 3.

2.1 Conductance

Typically, a vacuum system consists of a chamber to be evacuated that is connected to one or more vacuum pumps by an arrangement of tubes, valves, and traps (e.g. see Fig. 2.4, Section 2.5, p. 40). Both the rate of evacuation and the ultimate vacuum attained in the system are strongly dependent on the speed with which the gas molecules can flow from the chamber to the pumps. The relative ease with which the components of a vacuum system accommodate this flow of gas molecules is described in terms of their conductance C, which is defined as:

$$C = \frac{Q}{P_{hi} - P_{lo}} \tag{2.1}$$

In this equation P_{hi}–P_{lo} is the pressure difference (also often designated as ΔP) across the component, and Q is the *throughput*, or the quantity of gas flowing per unit time, in Pa-l/s or Pa-l/min. The calculation in Section 1.3 (p. 15) showed that 1 Pa-l of gas consists of about 2.47×10^{17} molecules, and so a flow of 1 Pa-l/s represents the movement of 2.5×10^{17} molecules/s, regardless of the pressure at which the flow occurs. Furthermore, the conservation of matter requires that Q be the same for all components of a pumping system that are in series with one another, provided there are no large leaks in any of them. Since conductance is obtained by dividing Q by a pressure, it has units of litres per min (l/min) or litres per second (l/s), the former having been traditionally used in the rough vacuum range, and the latter in the high and ultra-high vacuum ranges, in the past. Now, however, some manufacturers are beginning to use units of cubic metres per hour (m^3/h) generally, while conductances for large industrial apparatus are often expressed in units of cubic feet per minute (for which the unusual abbreviation of 'cfm' is commonly used). Only litres per minute and litres per second will be used in this book; however, factors for converting among these different units are given in Table 2.1.

Table 2.1 Conversion factors for units of flow rate

		litre/s	litre/min	cfm	m^3/h
1 litre/s	=	1	60	2.12	3.60
1 litre/min	=	0.0167	1	0.0353	0.060
1 cfm	=	0.472	28.3	1	1.69
1 m^3/h	=	0.278	16.7	0.589	1

The total conductance of a number of components connected in series is related to their individual conductances C_1, C_2, C_3 by:

$$\frac{1}{C_{total}} = \frac{1}{C_1} + \frac{1}{C_2} + \frac{1}{C_3} + \ldots \qquad (2.2)$$

while the equation for several components connected in parallel is:

$$C_{total} = C_1 + C_2 + C_3 \qquad (2.3)$$

The reciprocal of conductance is impedance Z (i.e. $Z = 1/C$).

2.2 Viscous flow

During the initial stages of an evacuation process, when the pressure is in the rough vacuum range, the mean free path of the gas molecules λ is so short that they collide much more frequently with each other than with the walls of the tube through which they are flowing, and the flow of the gas is affected by both its viscosity and by the pressure difference $\Delta P = P_{hi} - P_{lo}$ that exists between the two ends of the tube. Under these conditions the flow process is said to be viscous in character, and the movement of the gas molecules occurs by a cooperative mass flow phenomenon in which the motion of any particular molecule is strongly influenced by the numerous collisions (more than 10^6/s) it experiences with the other molecules that surround it. The overall result is that the molecules move along together in the general direction of the low pressure end of the tube. While it is certainly true that every molecule undergoes numerous collisions that cause it to move back upstream towards the high pressure end of the tube, collisions in the opposite direction are more compelling, and so it is virtually impossible for any molecule or group of molecules to move an appreciable distance in the direction opposite to that of this general mass flow process. Insofar as the operation of vacuum apparatus is concerned, this is one of the most important characteristics of this mode of gas flow, because we are so often concerned about oil molecules from vacuum pumps moving up the pumping lines and into the vacuum system. Practically, *backstreaming* or *back diffusion* processes do not occur during pumping operations in which the pressure is in the range of viscous flow.

As noted in Section 1.7 (p. 22) it is generally accepted that viscous flow occurs if the Knudsen number $K_n = \lambda/D$ has a value less than 0.01. Here λ is the mean free path of the gas molecules (mm) and D is the internal dimension of the tube (mm) through which the gas is flowing. This is equivalent to saying that the mean free path of the molecules must be less than one-hundredth of the diameter of the tube (i.e. $\lambda < 0.01D$). Using the relationship that $\lambda = 6.6/P$ mm (eqn. 1.4c, Section 1.7, p. 21), this requirement can be converted to $P > 660/D$, when P is measured in pascals and D is in mm ($P > 5/D$ for P in Torr and D in mm). The upper curve in Fig. 2.1 is a plot of this equation for tubes up to 200 mm in diameter. According to the above relationship, viscous flow will occur at all pressures above this curve. Intuitively, there is a tendency to associate viscous flow with small tubes, and to assume that it will prevail at lower pressures in small tubes than in large ones. Fig. 2.1 shows that this is not true, however. The lowest pressure at which viscous flow will occur in a tube 25 mm in diameter is about 30 Pa, while for a tube 200 mm in diameter it is near 3 Pa. In later discussions we will be concerned with carrying out certain operations before the pressure drops below the range of viscous flow. Fig. 2.1 shows that we will always be safe if such operations are performed at 100 Pa (1 Torr), and that we might reasonably extend this limit down to 10 Pa (10^{-1} Torr) because, as we will see shortly, there is considerable viscous character to flow in the upper part of the transitional range.

The conductance C_{vt} of cylindrical tubes for the flow of gases in the viscous range can be calculated with sufficient accuracy for our purposes in this book from the equation:

$$C_{vt} = 1.5 \times 10^{-9} \left(\frac{D^4}{\eta L} \right) P_{av} \quad \text{l/min} \qquad (2.4a)$$

This expression is based on Poiseuille's classical equation of fluid mechanics. The pressure dependence is given by the average pressure (Pa) in the tube $P_{av} = (P_{hi} + P_{lo})/2$. D is the diameter of the tube (mm), L is its length (m), and η is the gas viscosity. For air at 20°C, $\eta \approx 1.82 \times 10^{-5}$ Pa·s, and so this equation simplifies to:

$$C_{vt} \approx 8.2 \times 10^{-5} \left(\frac{D^4}{L} \right) P_{av} \quad \text{l/min} \qquad (2.4b)$$

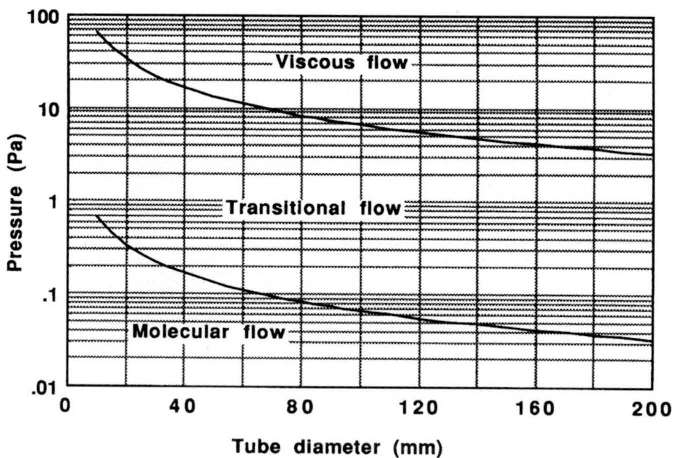

Fig. 2.1 The pressure ranges for the different types of gas flow.

The constant in this equation becomes 1.45×10^4 if you prefer to measure P in Torr, D in inches, and L in feet. One important thing to note here is the fact that conductance depends on the fourth power of the diameter of the tube; therefore, *small* differences in diameter make *very large* differences in conductance. A tube with a diameter of 50 mm has a conductance 16 times greater than one of the same length with a diameter of 25 mm.

Fig. 2.2 is a plot of the viscous conductance values for tubes 1 m long with diameters from 10 to 150 mm at average pressures ranging from 1 atm down to 1 Pa. Since viscous conductance varies inversely with length, the conductance of a piece of tubing of any arbitrary length can be calculated simply by dividing the value read from this graph by the tube's actual length in metres. For example, Fig. 2.2 gives the conductance of a piece of tubing 50 mm in diameter and 1 m in length as about 5×10^4 l/min at an average pressure of 100 Pa (1 Torr). A piece of this tubing 2.5 m long would have a conductance of about $C \approx 5\times10^4/2.5 = 2\times10^4$ l/min at the same average pressure. Although it is a little difficult to read numbers from a log–log graph such as this, values good enough for most purposes can be obtained from Fig. 2.2 by enlarging it with a photocopier and then using a bit of care in reading it.

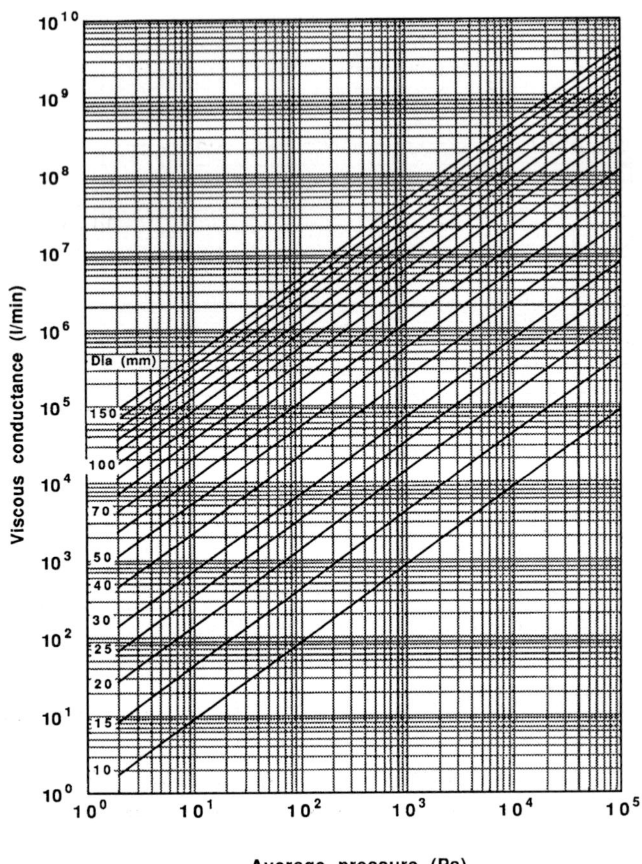

Fig. 2.2 The viscous conductance of tubes one metre in length.

The viscous conductance of a small thin circular orifice or aperture is approximately:

$$C_{va} \approx 9.42\, D^2 \quad \text{l/min} \tag{2.5}$$

if its diameter D is measured in millimetres.

2.3 Molecular flow

At sufficiently low pressures the mean free path of the gas molecules exceeds the internal dimensions of most tubes and other components through which they must flow, and the molecules collide with the walls of these components more frequently than with each other. Under these conditions the passage of some gas molecules from the vacuum chamber into the pumping line will have little influence on the motion of either the molecules remaining in the chamber or those already in the line. As a consequence, the flow of the gas is independent of pressure and viscosity, and depends primarily on the dimensions of the tube and the speed of the molecules as determined by their temperature. Flow under these conditions is called molecular flow.

The fact that conductance is independent of both pressure and viscosity in the range of molecular flow can be demonstrated in a straightforward manner for an aperture of area A (mm^2) separating two large chambers at pressures P_{hi} and P_{lo}. Using eqn. 1.5 (Section 1.8, p. 23), the number of air molecules entering the aperture, and passing through it each second from the high pressure side will be:

$$N_{hi} = A\gamma_{hi} = 2.64 \times 10^{18}\left(\frac{A}{\sqrt{MT}}\right)P_{hi}$$

A similar expression involving P_{lo} can be written for N_{lo}, the number of molecules passing through the aperture per second from the low pressure side. The net number of molecules flowing through the aperture per second will be the difference between these two:

$$N_{flow} = N_{hi} - N_{lo} = 2.64 \times 10^{18}\left(\frac{A}{\sqrt{MT}}\right)(P_{hi} - P_{lo})$$

Division by Avogadro's number gives the number of moles of gas flowing per second (i.e. $n_{flow} = N_{flow}/N_A$). According to eqn. 1.1 (Section 1.3, p. 14), the ideal gas equation, multiplication by RT gives the quantity of gas flowing per unit time, i.e. the throughput, $Q = PV_{flow} = n_{flow}\,RT$ in pascal-litres per second, as:

$$Q = \frac{2.64 \times 10^{18}\ RTA}{N_A\sqrt{MT}}\left(P_{hi} - P_{lo}\right)$$

Utilizing the definition for conductance in eqn. 2.1 gives the following expression for the molecular conductance of an aperture:

$$C_{ma} = \frac{Q}{P_{hi} - P_{lo}} = 3.64 \times 10^{-2} A\sqrt{\frac{T}{M}} \quad \text{l/s} \tag{2.6a}$$

Thus, molecular conductance does not depend on either viscosity or pressure, but only on the area of the aperture A and the variables T and M that determine the speed of the gas molecules. For air at 20°C, $M = 29$ g/mol and $T = 293$ k, giving a value of

$$C_{ma} = 0.12A \quad \text{l/s} \tag{2.6b}$$

for an aperture with an area of A mm². The area of a circular aperture is related to its diameter D by $A = \pi D^2/4$, and so this equation can also be written as:

$$C_{ma} = 0.12\pi D^2/4 = 0.09\ D^2 \quad \text{l/s} \tag{2.6c}$$

for a circular aperture with a diameter of D mm.

The molecular conductance of circular tubes for air at room temperature can be calculated with acceptable accuracy for present purposes from:

$$C_{mt} = \frac{0.0121\ D^3}{100L + 0.133D} \quad \text{l/s} \tag{2.7}$$

where, again, D is the tube diameter in millimetres and L is its length in metres. This equation becomes $C_{mt} = 78\ D^3/(12L+1.33D)$, if units of inches are used for D, and feet for L. Molecular conductance values for tubes having diameters ranging from 20 to 200 mm and lengths up to 1 m are plotted in Fig. 2.3 which can be enlarged by photocopying for easier use. Note again the strong dependence of the conductance on the diameter of the tube. Changing the diameter of a tube by a factor of two changes its conductance by a factor of eight. For example, a tube 0.5 m long with a diameter of 100 mm has a conductance of about 200 l/s, while the conductance of a tube of the same length with a diameter of 50 mm is only about 25 l/s.

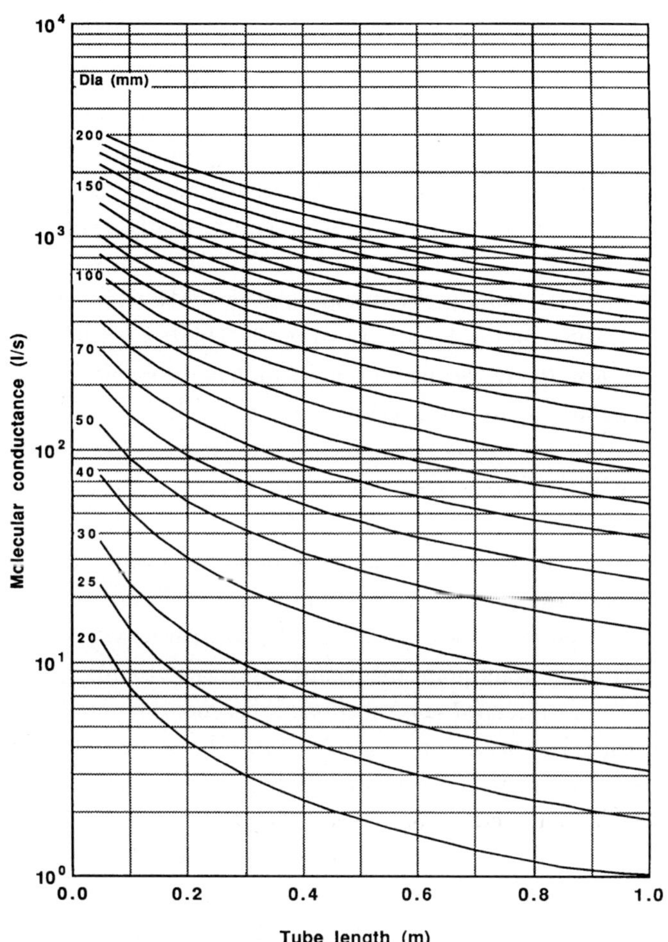

Fig. 2.3 The molecular conductance of circular tubes.

A bend or an elbow in a tube will cause a small decrease in its molecular conductance. This can usually be taken into account with sufficient accuracy by using a value for L in eqn. 2.7 that equals the maximum length of the tube as measured around the outside of the bend or elbow.

One very important characteristic of molecular flow can be illustrated by considering the aperture connecting the two large chambers described above. Again using eqn. 1.5, we find that during each second $N_{hi} = A\gamma_{hi} = 2.86{\times}10^{16}\ AP_{hi}$ molecules enter the aperture from the high pressure side, while $N_{lo} = A\gamma_{lo} = 2.86{\times}10^{16}\ AP_{lo}$ enter it from the low pressure side. Since the molecules collide with one another so infrequently there is free movement of molecules in both directions, and so the ratio is simply equal to the ratio of the two pressures (i.e. $N_{lo}/N_{hi} = P_{lo}/P_{hi}$). If $P_{lo} = 0.1P_{hi}$ then one-tenth as many molecules move through the aperture from the low pressure side as from the high pressure side. This *back diffusion* might seem insignificant, but it would amount to nearly 3×10^{10} molecules/s/mm^2 even if P_{lo} is 10^{-6} Pa (10^{-8} Torr), about the lowest pressure likely to be encountered in most electron microscopes. Also, note that if P_{hi} approaches P_{lo}, then N_{lo} become nearly equal to N_{hi}. This has rather interesting implications. For example, consider the situation in the manifold leading from the high vacuum pump to the electron gun of an electron microscope after a day or two of evacuation. Then, the pressure at the top of the manifold will be very little different from that at the bottom near the entrance to the pump, and nearly as many gas molecules will be moving up the manifold toward the electron gun each second as move down it towards the pump. Thus, whereas backstreaming cannot occur under conditions of viscous flow, it cannot be prevented under conditions of molecular flow.

This is a very important point that is not appreciated by many people who operate vacuum systems. I remember one colleague who purchased his own vacuum evaporator because he maintained that the one in our electron microscopy laboratory was contaminated with pump oil. When his new evaporator arrived, he instituted the policy of keeping it always under vacuum as a means of ensuring that it would stay clean. This might have been a good policy, except that when the evaporator was not actually being used to carry out an evaporation process he turned off the oil diffusion pump and used only the oil-sealed rotary-vane roughing pump to do the evacuation. This pump easily held the pressure at a value near 5 Pa (0.05 Torr), and this is close enough to being in the range of molecular flow for the 25 mm diameter tubing used in the pumping line to allow backstreaming to occur there. Unfortunately, oil-sealed rotary-vane pumps are noted for breaking the molecules of the sealing oil used in them down into low molecular mass fragments which are sufficiently volatile to readily diffuse into the vacuum line leading into the pump if conditions there permit it, as they did in this instance (Section 4.1.5, p. 144). As a result, within a few weeks the interior of his 'clean' evaporator was thoroughly coated with these hydrocarbon products, and his new evaporator was actually in worse condition than the old one in our laboratory which he refused to use.

As was noted in Section 1.7 (p. 22), molecular flow is usually considered to occur if the Knudsen number $K_n = \lambda/D$ is greater than one; that is, when the mean free path of the gas molecules λ exceeds the internal diameter D of the tube through which the gas is flowing (i.e. $\lambda/D>1$ or $\lambda>D$). Using the relationship $\lambda = 6.6/P$ mm of eqn. 1.4c again, this can be restated in terms of pressure as $P<6.6/D$ when P is measured in pascals and D in mm (or $P<0.05/D$ for P in Torr and D in mm). The lower curve in Fig. 2.1 is a plot of the equation $P = 6.6/D$ for tubes of various diameters. According to the concepts just discussed, molecular flow will occur at all pressures below this curve. This curve shows that molecular flow can occur at higher pressures in small tubes than in large ones. For example, the highest pressure at which molecular flow will occur in a tube 100 mm in diameter is about 7×10^{-2} Pa (5×10^{-4} Torr), while in a tube 25 mm in diameter it can occur at pressures as high as 2.7×10^{-1} Pa (2×10^{-3} Torr). If a leak consists of a hole 10^{-3} mm in diameter through the wall of a vacuum vessel, flow through it will be molecular in character at pressures up to nearly 7000 Pa (50 Torr).

This trend could lead to unexpected problems in some circumstances because of the back-diffusion phenomenon just described. Consider, for example, the specimen airlock of a high resolution electron microscope that is evacuated by an oil-sealed rotary-vane pump through a tube 10 mm in diameter, a typical arrangement for many instruments. During the long periods between specimen changing operations, when the airlock is not being used, the pump can easily bring the pressure in this small tube down into the range of molecular flow, whereupon oil molecules from the pump can diffuse into the pumping line and ultimately reach the valve that leads into the airlock chamber. If the pressure in the airlock chamber approaches the molecular range when this valve is opened to evacuate the airlock chamber during specimen exchange operations, the oil molecules can then diffuse into the airlock chamber. This is one of the worst places such contamination could occur, because then each time a specimen is introduced into the instrument it is bathed in oil vapours. Over a few months of operation the oil molecules will ultimately find their way into the specimen chamber of the microscope, making the situation even worse. Specimen airlock systems on electron microscopes should not be pumped out directly in this manner with an oil-sealed pump.

2.4 Transitional flow

The two curves in Fig. 2.1 that define the limits of the pressure ranges for viscous and molecular flow are separated by a considerable pressure interval. In this interval the Knudsen number varies from 0.01 to 1.0 and the flow mechanism is intermediate between viscous and molecular in character. Theoretical analyses (Roth, Section 3.6) show that in this range of transitional or intermediate flow the conductance of circular tubes can be treated as having both a viscous and a molecular component, and can be calculated with reasonable accuracy using the following equation:

$$C_{tt} \approx \frac{C_{vt}}{60} = \left(\frac{1 + 0.19 DP_{av}}{1 + 0.23 DP_{av}} \right) C_{mt} \ \text{l/s} \qquad (2.8)$$

Here, C_{vt} is calculated with eqn. 2.4, C_{mt} is calculated with eqn. 2.7, D is the diameter of the tube (mm), and $P_{av} = (P_{hi}+P_{lo})/2$ is the average gas pressure in the tube (Pa). Hablanian (Section 4.5.3) presents graphs that show the variation of the conductance of tubes with diameters ranging from 10 to 80 mm over the transitional range. Note that the above equation suggests that flow characteristics are predominantly viscous in character in the upper part of the transitional range, and predominantly molecular in the lower part of the range. Therefore, when making calculations of pumpdown times for most practical vacuum systems it is usually acceptable to assume that viscous flow occurs down to a pressure of about 1.0 Pa (10^{-2} Torr), and that molecular flow prevails below that, in effect ignoring the transitional region. The inaccuracies in conductance values this approach introduces do not seriously affect the overall results obtained for most practical purposes because of the relatively narrow range of time and pressure over which transitional flow prevails. Furthermore, the physical characteristics of most systems are not well defined over the range of transitional flow because valve operating procedures are usually performed at about the time transitional flow is encountered, and this interrupts the pumpdown process momentarily and drastically changes the system's configuration. If, however, calculations need to be made for a system that operates steadily at pressures in the range of transitional flow, it would be necessary to use conductance values derived from the above equation or from the graph presented by Hablanian.

Berman (Chapters 4, 5, 6, and 7), Hablanian (Chapter 4), O'Hanlon (Chapter 2), Roth (Chapter 3) and Van Atta (Chapter 2) discuss the derivations, applications, and limitations of these conductance equations in some detail, and also present equations for the conductance of traps, baffles, and non-circular tubes.

2.5 *Pumping speed*

The pumping speed S_p of a vacuum pump is defined as the volume of gas entering the inlet of the pump per unit of time. Pumping speeds have traditionally been expressed in litres per minute for rough vacuum pumps, and in litres per second for high vacuum pumps. If P_p is the pressure at the inlet plane of the pump (Pa), and Q is the throughput (in Pa-l/min or Pa-l/s), then the speed of the pump is given (in l/min or l/s) by:

$$S_p \approx \frac{Q}{P_p} \tag{2.9}$$

If a pump is connected to a vacuum chamber by a pumping line consisting of tubes, traps, baffles, and valves, as indicated schematically in Fig. 2.4, the rate at which the pump removes gas from the vacuum chamber will be influenced by the ease with which the gas molecules can flow through this line as given by its conductance C. As a consequence, there will be an effective pumping speed S_e at the outlet of the chamber, often called the *speed of evacuation*, equal to:

$$S_e \approx \frac{Q}{P_c} \tag{2.10}$$

where P_c is the pressure in the chamber. This can be related to C and S_p as follows:

$$\frac{1}{C} = \frac{P_c - P_p}{Q} = \frac{P_c}{Q} - \frac{P_p}{Q} = \frac{1}{S_e} - \frac{1}{S_p} \tag{2.11}$$

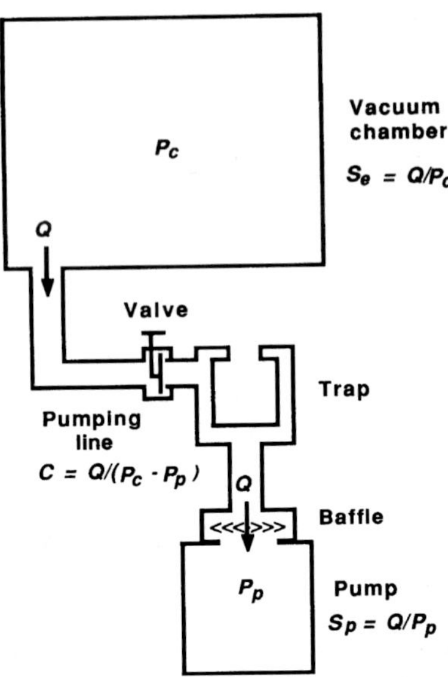

Fig. 2.4 The components of a vacuum system and the basic relations for evacuation processes. P_c, the pressure in the vacuum chamber; P_p, the pressure at the pump; S_e, the speed of evacuation at the vacuum chamber; Q, the gas throughput; and C, the overall conductance of the pumping line which consists of the trap, valve, baffle and tubing between the pump and the vacuum chamber.

This fundamental relationship can be rearranged into the following convenient form:

$$S_e = \frac{S_p C}{S_p + C} \tag{2.12}$$

This is a very important and informative equation because it shows the critical influence that the conductance of the pumping line has on the speed of evacuation. In effect, the conductance of the pumping line can be considered as degrading the pumping speed of the evacuation system. If there were no such effect then the speed of evacuation would equal the speed of the pump (i.e. $S_e = S_p$), and the ratio of these two quantities would be unity (i.e. $S_e/S_p = 1$). The extent of the degradation is then indicated by the extent to which this ratio is less than one. The value of this ratio is plotted as a function of the ratio of the conductance of the pumping line to the speed of the pump C/S_p in Fig. 2.5. Note that even when the conductance of the line is three-times the speed of the pump

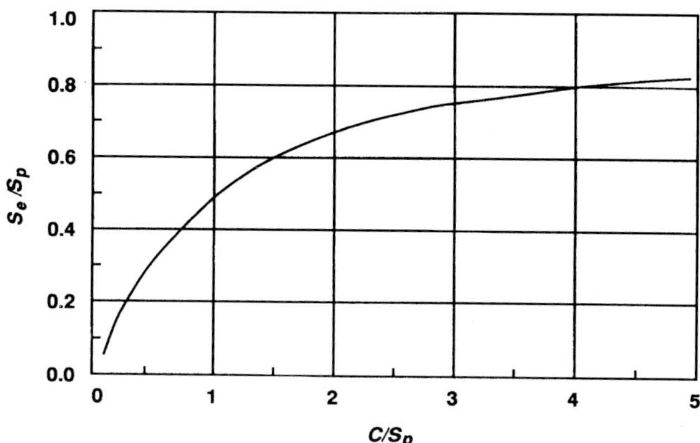

Fig. 2.5 The effect of conductance C in reducing the speed of evacuation S_e relative to the speed of the pump S_p.

(i.e. $C/Sp = 3$), the speed of evacuation is only three-quarters of the speed of the pump (i.e. $S_e/S_p = 0.75$).

The viscous conductance of ordinary pumping lines can be quite large, giving correspondingly large values for C/S_p. As a consequence, S_e can be nearly equal to S_p for rough pumping operations. For example, the vacuum system of a scanning electron microscope, such as the one shown in Fig. 2.6, might typically use a roughing pump with a speed of about 100 l/m with a roughing line about 2.5 m long and 25 mm in diameter. From Fig. 2.2 we find that the conductance of this line at an average pressure of 10^2 Pa (1 Torr) is about $C_{line} \approx 3500/2.5 = 1400$ l/m, giving $C/S_p \approx 14$. Then $S_e \approx 100(1400)/(100+1400) \approx 93$ l/m, or about 90 percent of the speed of the pump.

The molecular conductance of the pumping line for the high vacuum pump on most vacuum systems is usually considerably less than the speed of the pump (i.e. $C/S_p<1$), however, and this causes the speed of evacuation in the high vacuum range to drop to a small fraction of the speed of the pump. Consider again the vacuum system illustrated by Fig. 2.6, for example. Here the specimen chamber and electron gun are evacuated through a manifold about 0.75 m long and 80 mm in diameter by a high vacuum pump working through the high vacuum valve V1 and providing a pumping speed of about 600 l/s. From Fig. 2.3 we find that the molecular conductance of this manifold is about $C_{mnf} \approx 70$ l/s ($C/S_p = 70/600 = 0.12$). The electron gun is connected to this manifold through a tube 25 mm in diameter and 0.15 m long

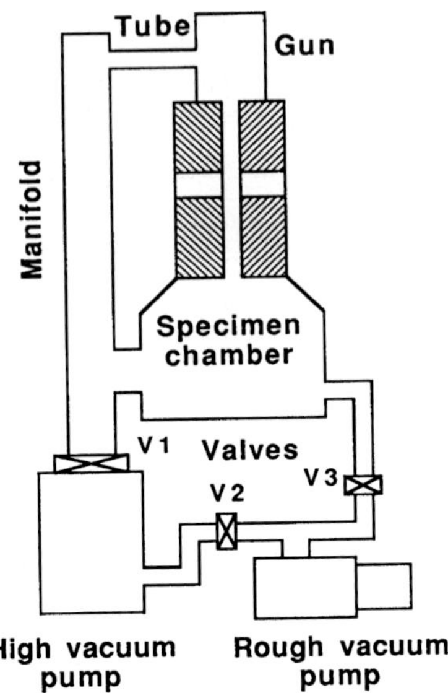

Fig. 2.6 A schematic diagram of the vacuum system for a scanning electron microscope.

with a conductance of only about $C_{tube} \approx 10$ /s ($C/S_p = .02$). According to eqn. 2.2, the resulting conductance to the electron gun is:

$$C_{gun} = \frac{1}{1/C_{tube} + 1/C_{mnf}} = \frac{1}{1/10 + 1/70} \approx 9 \quad \mathrm{l/s}$$

so that overall C/S_p for the electron gun has a value of only 9/600 = 0.015.

It is important to note that the conductance for evacuation of the gun is totally dominated by the small value of the conductance of the tube between the manifold and the gun. Using eqn. 2.12 we find that the conductance of this tube also severely limits the speed of evacuation at the gun:

$$S_{gun} = \frac{600 \times 9}{100 + 9} \approx 9 \quad \mathrm{l/s}$$

If the diameter of this tube were increased slightly to 40 mm its conductance would increase to about 37 l/s, whereupon both the overall conductance to the gun and the speed of evacuation at the gun would increase to about 25 l/s, a very significant improvement for a rather modest change in design.

Even the speed of evacuation of the specimen chamber is severely limited by the conductance of the portion of the manifold leading to it. This consists of a tube 80 mm in diameter and about 0.15 m long. The conductance of this tube is approximately 240 l/s, giving a speed of evacuation for the specimen chamber of:

$$S_c = \frac{600(240)}{(600 + 240)} \approx 170 \text{ l/s}$$

which is less than one-third the speed of the pump. So great is the influence of the conductance of the manifold that even if the pumping speed were doubled the speed of evacuation for the sample chamber would increase by only about 20 percent to:

$$S_c = \frac{1200(240)}{(1200 + 240)} \approx 200 \text{ l/s}$$

The overall pumping characteristics of this system in the high vacuum range could be improved considerably if it were reconfigured as shown in Fig. 2.7. Here the high vacuum pump and high vacuum valve are mounted directly onto the specimen chamber so that the speed of evacuation of the chamber becomes essentially equal to the pumping speed of 600 l/s, giving a 3-fold increase over the original design. The electron gun is still evacuated through a manifold 80 mm in diameter, which now leads from the back of the specimen chamber, but the diameter of the tube connecting into the gun itself is increased to 50 mm. Interposing the specimen chamber into the pumping line to the gun would probably have negligible effect on the net conductance to the gun, whereas the new manifold with the larger diameter tube leading into the gun increases the speed of evacuation at the gun to about 35 l/s, nearly a 4-fold increase over the original design.

In considering the effect of the sizes of tubes and pipes on the pumping characteristics of vacuum systems, it is of the utmost importance to remember that conductance increases with the fourth power of the diameter in the viscous flow range (eqn. 2.4) and with the third power of the

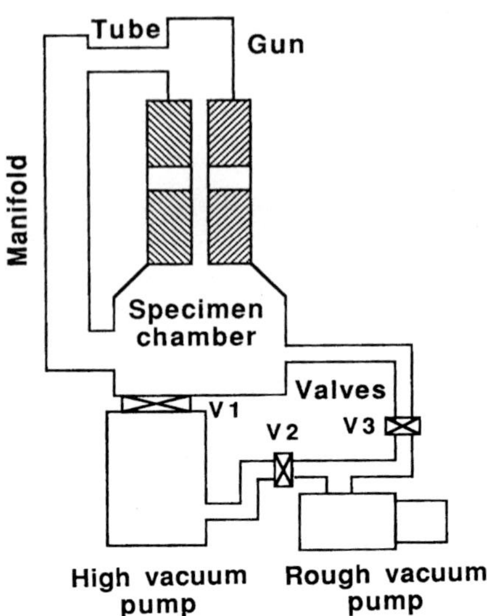

Fig. 2.7 A rearrangement of the vacuum system for the scanning electron microscope shown in Fig. 2.6.

diameter in the range of molecular flow (eqn. 2.7). Therefore, seemingly small differences in diameters can make surprisingly large differences in conductances and pumping speeds. While the effect of length is important, it is only to the reciprocal first power, and is therefore considerably less drastic than that of diameter, and so a modest increase in the diameter of a pumping line can often offset a rather considerable increase in its length.

These principles are illustrated by modifications made on an instrument purchased several years ago which had a 300-litre vacuum chamber, and which, as purchased, was evacuated from atmospheric pressure to about 10 Pa (10^{-1} Torr) by a large mechanical pump ($S_p \approx 400$ l/min) through about 1 m of rubber tubing 25 mm in diameter. Fig. 2.2 shows that the viscous conductance of this tube, evaluated at $P_{av} = 100$ Pa, is about 3000 l/min. During the evacuation process the mechanical pump discharged large quantities of oil mist into the laboratory, creating a situation that was both unpleasant and unhealthy. To eliminate this, the pump was moved to an adjacent well-ventilated and little-used room, and connected to the chamber by 6 m of copper tubing 50 mm in diameter. Fig. 2.2 shows that a tube of this diameter with a length of 1 m has a conductance of about 5×10^4 l/min at an average pressure of 100 Pa. Because viscous conductance varies with the inverse of the tube length, the conductance of this new pumping line was about $(5\times10^4)/6 \approx 8000$ l/min, which is more than twice the conductance of the original tubing. Thus, it was possible to move the pump six times as far from the chamber without significantly decreasing the speed of evacuation by using a new pumping line with a larger, but not impracticably large, diameter. Note that the conductance of this pumping line is much larger than the speed of the pump; that is, $C/S_p = 8000/400 = 20$, giving $C = 20S_p$. Because of this, the speed of evacuation is not appreciably less than the speed of the pump [using eqn. 2.12, $S_e = S_pC/(S_p+C) = 400\cdot(8000)/(400+8000) = 380$ l/min $= 0.95\ S_p$]. As mentioned above, this is typical for many rough pumping lines.

As a general rule, the diameters of pumping lines should be as large as possible, while their lengths should be kept as short as is convenient. We will see in the following sections that the highest possible speed of evacuation is always to be desired to ensure rapid pumpdown and the best ultimate vacuum. Electron microscopes with small diameter tubes for the evacuation of the electron gun and, particularly, the specimen chamber, as in Fig. 2.6, are not well designed.

2.5.1 *The theoretical maximum pumping speed*

In closing this section, it may be useful to emphasize again the fact that pumps do not draw, suck, or attract gas molecules out of vacuum systems. Instead, pumps are basically devices that capture those gas molecules that happen to wander into them under the influence of the random, thermally-activated motion of the gas molecules. It is also interesting to note that based on this concept it is possible to derive a theoretical maximum pumping speed S_{tm} for high vacuum pumps operating in the range of molecular flow. To do this let us assume there is a material which might be called 'molecular flypaper' that is so 'sticky' that every gas molecule that strikes it sticks to it permanently. If a package containing a piece of this material were quickly opened up inside a vacuum system, eqn. 1.5b (Section 1.8, p. 23) tells us that each square millimetre of the molecular flypaper would capture about $N_{cap} = 2.86\times10^{16}P$ molecules of air per second at 20°C. Dividing by Avogadro's number converts this to $n_{cap} = N_{cap}/N_A = 4.75\times10^{-8}P$ moles of air captured per second per mm^2, and using the ideal gas equation (eqn. 1.1, Section 1.3, p. 14) gives the volume of gas captured as $V_{cap} = n_{cap}RT/P = (4.75\times10^{-8}P)\times(8314\times293)/P = 0.12$ l/s per mm^2. Note that, contrary to what one might intuitively conclude, the pressure term cancels out in this calculation. Whereas the *number* of air molecules that strike a square millimetre of area per second at a given temperature varies linearly with the pressure, the *volume* of air striking a square millimetre of a surface per second at a given temperature is independent of pressure. On this basis a piece of molecular flypaper with an area of A mm^2 would have a pumping speed of $S = V_{cap}A = 0.12A$ l/s for air at 20°C. Air molecules cannot reach any other device faster than they reach this unobstructed surface, and no other device can capture them more efficiently, and so no other device can remove them from the volume of a vacuum system faster. Therefore, this value of:

$$S_{tm} = 0.12\text{A} \quad \text{l/s} \tag{2.13a}$$

represents the maximum speed a pump with an acceptance area of A mm^2 can possibly develop when pumping air at room temperature under conditions of molecular flow. This is determined only by the rate with which the gas molecules enter the acceptance area of the pump because of their random thermally-activated motion, and is not dependent on any force exerted on the molecules by the pump nor on the pumping mechanism involved. If the pump has a circular inlet, as most pumps do, and if its size is expressed in terms of its inlet diameter D (mm), a common practice, this equation becomes:

$$S_{tm} = 0.12\pi D^2/4 = 0.09D^2 \quad \text{l/s} \tag{2.13b}$$

which is the same as eqn. 2.6c for the molecular conductance of a circular aperture C_{ma}. Roth (Sections 3.2 and 3.3) gives a more detailed treatment of this topic for both viscous and molecular flow.

As noted, the above equations apply only to the pumping of air at room temperature; the more general equation is:

$$S_{tm} = 3.64 \times 10^{-2} A\sqrt{\frac{T}{M}} \quad \text{l/s} \tag{2.13c}$$

in which T is the temperature (K) of the gas being pumped, M is its molar mass (g/mol), and A is the acceptance area (mm^2) of the pump. Again, for a pump with a circular inlet of diameter D (mm) this equation can be written as:

$$S_{tm} = 2.86 \times 10^{-2} D^2 \sqrt{\frac{T}{M}} \text{ l/s} \tag{2.13d}$$

These equations suggest that the speed of a pump can be expected to be different for gases which have different molar masses. For example, $M = 4$ for helium, compared with $M = 29$ for air, and so the pumping speed of a pump for helium could be:

$$\frac{S_{He}}{S_{Air}} = \sqrt{\frac{M_{Air}}{M_{He}}} = \sqrt{\frac{29}{4}} = 2.7$$

times greater than for air, providing both gases are pumped at the same temperature and both are captured by the pump with the same efficiency, simply because helium molecules travel faster and have longer mean free paths than air molecules.

2.6 The rate of evacuation

One aspect of working with the vacuum systems of electron microscopes that is inevitably annoying to operators is the need to wait for what may seem like an unreasonably long time for the instrument to pump down to an operating vacuum after changing specimens, film, or a filament. It will therefore be instructive to examine the factors that govern this process.

Physically, the rate of evacuation is the rate at which the pressure P_c inside a vacuum chamber of volume V_c decreases with time. Mathematically, this is given by the differential expression dP_c/dt, in which dP_c is the incremental amount by which the pressure decreases in an increment of time dt. For an ideal system this is related to the speed of evacuation S_e by the equation:

$$\frac{dP_c}{dt} = -\left(\frac{P_c}{V_c}\right) S_e \tag{2.14}$$

The rate of evacuation is usually given in units of pascals per minute in the rough vacuum range and in pascals per second in the high and ultra-high vacuum ranges, corresponding to the units used for S_e. Note particularly that this equation shows that the rate of evacuation must decrease as P_c, the pressure inside the chamber, decreases, assuming the speed of evacuation remains constant. This means that the lower the pressure gets, the slower the evacuation process becomes. This accounts at least in part for the fact that it often seems to take forever for the pressure to drop that last little bit before the filament can be turned on.

Actual vacuum systems, particularly those as complex as are found in electron microscopes, are not ideal because they are neither free from leaks nor perfectly clean. Therefore, the decrease in pressure being effected by the pump is inevitably opposed by an influx of gas into the system from such sources as leaks, the desorption of gas molecules from the interior walls of the system, the diffusion of gas through gaskets and even the walls of the system, the back-diffusion of gas through the pump itself, and the evolution of gas by gaskets, plastics, and other materials in the system. This influx of gas q (expressed in Pa-l/min or Pa-l/s, depending on the units used for S_e) opposes the evacuation process, in effect contributing a pressure rise equal to q/V_c. When this is taken into account the above equation for the rate of evacuation becomes:

$$\frac{dP_c}{dt} = \frac{q}{V_c} - \left(\frac{P_c}{V_c}\right)S_e \tag{2.15}$$

Because, as explained above, the rate at which gas molecules are removed from the system $(P_c/V_c)S_e$ decreases as P_c decreases, it ultimately drops to the same level as the rate of gas influx, whereupon $q/V_c = (P_c/V_c)S_e$. This makes $dP_c/dt = 0$, and so no further decrease in pressure occurs. The pressure existing in the chamber at this point is called the *ultimate pressure* P_u. Substituting P_u for P_c in the equation gives $q/V_c = (P_u/V_c)S_e$ and rearranging the terms yields the very important relationship:

$$P_u = \frac{q}{S_e} \tag{2.16}$$

This shows that the ultimate pressure attainable in a vacuum system is fundamentally determined by the level of gas influx and the speed of evacuation. A low ultimate pressure requires a clean, tight vacuum system to ensure a low level of gas influx and a well designed pumping system to provide a high speed of evacuation. Now, by substituting $q = P_u S_e$ into eqn. 2.15 we obtain:

$$\frac{dP_c}{dt} = \left(\frac{P_u}{V_c}\right) S_e - \left(\frac{P_c}{V_c}\right) S_e \qquad (2.17)$$

as the final required form of the equation for the rate of evacuation.

2.7 Pumpdown time

It would be expected intuitively that the time required to pump an instrument down to the operating level would be directly proportional to the rate of evacuation and inversely proportional to the volume of the system. An approximate equation for this relationship can be obtained by rearranging eqn. 2.17 to separate the variables, and formatting it for integration as follows:

$$\int_{P_0}^{P_t} \frac{dP_c}{P_c - P_u} = -\left(\frac{S_e}{V_c}\right) \int_0^t dt \qquad (2.18)$$

As indicated, the integration is carried out over the range from the initial pressure P_o at the start of the process when time $t = 0$ to the pressure P_t that will be attained at the pumpdown time t. It is very important to note that inherent in this equation is the assumption that S_e is independent of pressure, and therefore does not change during the pumpdown process. When the integration is performed and the resulting equation is rearranged to solve for the pumpdown time, we obtain:

$$t = \left(\frac{V_c}{S_e}\right) \ln\left(\frac{P_o - P_u}{P_t - P_u}\right) \qquad (2.19)$$

where ln denotes the natural or Naperian logarithm (base $e = 2.71828$) of the quantity in parentheses that follows.

In many situations the final pressure required for operation is considerably greater than the ultimate pressure the system is capable of achieving (i.e. $P_t > P_u$). For example, the ultimate pressure attainable in an electron microscope might be in the low end of the 10^{-4} Pa (10^{-6} Torr) range, whereas it would be perfectly acceptable to turn a tungsten filament on when the pressure P_t reaches a value near 10^{-2} Pa (10^{-4} Torr). Under such circumstances $P_t - P_u \approx P_t$ and $P_o - P_u \approx P_o$, and it is possible to eliminate P_u from eqn. 2.19, which then simplifies to:

$$t = \left(\frac{V_c}{S_e}\right)\ln\left(\frac{P_o - P_u}{P_t - P_u}\right) \tag{2.20}$$

2.8 The pumpdown process

Examination of these equations provides considerable insight into the behaviour of practical vacuum systems. First of all, they show the overwhelming importance of a high speed of evacuation S_e on the performance of a vacuum system. Eqn. 2.16 shows that, for a system with a given level of gas influx q, the higher the speed of evacuation is, the lower the attainable ultimate pressure will be. Eqns. 2.19 and 2.20 likewise show that a high speed of evacuation leads to a shorter evacuation time. The discussion in Section 2.5 (p. 40) relating to eqn. 2.12 showed clearly that high speeds of evacuation require high values of both S_p and C. A well-designed vacuum system will therefore have high-speed pumps connected to the system through short pumping lines of large diameter. Unfortunately, the pumping characteristics of an instrument are pretty much determined by the manufacturer and it is usually very difficult to alter them significantly once an instrument has been purchased, although some instruments may have extra ports that make it possible to make some modifications without undue difficulty. For example, Fig. 2.8 is a photograph showing ion pumps added to ports in the specimen chamber and pumping manifold of an analytical transmission electron microscope to increase the speed of evacuation of the specimen chamber for critical analytical applications.

It is surprisingly instructive to apply eqn. 2.20 to the process of evacuating an instrument, such as the scanning electron microscope depicted

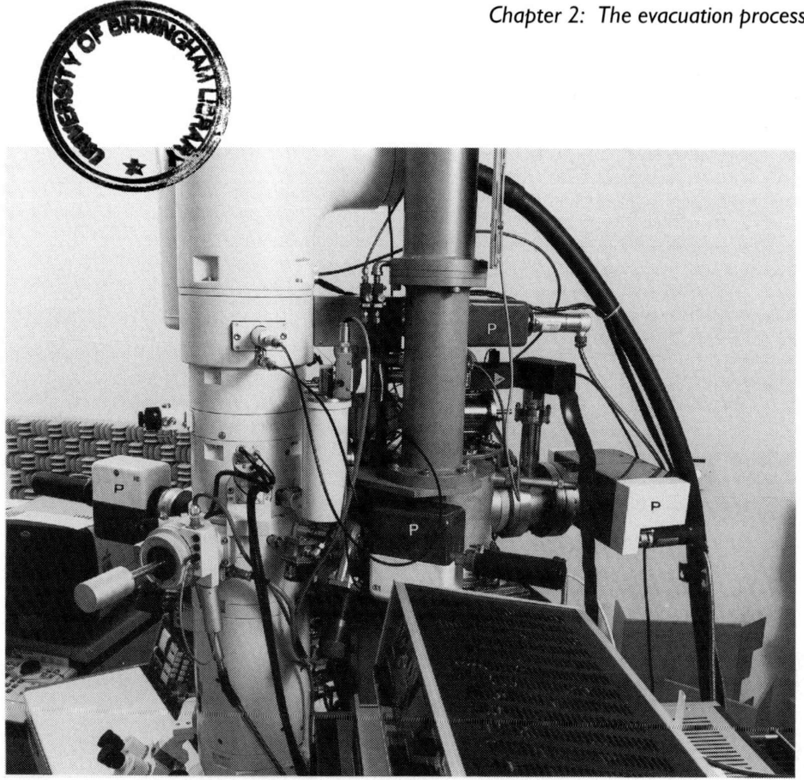

Fig. 2.8 Supplementary ion pumps (P) added to an analytical electron microscope. (Courtesy of A.J. Mardinly, Intel Corporation.)

schematically in Fig. 2.6, after exchanging specimens. Assuming air is admitted to the whole vacuum system in this exchange process, as is common practice for many scanning electron microscopes (SEMs), the first stage of evacuation would involve rough pumping the system from atmospheric pressure down to about 10 Pa (0.1 Torr) using the rotary-vane roughing pump, with valves V1 and V2 closed and V3 open. We now encounter a basic shortcoming of eqn. 2.20. In carrying out the integration of eqn. 2.18 it was assumed that the speed of evacuation S_e remains constant. In practice, however, both the speed of the pump and the conductance of the pumping line usually vary as the pressure is reduced, possibly causing a significant variation in S_e. This shortcoming can be largely offset by dividing the pressure range into small increments i over which these variables are 'practically' constant, calculating the time of evacuation for each increment t_i, and summing to obtain a total time.

Table 2.2 summarizes the application of this incremental procedure to the evacuation of the SEM of Fig. 2.6 from atmospheric pressure down to

Table 2.2 Calculation of the pumpdown time for a rough pumping operation

i	P_o	P_t	P_{av}	C	S_p	S_e	t_i
1	10^5	10^4	5×10^4	600 000	100	100	0.6
2	10^4	10^3	5×10^3	60 000	100	100	0.6
3	10^3	10^2	5×10^2	6 000	100	98	0.6
4	10^2	10^1	5×10^1	600	100	85	0.7

Total pumpdown time - - - - - - - - 2.5 min

P_o, P_t, and P_{av} are, respectively, the initial, final and average pressures (Pa) for the i incremental steps in the pumpdown process. C is the viscous conductance of the tube (in l/min from Fig. 2.2), and S_p is the speed of the pump (l/min), both at P_{av}. S_e is the speed of evacuation (in l/min from eqn. 2.12) and t_i the pumpdown time for the increment calculated using eqn. 2.20.

10 Pa (0.1 Torr), assuming that the volume of the SEM is about 25 l, and that it is being evacuated by a rotary-vane pump with a rated speed of 100 l/min through a roughing line 25 mm in diameter and 2.5 m long. Manufacturers' literature indicates that the speed of such a pump remains essentially constant over this entire pressure range. The conductance of the tubing varies considerably, but is so great at all pressures that its variation causes only a modest decrease in the speed of evacuation in the lowest pressure increment.

This calculation predicts a rough pumping time of about two and a half minutes, and it would probably be found that this would correspond to the actual pumpdown time within a factor of less than two. This is an acceptable outcome considering all the uncertainties and limitations involved in calculations and measurements of this kind.

When the same approach is used to calculate the time to pump the system on down to an operating pressure of 10^{-3} Pa (10^{-5} Torr) using the diffusion pump, with valves V1 and V2 open and V3 closed, the results shown in Table 2.3 are obtained. In this range, conditions of molecular flow prevail and so the conductance of the pumping line remains constant, in accordance with eqn. 2.7, but the speed of the pump increases considerably from the first pressure increment to the last, giving a welcome 2-fold increase in S_e. However, anyone familiar with operating an electron micro-

Table 2.3 Calculation of the pumpdown time for an evacuation process in the high vacuum range

i	P_o	P_t	P_{av}	C	S_p	S_e	t_i
1	10^1	10^0	5	240	100	70	0.82
2	10^0	10^{-1}	5×10^{-1}	240	300	130	0.43
3	10^{-1}	10^{-2}	5×10^{-2}	240	500	160	0.35
4	10^{-2}	10^{-3}	5×10^{-3}	240	600	170	0.34

Total pumpdown time - - - - - - - - - - - - 2 s

P_o, P_t, and P_{av} are, respectively, the initial, final and average pressures (Pa) for the i incremental steps in the pumpdown process. C is the molecular conductance of the manifold (in l/s from Fig. 2.3) and S_p is the speed of the pump (l/s), both at P_{av}. S_e is the speed of evacuation (l/s from eqn. 2.12) and t_i is the pumpdown time for the increment calculated using eqn. 2.20.

scope will immediately comment that the calculated pumpdown time of only 2 s is totally unrealistic. In practice the actual pumpdown time would probably be in the range from 10 to 20 min, a factor of several hundred times greater than this calculated time. This is a major discrepancy and it arises because the effect of q, the influx of gas into the vacuum system during evacuation, that was introduced into eqn. 2.15 was subsequently neglected in the simplifying assumption used to obtain eqn. 2.20 from eqn. 2.19. This did not cause a serious error in the calculations for the rough pumping operation summarized in Table 2.2, because over the pressure range involved there q is negligible compared to Q for any high vacuum system in good operating condition. For example, $Q = P_c \cdot S_e \approx 10\times87 \approx 870$ Pa-l/min (9 Torr-l/min) at 10 Pa (0.1 Torr) in this SEM. It would take a phenomenal leak (or perhaps a big dab of wet silver paint on the specimen stub) to produce a value of q of comparable magnitude. In the high and ultra-high vacuum ranges, however, Q is much smaller (e.g. 1.7×10^{-1} Pa-l/s at 10^{-3} Pa) and q becomes a significant factor even in the best vacuum systems. Actually, our calculated pumpdown time is probably a pretty good estimate of the time that would be required to remove the free atmospheric gas from our SEM. The seemingly endless wait we are all accustomed to enduring while the last increment of vacuum is being established is due largely to gas influx q. We will examine gas influx in more detail in the sections that follow, and make a more realistic calculation of pumpdown time by taking it into account in Section 2.10.3 below. Hablanian (Section 4.3.4) gives an equation for pumpdown time which is equivalent to eqn. 2.19, but which includes terms that allow for an outgassing rate that decreases exponentially with time.

2.9 Managing vacuum systems

We have just seen that gas influx is *the* critical factor that determines two of the most important operating characteristics of a given vacuum system: the time required to pump it down to an operating vacuum, and the ultimate level of vacuum that can be attained in it. Shortly, we will also see that gas influx can be a major factor contributing to specimen contamination in an electron microscope. It is obviously desirable to manage a vacuum system so as to minimize the deleterious effects of gas influx. In fact, unless you are willing to make physical modifications in your vacuum apparatus, about all you can do to optimize its performance is to control gas influx processes. It will therefore be of considerable practical value to examine in some detail the characteristics of the phenomena that contribute to gas influx, and to consider their operational implications carefully, because this will provide useful, practical guidelines for managing vacuum systems to optimize their performance.

Before proceeding with this discussion I would like to confess that the heading ‘Managing vacuum systems’ was inserted above primarily to trap impatient readers (i.e. those who are trying to get some practical information without first putting in the effort needed to develop an understanding of the underlying principles) into reading the material presented in this chapter. Otherwise, I am sure that many of you would not have bothered to read this section. If the title were ‘Gas influx’, as it probably should be, many of you would have said, “Why bother to read about such a far-out topic when all I want to do is to find out how to get a better vacuum in my evaporator?” As asserted above, and as will be demonstrated below, many important practical operating principles are related to this phenomenon. However, I think you will end up with a much better understanding of these principles if you are reasonably familiar with the basic relationship between gas influx and the evacuation process that was developed in the preceding four sections. You will also be better able to appreciate the discussions given in Chapter 9 of the operation of the valves in various types of vacuum systems if you have a reasonable understanding of the differences in the characteristics of gas flow in the rough and high vacuum ranges that was presented in Sections 2.2 (p. 29) and 2.3 (p. 32). If you are not fairly confident of your understanding of the topics covered in the preceding sections of this chapter, please go back and read them now. They are basically descriptive and require only a very minimum of mathematical acuity.

2.10 Gas influx

The influx of gas into vacuum systems is of great importance to electron microscopists for two reasons. First, it is a major source of the organic materials that interact with the electron beam to produce carbonaceous contamination of the specimen. Secondly, gas influx has an overwhelming effect on the pumpdown rate for a given vacuum system and the ultimate vacuum that can be attained in it.

Gas influx is a highly complex process that is generally considered to involve contributions from at least four separate phenomena, and therefore to consist of four more or less independent components:

q_p — gas entering the system by permeation through gaskets and the walls of the system;
q_l — gas entering the system through leaks;
q_d — gas that desorbs from surfaces inside the system; and
q_e — gas evolved from materials inside the system by physical and chemical processes other than desorption.

The total influx q is, of course, the sum of these individual components. That is:

$$q = q_p + q_l + q_d + q_e \tag{2.21}$$

Note that it is the magnitude of this total influx q relative to the value of the throughput Q for the evacuation process that is of importance. This is because q gives the total number of gas molecules entering the volume of the system per second by these various influx processes, while Q is the number being removed from the system per second by the pumps. Both q and Q are normally given in units of Pa-l/s, a flow of 1 Pa-l/s being equivalent to the movement of 2.47×10^{17} molecules/s, as shown in Table 1.2 (Section 1.3, p. 16). We can calculate Q for an evacuation process using eqn. 2.10; that is, $Q = S_e P_c$. For example, using the data given in Table 2.3 for the evacuation of our SEM at 10^{-3} Pa (10^{-5} Torr), the lowest pressure considered there, we find that $Q = 170\times10^{-3} \approx 1.7\times10^{-1}$ Pa-l/s. It would decrease to 1.7×10^{-3} Pa-l/s at 10^{-5} Pa (10^{-7} Torr), the bottom of the high vacuum range, and the value of q would have to be less than this for the pressure to reach this level. To reach a pressure of 10^{-6} Pa (10^{-8} Torr), entering the ultra-high vacuum range, the level of gas influx would have to be less than 1.7×10^{-4} Pa-l/s. Again we can see the overwhelming importance of having high-speed pumps and pumping lines with large conductance to provide a high speed of evacuation. For example, we estimated that reconfiguring our SEM as shown in Fig. 2.7 would provide a speed of evacuation approaching 600 l/s. This would give a value for the throughput at 10^{-6} Pa of $Q = 600\times10^{-6} = 6\times10^{-4}$ Pa-l/s, and make a 3-fold higher level of gas influx tolerable. It is very difficult and quite expensive to increase S_e by a factor of 10 or 100, however, and so in the end it is q that must be controlled for systems that must reach very low pressures.

Berman (Chapter 8), O'Hanlon (Chapter 6), Roth (Chapter 4), and Weston (Chapter 2) describe the theoretical aspects of gas influx in some detail, and Berman (Chapter 8) and O'Hanlon (Appendix C) give tables of outgassing rates for a variety of materials. The following discussion is more concerned with those practical aspects of the phenomena which apply to the operation of electron microscopes and related equipment. In this discussion values of q will be given in units of pascal-litres per second per square millimetre of surface area. Factors for converting to some other commonly-used units are: 1 Pa-l/s·mm² = 10^6 Pa-l/s·m² = 10^3 Pa-m³/s·m² = 1 mbar-l/s·cm² = 0.75 Torr- L/s·cm² = 0.92 SCC/s·cm².

2.10.1 Gas permeation

Gas permeation refers to the diffusion of gas molecules through the solid materials that comprise the envelope of the vacuum system. The values given for the permeation of atmospheric gases at room temperature through most metals used in the construction of vacuum systems are well below 10^{-10} Pa-L/s per mm^2 of area for a wall thickness of 2 mm. This influx is many orders of magnitude too small to be of significance in even the best vacuum systems likely to be encountered in electron microscopy. However, since the rate of gas permeation increases exponentially with temperature, it can become significant in some special types of apparatus that operate at high temperatures in the ultra-high vacuum range.

The rate of permeation of atmospheric gases at room temperature through the rubber gaskets and O-rings used in easily demountable vacuum seals, and in seals that transmit motion through the vacuum wall (stage controls, aperture centring devices, etc.), is generally given as being in the range from 10^{-8} to 10^{-6} Pa-l/s per mm^2 of area for a thickness of 1 mm. Such values become significant at pressures below the high vacuum range. Consequently, rubber O-rings and gaskets are used only very sparingly in electron microscopes that are designed to attain pressures in the 10^{-5} Pa (10^{-7} Torr) range, and are not used at all in surface analysis instruments and similar apparatus designed to operate in the ultra-high vacuum range.

The only factor that has an appreciable effect on the rate of gas permeation is the difference between the gas pressure inside and outside the system. Since this difference does not change significantly once a modest vacuum is established, q_p does not vary with time during the evacuation process. Because of this, and because q_p is usually much smaller than Q, especially in the rough vacuum range, it usually does not have a noticeable effect on the rate of pumpdown, but fundamentally serves to limit the ultimate pressure attainable. Operationally, the rate of gas permeation is determined by the design and construction of an instrument and usually cannot be altered significantly by the operator or owner of it.

2.10.2 *Gas leaks*

Although gas leakage and gas permeation might seem to be the same phenomenon, they are fundamentally different. Gas permeation is determined by the characteristics of the materials used in constructing a vacuum system, while gas leakage is determined by how well the system is designed and constructed and how carefully it is used.

There are numerous ways in which leaks can arise during the manufacture of a vacuum system. One of the most common is for a pinhole to form during the casting or welding of a metal part. One of my graduate students once spent several days trying to get a good vacuum in a device he was constructing, working on the assumption that he had assembled it improperly. Finally, a few hours of systematic testing revealed the source of the trouble to be a pinhole in the body of one of the valves. Leaks can also arise from poor design. On one fairly sophisticated instrument we found a tremendous leak around a screw that had been threaded through the wall of the main vacuum chamber. An attempt had been made to achieve a vacuum seal by placing an O-ring around the threads of the screw on the outside of the wall and tightening a nut down against it. By stopping this and 22 other smaller leaks we decreased the pumpdown time by a factor of two, improved the ultimate vacuum by nearly two orders of magnitude, and obtained a 3-fold increase in filament life. In all fairness, it must be acknowledged that manufacturers of electron microscopes presently follow sound design principles and employ good manufacturing methods, and so problems of this kind are now rare. Even so, the first few units of one recent model high-resolution electron microscope developed a small leak when the current to the lenses was turned off for a few hours, because one of the joints in the microscope column opened slightly due to thermal contraction caused by cooling of the lenses. Once the problem was detected and its source located, the manufacturer quickly changed the design to eliminate it.

It is not at all difficult to generate leaks while using or servicing an instrument, however, and it is here that the operator can have a significant influence. Devices such as aperture controls, specimen exchange mechanisms, and stage drives are prominent sites for leaks. The O-rings and gaskets in these devices must be lubricated to maintain a proper seal, and are subject to wear and deterioration. When such devices are used they should be used with sensitivity. Above all, they should be operated slowly to allow time for the gaskets to adjust to the motion and to maintain a good vacuum seal. Driving them rapidly can cause wear and generate leaks, particularly if the gaskets are only marginally lubricated. Such devices should be taken apart periodically (perhaps yearly, or more often if use is heavy), and the O-rings in them should be cleaned or replaced, and properly lubricated (see Section 10.11, p. 455).

When airlock doors, electron guns, and similar devices are opened, great care should be exercised to keep their gaskets clean. Before such devices are closed or reassembled, the gaskets should be inspected with a 5× or 10× magnifier to be sure they are clean. A single hair lying across a gasket can prevent the vacuum from going below about 10^{-1} Pa (10^{-3} Torr). Particles of dust and tiny fibres of lint, which can also generate leaks, almost seem to be attracted to lubricated gaskets that are exposed to the atmosphere. I have often wondered if this may not be a manifestation of the 'fifth force' physicists have been looking for to complete the 'Grand Unified Theory'. This can be a particularly serious problem with the specimen rods of side-entry specimen stages, because the airlock systems of these stages are generally very difficult to disassemble and clean if lint fibres and dust particles get transferred to the gaskets inside them. The entire specimen rod should be inspected with a 10× magnifier or a binocular microscope to be sure it is free of such materials each time before it is inserted into the stage.

Like the rate of gas permeation, the leak rate is usually independent of the pressure inside the system, and thus does not vary with time during pumpdown. Consequently, if the leak rate is very small it will only cause a small increase in the ultimate level of vacuum attainable, and will not cause an operationally annoying increase in the pumpdown time. If, however, the leak rate becomes high enough so that P_u approaches the desired operating pressure P_t, the pumpdown time can increase noticeably, as indicated by eqn. 2.19.

Users should be aware at all times of the fact that it is possible to generate a leak in an instrument by improper operating and maintenance practices. It is helpful to equip an electron microscope with a Bayard–Alpert vacuum gauge that gives a numerical display of the pressure (see Section 3.2.1, p. 93), and to record periodically both the pumpdown time and the ultimate vacuum attainable under standard conditions. This will allow any deterioration in performance to be detected early, whereupon the leak detection methods described in Chapter 10 can be used to obtain a reliable diagnosis.

2.10.3 *Gas desorption*

When gas molecules collide with the surface of a solid they can become attached to it by weak secondary bonding forces. Since a well-defined chemical reaction product is not formed between the gas and the solid, this is usually considered to be a physical process which is called *gas adsorption*.

As pointed out in Section 1.9 (p. 23), the surfaces of most solids become completely covered with adsorbed molecules in a fraction of a second at pressures above the high vacuum range. The term 'gas desorption' refers to the release of these adsorbed gas molecules from surfaces inside the vacuum system while the system is being evacuated. Of the common atmospheric gases, water is the most strongly adsorbed. Because water molecules are highly polar they not only bind strongly to most solid surfaces, but once a surface itself is covered they also bind to the water molecules already attached to it. As a consequence, as many as 50 to 100 layers of water molecules can be adsorbed onto the surface of a solid that has been exposed to the atmosphere at moderate levels of humidity.

The adsorption of gas molecules occurs on all surfaces inside a vacuum system whenever it is opened to the atmosphere. When it is subsequently pumped down, the free gas molecules are pumped out first and then the adsorbed molecules are slowly released into the system. The calculations summarized in Tables 2.2 and 2.3 indicate that removal of the free gas molecules can be accomplished surprisingly quickly. Removal of the adsorbed molecules is a much slower, and a much more difficult process, however. The data available generally indicate that the rate of desorption of gases from unbaked metal surfaces varies approximately linearly with time after the first few minutes of evacuation, reaching values after 1 h of evacuation at room temperature that range from 10^{-9} Pa-l/s per mm^2 of surface area for polished stainless steel to 10^{-6} Pa-l/s per mm^2 of surface area for slightly rusty mild steel. Mathematically the variation in the rate of desorption with pumping time can be described by an equation of the form:

$$D_t \approx D_u + D_r \left(\frac{t_r}{t} \right)^x \tag{2.22}$$

Here D_t is the rate of desorption to be expected after pumping time t, D_r is a reference value known to apply after pumping time t_r (usually the value after 1 or 10 h of pumping), D_u is the ultimate rate, and x is usually taken as having a value of one for metals and 0.5 for non-metallic materials.

By using this equation with the approximate desorption rates given above it is possible to make a rough estimate of the contribution of gas desorption to a pumpdown process, such as the one involved in the calculations of Table 2.3. In that example the speed of evacuation S_e at the desired operating pressure of 10^{-3} Pa (10^{-5} Torr) was estimated to be about 170 l/s, giving a rate of gas removal from the system at this pressure of $Q = S_eP = 170\times10^{-3} = 1.7\times10^{-1}$ Pa-l/s. Clearly, the desired operating pressure cannot be attained until the rate of gas desorption q_t decreases to this level; that is, until $q_t = D_t A_i < 1.7\times10^{-1}$ Pa-l/s. Assuming the internal surface area A_i of the SEM to be about 10^6 mm², this requires that D_t decreases to a value of $D_t = q_t/A_i \approx 1.7\times10^{-1}/10^6 \approx 1.7\times10^{-7}$ Pa-l/s per mm². Using eqn. 2.22, assuming a median value for D_r of 5×10^{-8} Pa-l/s per mm² when t_r = 60 min, considering D_u to be negligibly small, and taking $x = 1$, gives:

$$t \approx \frac{D_r \cdot t_r}{D_t} = \frac{(5 \times 10^{-8}) \cdot 60}{1.7 \times 10^{-7}} \approx 18 \text{ min}$$

as the pumping time required for the desorption rate to decrease to this level. This is a new estimate of the time needed to pump our SEM down to the desired operating vacuum of 10^{-3} Pa (10^{-5} Torr). When compared to the time of 2 s calculated in Table 2.3 for removing the free gas molecules from the system, this shows the overwhelming influence that gas desorption can have on the pumpdown process. It now becomes quite clear why it can seem to take forever to pump the bell jar of a vacuum evaporator down after it has been standing open for an hour or so on a humid summer day, especially if its interior is coated with a thick layer of evaporated carbon which can adsorb truly tremendous quantities of water and other gases.

It must be appreciated that calculations of the type just carried out are likely to be very imprecise for at least two important reasons. First, it is almost impossible to estimate the internal surface area of an instrument such as an electron microscope. Even if the apparent area could be measured with some accuracy, the true effective area for gas adsorption would not be known since this depends strongly on such factors as surface roughness and surface treatment. Data available indicate that the effective area for gas adsorption can be from 10 to 100 times greater than the apparent area. Secondly, the capacity of surfaces to adsorb gases and the rate at which they subsequently release them during evacuation both depend strongly on the preparative treatment the surfaces received before being put into use. Because of this, it is now common practice to polish the interior surfaces of ultra-high vacuum apparatus, using chemical or electrolytic methods that have been developed especially for this purpose, to reduce gas desorption effects.

In addition to having a strong effect on the pumpdown time, gas desorption can be the principal factor determining the ultimate pressure attainable in a vacuum system. This is so because of the large amount of gas, particularly water, that can adsorb onto most metallic surfaces, the tenacity with which it adheres, and the low rate at which it desorbs. In order to remove adsorbed molecules from solid surfaces it is necessary to overcome the secondary bonding forces that hold them there. One way to do this is to increase their thermal vibrational energy by heating the surface. Since the rate of desorption increases exponentially with temperature, a few hours of pumping on a system at an elevated temperature will be equivalent to a great many hours of pumping at room temperature insofar as removing adsorbed gases and reducing the overall outgassing rate is concerned. For example, the gas desorption rate for clean, polished stainless steel decreases from about 10^{-7} Pa-l/s per mm^2 in the unbaked condition to below 10^{-10} Pa-l/s per mm^2 after baking for one day at 250°C. Therefore, it is common practice to *bake out* vacuum systems to improve their ultimate attainable level of vacuum. Baking is usually accomplished by placing heating tapes, an oven, or a heating mantle around the outside of the system, although heating elements and high intensity lamps are sometimes installed inside the system.

Even modest heating can be beneficial. Some models of electron microscopes are designed so that high currents can be run through their lens coils for a limited time while the cooling water to the lenses is shut off. High intensity lamps are currently being installed in the specimen chambers of some electron microscopes. These procedures warm the column enough to measurably reduce the pumpdown time and the ultimate pressure attainable. However, it is generally considered to be virtually impossible to attain pressures much below 10^{-5} Pa (10^{-7} Torr) in a practical vacuum system without baking at temperatures above the boiling point of water. Baking for several hours at temperatures in the range from 100 to 200°C will usually make it possible to achieve pressures in the low end of the 10^{-6} Pa (10^{-8} Torr) range. Viton rubber will tolerate these temperatures, so such a treatment can be used on properly designed systems that employ Viton O-rings sparingly for static seals. O-rings in seals that transmit motion usually cannot be baked at these temperatures because it is difficult to maintain proper lubrication. An extended bakeout at temperatures above 250°C is usually required to reach lower pressures, and temperatures as high as 450°C are often used when pressures near 10^{-9} Pa (10^{-11} Torr) must be reached. Vacuum systems must be especially designed and constructed to withstand such treatments. Stainless steel is the preferred material of construction. Rubber O-rings cannot be used; instead, metal (usually copper) gaskets are employed for static seals, and metal bellows to seal devices that transmit motion through the vacuum walls. O'Hanlon (Chapter 11), Van Atta (Chapter 9), and Weston (Chapter 6) discuss the design and construction of such systems. Most dedicated scanning transmission electron microscopes and surface analysis instruments have vacuum systems of this kind. The electron guns and the upper parts of the columns of some newer transmission and scanning electron microscopes that have field emission guns are also designed to tolerate high-temperature bakeout treatments.

Photons of ultraviolet light can also provide the energy needed to desorb gas molecules from solid surfaces. Danielson Associates offer devices that produce ultraviolet light with an optimal distribution of wavelengths for this purpose. This method can be used on systems that employ rubber O-rings and gaskets or contain other non-metallic materials, because it does not cause an appreciable increase in temperature. It is also highly effective because the ultraviolet photons are reflected around inside the system and thus tend to reach most exposed surfaces.

There are several very practical measures that the users of vacuum apparatus can take to reduce the deleterious effects of gas desorption. One obvious approach is to keep the apparatus evacuated except when it is absolutely necessary to let it up to atmospheric pressure. In this way the value for the pumping time t will become very large, the second term in eqn. 2.22 will become negligibly small, D_t will approach the ultimate value D_u, and the pressure in the system can approach its ultimate value.

Water is a very troublesome substance because, as mentioned above, it adsorbs so tenaciously and in such large quantities to most types of materials. Every reasonable effort should therefore be made to keep it from entering a vacuum system. One way to do this is to admit a dry gas, rather than ordinary moist air, into the vacuum system whenever it is necessary to bring it up to atmospheric pressure. This will minimize the introduction of water; in addition, the internal surfaces become coated with adsorbed molecules of the dry gas which then impede adsorption of any water that may get into the system while it is open to the atmosphere. The entry of water can be further reduced by covering all openings with aluminum foil, and by maintaining a slow flow of dry gas through the system, while it is at atmospheric pressure. In a relatively confined system the use of dry gas in this way can be highly effective in reducing pumpdown time when it is subsequently evacuated. However, it is of doubtful value to admit dry gas into a vacuum evaporator if the bell jar is subsequently set aside and left open to the atmosphere, or into a scanning electron microscope if the specimen chamber door is left standing open for a long period of time.

There are several ways to obtain the dry gas needed for this purpose. It is, of course, possible to dry room air by passing it slowly through a large drying tube filled with a desiccant. In practice, however, the tendency to neglect replacing the desiccant as it becomes saturated with water generally renders this method ineffective in the long run. One common approach is to use dry nitrogen purchased from a supplier of compressed gases. Care must be taken when ordering gas for this purpose to be sure that oil-free, dry nitrogen, not the ordinary commercial variety, is obtained. This is often not a simple matter to accomplish. Suppliers of compressed gases may not reserve tanks specifically for oil-free gases, but may put oil-free gas in them one week and dirty gas the next. The safest procedure is to purchase a tank and have it reserved for oil-free gas, after it has been thoroughly cleaned to remove existing oil contamination. It is necessary to use a pressure-reducing valve to control the pressure of the gas fed to the instrument. This valve must be one that provides sensitive control of its output pressure in the range under 100 kPa (10 psi) so that the pressure of the gas fed to the vacuum system can be accurately regulated not to exceed atmospheric pressure by more than about 10 kPa (2 psi) to avoid damaging the system. A needle valve should also be used to control the rate of flow of the gas so that it takes at least 5 min to bring an apparatus as large as an electron microscope up to atmospheric pressure. Special precautions must be taken with instruments that are equipped with recent models of energy dispersive X-ray detectors which have ultra-thin windows to avoid damage to these delicate devices. Manufacturers of these detectors usually specify the use of a demand valve that will shut off the flow of nitrogen as soon as the pressure inside the instrument equals the outside atmospheric pressure.

Since the pressure in these tanks of compressed gas may be greater than 15 MPa (2000 psi) great care must be exercised, and all prescribed safety procedures must be observed, in handling them. In particular, every tank containing compressed gas must be firmly fastened to a wall or a strong table, using a strap or chain harness specifically designed for this purpose, in such a manner that it cannot possibly tip over. If this should happen, the valve on the top of the tank can be broken off, whereupon the rapidly escaping high pressure gas can drive the tank around with great speed and force, making it a potentially destructive and lethal projectile.

The vapour pressure of water at the temperature of liquid nitrogen is below 10^{-18} Pa (10^{-20} Torr), and so the gas that is constantly boiling off any container of liquid nitrogen is about as dry as any gas can be and is well suited for use in back-filling a vacuum system. The liquid nitrogen used to cool the solid-state X-ray detectors on many SEMs and TEMs, for example, can be used in this way. This can be done by fitting a one-hole rubber stopper snugly into the Dewar flask of the detector and connecting it to the gas inlet of the vacuum system with flexible, non-collapsible tubing (ordinary polyethylene tubing will do). A flexible, inflatable, plastic container, such as a large inflatable plastic toy ball, is connected to a T-junction in this tubing by a short length of soft, highly-flexible surgical rubber tubing that has a clean slit about 100 mm long made with a sharp razor blade or a scalpel. This slit will ordinarily close tightly enough so that the nitrogen from the Dewar will flow into the plastic ball. When the ball becomes full, however, the slit will serve as a primitive pressure release valve by opening slightly and allowing the gas to escape, thereby preventing the ball from rupturing. When the gas inlet to the vacuum system is opened, the dry nitrogen in the ball will flow into the system under atmospheric pressure, and so there is no danger of over-pressurization. A small weight can be placed on the ball to sustain a slow flow after atmospheric pressure is reached, if desired. A ball 0.6 m in diameter will contain about 100 l of nitrogen, enough to fill most laboratory vacuum systems several times. This arrangement is simple and inexpensive to construct and operate, it makes double use of expensive liquid nitrogen, and a colourful plastic toy ball used in this manner will provide a topic of conversation for all visitors to your laboratory.

Items that are to be placed into a vacuum system should be kept as clean and dry as possible, again to minimize the amount of water on them. Storing such items in a clean desiccator and warming them gently with a hair dryer just before they are introduced will help considerably in accomplishing this. Handling them with clean, dry, lint-free nylon or cotton clean-room gloves will also prevent them from picking up moisture and grease from your fingers.

2.10.4 Gas evolution

The term 'gas evolution' is usually used to refer to the release of gases from objects inside a vacuum system by physical and chemical processes other than the release of physically adsorbed molecules of atmospheric gases from surfaces. Gas evolution processes are very important in the design and operation of electron microscopes, and most other vacuum apparatus. We are all familiar with the evolution of water from photographic film, and are accustomed to pre-pumping film before putting it into an electron microscope to reduce this source of gas influx (Section 9.2.3, p. 386). Even with pre-pumping, however, gas evolution from film limits the attainable vacuum to about 10^{-4} Pa (10^{-6} Torr). This problem is so serious that most electron microscopes now have a pumping system for the photographic chamber that is effectively separate from the one used for the gun and specimen chambers (Section 9.3, p. 392). The back diffusion of oil vapours from the pumping system into the specimen chamber is another serious problem involving gas evolution, because the oil molecules eventually find their way onto the specimen where they are bombarded by electrons and converted to a carbonaceous lacquer that seriously degrades image quality. This has led to major changes in the design of the vacuum systems of electron microscopes in recent years (Chapter 9).

2.10.4a Gas evolution from construction materials

Gas molecules penetrate into the body of many solids when they are exposed to atmospheric pressure, and are subsequently released if these materials are placed under vacuum. The amount of gas absorbed and subsequently desorbed in this way by metals is generally negligible insofar as most vacuum systems involved in electron microscopy are concerned. However, the evolution of gases by vacuum grease, rubber gaskets, and other polymeric materials, such as insulation on wires, can be very significant. For Viton, the type of rubber most commonly used for the O-rings in electron microscopes, this 'outgassing rate' can be less than 5×10^{-9} Pa-l/s per mm^2 of surface if the Viton has been baked for several hours at 200 to 300°C immediately before use, but can approach 10^{-6} Pa-l/s per mm^2 of surface if it is unbaked. Most other polymers have outgassing rates in the range from 10^{-7} to 10^{-5} Pa-l/s per mm^2. Harris (p. 240) gives outgassing rates for a variety of these materials (note:1 mbar-l/s per cm^2 = 1 Pa-l/s per mm^2). The principal substances usually released by polymers are water, carbon monoxide, carbon dioxide, oxygen, and nitrogen, but if they are heated to high enough temperatures plasticizers, stabilizers, and other additives may also be released, depending on the formulation of the polymer and its method of manufacture. The outgassing rate for polymers is usually considered to decrease with the time of evacuation roughly in the manner described by eqn. 2.22, with the exponent x having a value of 0.5. The use of O-rings and other items made of polymeric materials is now kept to a minimum in constructing the gun and electron optical column of most models of electron microscopes to control this source of gas influx.

2.10.4b *Gas evolution from specimens*

While it is necessary to alter the design of the instruments to counteract gas evolution from materials of construction, there are several other ways in which gas influx can be limited by appropriate day-to-day operating procedures. One of these consists of exercising care not to introduce volatile materials into an electron microscope on the specimen and specimen holder. Biological specimens present a particular problem here because they are basically organic in character and evolve volatile components during electron bombardment. In addition, they are usually embedded in a polymerized resin which may also degrade under the influence of the electron beam. These processes are particularly troublesome, because the materials evolved are mostly organic compounds which interact with the electron beam to form the carbonaceous contamination that has such a degrading effect on image quality and resolution (Section 1.1, p. 4). Biologists have to accept these problems as inherent in their work. However, these problems can be controlled well enough to obtain high quality micrographs by using cold stages (Butler and Hale, Chapter 4) and specimen anticontamination devices, and by minimizing the time the specimen is exposed to the electron beam.

Metals, ceramics, and minerals are completely non-organic in character, and so ideally it should be possible to work with specimens of these materials without encountering contamination problems involving organic materials. In practice, however, specimen contamination remains a problem due to organic materials from other sources. The specimen holder is one such source. Since it is the device in closest proximity to the specimen, every possible effort should be made to keep it scrupulously clean and free of organic materials. Fingerprints are a major source of trouble here. Various estimates have been given for the rate of gas evolution from a fingerprint, some of which are as high as 10^{-4} Pa-l/s per mm^2. These rates are comparable to the rates of gas desorption used in the calculation of pump-down time in the previous Section. Hablanian (p. 88) presents a classical calculation which suggests that it can take as long as a year to pump away a single fingerprint under typical operating conditions. Such estimates and calculations obviously have a high level of uncertainty because of the highly variable character of fingerprints. Nonetheless, it is generally accepted that fingerprints can produce significant levels of outgassing, and that they are a serious potential source of gas influx and organic contaminants. Therefore, never touch any part of the specimen manipulator mechanism that is inserted into the specimen chamber with your bare fingers. Similar care should obviously be exercised when handling the electron gun while changing a filament, and when working with other parts from inside the instrument. Always wear clean, grease-free, lint-free, cotton or nylon gloves, or the special grade of grease-free, powder-free, plastic gloves used in the electronics industry, when handling these objects. Latex or plastic gloves of the type used in medicine are not satisfactory, because they usually contain plasticizers, lubricants, and similar organic compounds which make them comfortable to wear, but which will be deposited on any item they touch. All tools used in such work should be thoroughly degreased. In fact, it is a good idea to reserve a special set of tools for this purpose, which are kept scrupulously clean and are handled only with gloves. Even when wearing appropriate gloves it is still necessary to exercise a certain amount of vigilance, because most of us have a natural tendency to unconsciously rub our nose, ears, and hair periodically, and this can quickly contaminate the gloves. Above all, when you are having difficulty getting something to fit together correctly, resist the temptation to pull off the gloves and work with your bare fingers.

2.10.4c Cleaning procedures and gas evolution

The use of grease-based polishing compounds to clean aperture holders, specimen holders, grid caps, and other parts from inside electron microscopes is a practice that is dear to the hearts of most electron microscopists, but which really makes little sense at all. In fact, it is somewhat analogous to taking a bath in a mud puddle. One of the basic objectives of the cleaning process is the removal of organic materials from internal parts of the microscope, and so it is hard to justify using a greasy substance as the primary cleaning agent. In doing so all surfaces of the parts being 'cleaned' become thickly covered with the greasy polishing compound, and it is highly likely that all screw holes, screw threads, cracks, and crevices on these parts will be filled with it. It then becomes necessary to subject all parts to several treatments with organic solvents in an ultrasonic cleaner to attempt to remove this polishing compound. Most of the grease-based polishing compounds are not readily soluble, and so it is very difficult to remove them completely. (If you insist on using these materials, it is important to check to be sure that the solvents being used are actually effective in dissolving them.) Unless the solvents are changed frequently, and unless a fresh batch, and a clean container for it, is faithfully used for the final cleaning treatment, it is quite likely that the parts will end up being covered with a residual layer of the greasy cleaning compound that, while possibly difficult to observe visually, is about as heavy as any contamination that might have been on them when they were initially removed from the microscope. In addition, there is a danger that traces of the polishing compound may remain in cracks, holes, screw threads, and crevices, from which it can slowly evolve into the microscope after the 'cleaned' parts are put back into service. A further disadvantage of this overall process is that it involves the use of large quantities of expensive organic solvents which are somewhat hazardous to use, and which present serious, and often expensive, disposal problems under prevailing stringent health, environmental, and safety regulations.

An alternative approach is to use a cleaning procedure that is entirely water-based. There are many water-soluble, liquid metal cleaners on the market that are very effective. Some care must be exercised in selecting one of these products for cleaning parts from a vacuum system, however. Most of those formulated for cleaning copper and brass are unsatisfactory because they contain a silicone compound that coats the metal surface to inhibit tarnish. Try the cleaner on a metal coin first; if the coin ends up being strongly water repellant, reject the cleaner. One product that has proven satisfactory for most metal parts found in electron microscopes, and that is widely available in hardware and department stores throughout the United States, is the 'Revere Ware' brand of stainless steel cleaner manufactured by Copper Clad Products. This company also manufactures liquid cleaners specifically formulated for copper and aluminium. In use, all areas of the part being cleaned are rubbed firmly with a swab or a pad of cloth or cotton wool moistened with the liquid cleaner, and then rinsed immediately with running hot water to remove the cleaner. Next, the part is similarly scrubbed with a hot detergent solution and is then treated ultrasonically in a strong detergent solution for 5 to 10 min to remove any particles of the polishing compound that may remain on it. If the part is made of copper, brass, or aluminum it should not be allowed to remain in contact with the metal cleaner nor kept in the detergent solution any longer than necessary or it is likely to tarnish or oxidize. (However, Noel Martin, of FEI Inc., reports that Branson's GP brand cleaning solution has minimal adverse effect on most metals.) The part is removed from the ultrasonic cleaner with clean, grease-free tweezers, forceps, or tongs, and then rinsed thoroughly in running hot water to remove the detergent solution. A sensitive criterion for cleanliness is that the part remains freely wetted by water after the detergent has been thoroughly rinsed away. Considerable care is needed, however, to prevent an initially clean part from becoming recontaminated by oil or grease from the tweezers or tongs (or fingers) that are used to handle it during these final rinsing and drying operations. Finally, the part is rinsed with a generous stream of clean isopropyl alcohol, or treated with isopropyl alcohol in an ultrasonic cleaner, to remove the water. The alcohol is not allowed to evaporate on the part but is forcibly blown off with a high-speed jet of clean, dry air or nitrogen so that it carries any dissolved material away with it rather than depositing this material on the part as it evaporates. When the cleaning is completed, the part is wrapped in aluminum foil or grease-free, lint-free tissue so that it remains clean. It is then heated with a hair drier to reduce surface moisture just before it is put back into the electron microscope. The hair drier must, of course, be kept clean and grease-free, and can only be used in a clean environment, or it will contaminate the specimen again. Parts should be sprayed with a jet of clean, dry air or nitrogen, and then inspected with a magnifying glass to be sure that they are free of dust particles, immediately before they are installed in the electron microscope.

The advantages of such a procedure are: it does not introduce extra greasy material into the cleaning process, it is fast, easy, and convenient to carry out, and it uses a minimum of organic solvents. It also involves a minimum of safety problems. It is as safe to use a detergent for this purpose as to use one for washing dishes at home, and isopropyl alcohol, which is the basic component of some grades of rubbing alcohol, is one of the safest organic solvents to work with. This cleaning method was suggested to me by Yoshio Noguchi of the JEOL Company in the 1960s, and has been used with very satisfactory results in my laboratories over the intervening years. Although I have recommended this procedure to service engineers from several companies, the devotion to the grease-based procedure is so strong that most laboratories are very reluctant to change. One engineer even tells of being told to revert to the old grease-based procedure or to leave the premises when he tried to introduce this new cleaning procedure at one customer's laboratory.

A number of very effective detergents are currently available in the United States for use in this procedure. One such is the 'Micro' brand laboratory cleaning solution produced by International Products Corporation. Another, 'Liqui-Nox', produced by Alconox, Inc., is stocked by many hospital and laboratory stores, while a third, 'Clear Magic', produced by Westley Products, can be found on the shelves of most department and automotive supply stores. Others have been developed for use in the semiconductor manufacturing industry, and can be obtained from suppliers there. Manufacturers of ultrasonic cleaners also sell cleaning solutions specifically formulated for use in ultrasonic cleaning procedures. Examples are the 'General Purpose', 'Industrial Strength', and 'Oxide Remover' cleaning solutions supplied by the Branson Corporation. All of these cleaning solutions are liquid formulations which contain wetting, emulsifying, sequestering, and suspending agents, and which rinse away readily in warm water, leaving negligible residue behind. Many of these, or substantially similar products, are available from the several companies that sell general supplies for electron microscopy. Some of these formulations are quite powerful, and will do an acceptable job of cleaning metal parts without the prior use of an abrasive. For example, the 'Clear Magic' brand cleaning solution mentioned above is capable of degreasing automobile engines, while the 'RBS-35' brand cleaning solution produced by Pierce Chemical Co. is advertised as being capable of removing silicone and Apiezon oils and greases, yet both are also used in the electronics industry. Noel Martin, of the FEI Company, reports having found ultrasonic treatment in the Branson 'General Purpose' cleaner, followed by thorough rinsing with deionized water and drying with hot air, to be an effective means of cleaning metal parts for ultra-high vacuum systems. While the cleaning solutions mentioned above are primarily those available in the United States, comparable products are certainly available in other countries, particularly from companies that deal in accessories for electron microscopy, such as Agar Scientific, Fisons, and Taab. It is definitely worth experimenting with a variety of these products to determine their properties and capabilities.

Some people have expressed a reluctance to use liquid metal cleaners of the type mentioned above because they think the abrasives in them are too coarse, and that they do not leave a sufficiently smooth and bright polish on the metal parts that are cleaned with them. It is not entirely clear that a bright shiny finish is an important requirement for parts inside an electron microscope, although smooth surfaces do generally produce lower outgassing rates than rough ones. If, however, you think a smooth bright finish is important then an alternative approach that will not affect the surface finish, but is nearly as fast, convenient, and effective, is to use a thick suspension of a fine abrasive powder in one of the detergent solutions instead of a liquid metal cleaner. Very fine calcium carbonate powder ($CaCO_3$), which can be obtained from most chemical supply companies, is often referred to as a 'soft abrasive'. Its hardness is only about 3 on the Mohs scale, and so it will not appreciably alter the finish on most metal parts, yet it does aid significantly in removing organic materials from them. Interestingly, the common kitchen cleaners that are formulated for use on fibreglass sinks and bathtubs (e.g. 'Comet Cleanser', made in the U.S.A. by the Procter and Gamble Company) do not scratch most metal parts, and are effective agents for cleaning the grid caps of electron guns and many other parts from electron microscopes. In fact, the grid caps of most electron guns can be cleaned without the use of any abrasive whatsoever. Peter B. Sewell, of LAB-6 Inc., recommends cleaning lanthanum hexaboride deposits from apertures and Wehnelt cylinders by soaking them for about a minute in a solution consisting of 1 part by volume of concentrated hydrochloric acid and 4 parts water, rinsing sequentially with water, dilute ammonia, deionized water, and finally with isopropyl alcohol, and then drying with a jet of clean, hot air. Fred Sheldon, of Agar Scientific, has found that the deposits on the Wehnelt cylinders of electron guns with tungsten filaments are readily dissolved by a solution made by diluting a 'concentrated' solution of NaOH or KOH to about 20% with ethyl alcohol. As noted above, ultrasonic treatment in some detergents will also remove tungsten deposits from grid caps.

If a somewhat more aggressive abrasive than the calcium carbonate is needed then a slurry of iron oxide (Fe_2O_3), zinc oxide (ZnO), or titanium dioxide (TiO_2) in a detergent can be used. All of these materials have Mohs hardness values near 6, and all are readily available in the form of very fine powders from chemical suppliers. Companies that handle metallographic supplies can provide aluminum oxide (Al_2O_3) powder in a range of particle sizes, down to 0.05 μm. This material has a Mohs hardness of about 9 and can be used to re-finish parts or to remove stubborn deposits, if necessary.

The use of a 'chamois leather' to clean the insulator of the electron gun and to polish other parts of an electron microscope is a practice that has been recommended in the past, but which also seems highly inappropriate. Chamois leathers are usually treated with an oil or a fatty compound to make them soft and flexible. If a chamois leather is used to wipe parts of an electron microscope it is highly likely that this greasy material will be deposited on the parts, leaving them as contaminated as when the cleaning process started, although possibly a bit more shiny in appearance. If an attempt is made to remove this greasy material from a chamois by washing it or by treating it with a solvent, it will become hard and stiff, and generally unsuitable for cleaning operations. This phenomenon is familiar to many bicycle riders who use chamois-lined riding shorts, and in recognition of it many bicycle shops sell 'chamois fat' for softening the liners again after the shorts have been washed. A better procedure for cleaning the ceramic insulator and the interior of the gun chamber of an electron microscope is to rub them firmly with lint-free, grease-free cloth pads that have been dipped in a thick slurry of isopropyl alcohol and either the soft, fine calcium carbonate powder or the very fine (0.05 μm) aluminum oxide powder described above. After all areas are cleaned in this way they are wiped repeatedly with clean pads moistened only with isopropyl alcohol to remove the polishing powder. Then all surfaces, corners, and joints are blasted with a jet of clean, dry air or nitrogen to remove any remaining loose particles. Finally, the insulator and the interior of the gun chamber should be heated with a hair dryer just before they are put back together to remove as much residual surface moisture as possible. This procedure has given satisfactory results with guns that operate at potentials up to 200 kV. The cleaning of guns for higher voltage instruments should probably be left to the manufacturer's service engineers. If they start using a chamois leather or a grease-based polishing compound, however, you have just cause for limited optimism about the outcome.

A word of *caution* is in order here. Pole pieces, magnetic shield tubes, and other parts that conduct magnetic fields are usually made of soft iron or an iron-based alloy, and will corrode very easily. Therefore, these items should not be cleaned with water or a water-soluble metal cleaner. Electron microscopes are usually constructed so that these critical parts are well shielded from bombardment by the electron beam to prevent contamination building up on them, and so they seldom need cleaning. If, however, cleaning should become necessary, first try a prolonged ultrasonic treatment in a mixture of equal parts of acetone with toluene or xylene, and then a shorter treatment in isopropyl alcohol. If this approach is not sufficient, rub the parts carefully with a slurry of fine calcium carbonate or aluminum oxide powder in isopropyl alcohol, as described above for cleaning electron gun insulators, and then use ultrasonic treatments in isopropyl alcohol to remove particles of the abrasive. When they are not in the electron microscope, pole pieces should be wrapped in grease-free, lint-free tissue and stored in a clean desiccator so that they do not corrode from exposure to atmospheric humidity. Above all, always handle pole pieces with the utmost care; the resolution of your microscope is determined by these devices.

Finally, it might be noted that cleaning procedures are a critical part of many vacuum-based processes in the electronics and plating industries, and that many other highly effective methods have been developed for use there. Guthrie (Chapter 9) and Stuart (Chapter 3) describe a number of these methods.

2.10.4d *Gas evolution problems in SEMs*

The type of work that is done with SEMs provides an abundance of opportunities for operator-induced gas evolution problems. The nice juicy insects that yield those exotic micrographs also evolve all sorts of interesting materials into the vacuum system. Many kinds of polymeric materials release solvents and plasticizers that lead to horrendous rates of specimen and aperture contamination, and may degrade the pumpdown rate and the ultimate vacuum attainable. It is a good practice to pump such kinds of specimens for several hours in a chamber of the type used to prepump film for transmission microscopes before putting them into an SEM. Polished and etched metal and ceramic specimens can also be serious offenders. These are usually embedded in a block of polymerized resin for ease of handling. They are often polished with an oil-based diamond polishing compound, and then are usually etched with a corrosive chemical of some type. If they are porous, as sintered ceramics usually are, or if there are small cracks between the specimen itself and the mounting resin, the oil or the etching reagent can enter these interstices and be carried into the SEM. The evolution of an etching reagent with hydrofluoric acid as a major constituent into your SEM is unlikely to have many beneficial effects. Before such specimens are placed in an SEM they should be subjected to extensive cleaning with appropriate solvents in an ultrasonic cleaner, pre-evacuated in the same manner as just described for polymeric specimens, and then examined by light microscopy for signs of residual amounts of polishing or etching reagents in pores and cracks. Stains on the surface near pores and cracks usually indicate that they still contain some reagent.

Even the mounting polymer itself can evolve significant quantities of organic materials into a vacuum system. This is particularly true for the 'cold mounting' resins that are formulated to cure at room temperature. While these materials may become physically hard enough to handle in less than an hour, they may evolve large quantities of materials when placed under vacuum for many hours after the recommended curing time has passed. In our initial encounter with extensive use of these materials, our SEM became so badly contaminated that there were visible deposits on the objective lens, the instrument was very slow to pump down to an operating vacuum, filament life dropped from 50 to below 10 h, and the specimen contamination rate became intolerable. We now require that specimens mounted in these cold mounting resins be allowed to stand for at least 24 h to ensure that complete curing has occurred, and then be prepumped for a similar period, before being placed in the SEM. Mounting resins that are prepared under pressure at temperatures above 100°C, such as the thermosetting diallyl phthalate, phenolic, and epoxy resins, and the thermoplastic acrylic resins, cure more completely during the mounting process and evolve much less material when subsequently evacuated. Nonetheless, we require several hours of prepumping before allowing them in the SEM.

In the hands of an unthinking user, silver and carbon paints can be surprising sources of gas evolution. On one occasion a user complained that something was badly wrong with the vacuum system of our SEM because it was not pumping down as fast as it should. This is the classical approach in which the user of an instrument puts the blame on the instrument and those who maintain it, rather than suspecting that he might be using it in an inappropriate manner. Investigation of this problem revealed that this user had smeared a mounting stub heavily with carbon paint, laid a large, flat metal specimen on it, waited a few minutes for the paint to 'dry', and then put it into the microscope. The problem arose because in the short time allowed, the paint only dried superficially around the edges of the specimen, but remained wet underneath it. In the vacuum system the solvent from this wet paint diffused out slowly through the dry surface layer giving a remarkably high level of gas evolution for a surprisingly long time. A better practice for mounting such a specimen is to place it on a clean, dry stub and to put a few small dabs of paint at well separated locations around the edge of it. This minimizes the trapping of paint underneath the specimen, and allows the solvent to escape all around each dab, promoting rapid, thorough drying. If the assembly is prepumped or heated gently, and preferably allowed to stand overnight, so that the dabs of paint are really dry, they will usually hold the specimen quite firmly to the stub, and there will be a minimum of gas evolution when it is placed in the SEM. It would, of course, be still better to use a mechanical clamping device to hold the specimen on the stub. All users of SEMs should be sensitive to sources of gas evolution such as these, and to their possible effects on the performance of the instrument. It usually takes very little extra effort to avoid them, while the rewards are shorter pumpdown times, lower levels of astigmatism, lower rates of specimen and aperture contamination, and longer filament life.

2.10.4e *Gas evolution from virtual leaks*

Virtual leaks constitute another potentially serious source of gas evolution in vacuum systems. A virtual leak occurs when a small amount of gas is trapped at atmospheric pressure in a crevice or cavity in such a way that it can escape only very slowly when the system is subsequently evacuated.

The classical example of a virtual leak is the entrapment of gas by a screw that is threaded into a blind hole. The small cavity at the bottom of the hole fills with gas at atmospheric pressure, and when the instrument is then evacuated this gas can escape from the cavity only through the tiny channel between the screw threads. Various estimates and measurements indicate that the level of gas influx from such a source can remain significant for as many as 10 or 20 h of pumping. In systems that contain specimen stages, aperture manipulators, beam deflecting coils, and other such devices, there are many opportunities for virtual leaks to occur. Only careful design can minimize this problem. If you design a device to be used in a vacuum system, be sure to exercise care to avoid virtual leaks. For example, the problem with the blind screw hole can be avoided by drilling the hole clear through the piece, by boring a small hole into it through the side of the piece, by boring a hole down the axis of the screw, or even by cutting a groove in the threads of the screw. A number of different sources of virtual leaks are illlustrated in the book on *Helium Mass Spectrometer Leak Detection* published by Varian (Chapter 6), while Hablanian (Section 4.3.4) discusses pumpout rates for different kinds of virtual leaks.

In general, it is difficult to characterize the overall rate of gas evolution accurately. In some systems it will decrease steadily with pumping time, in others it will seem to remain essentially constant for long periods. Consequently, it is almost impossible to give any general method for making meaningful calculations of the effect of gas evolution on pumpdown time and ultimate vacuum in a practical vacuum system. Qualitatively, however, it should be evident that significant gas evolution can have a very adverse affect on the performance of the vacuum system of an electron microscope. In addition, the evolution of organic compounds into an electron microscope can lead to serious optical problems due to contamination of specimens and apertures. Therefore, electron microscopists should always exercise great care to avoid practices, such as those described above, that can contribute to worsening these problems.

2.11 Concluding remarks

If Aristotle's statement that "Nature abhors a vacuum", which was given a somewhat unfavorable treatment in the first paragraph of Chapter 1, is taken in a broad, general sense, it may well be right; nature very well may be such that an absolute vacuum is very unlikely to exist. Atoms, ions, and molecules are such ubiquitous entities that it has proved to be virtually impossible to produce a significant volume of space here on earth that is totally devoid of them, and the best evidence available indicates that significant numbers of these particles exist even in the ultra-ultra-high vacuum of deep interplanetary space. In this chapter we have seen that there are a number of factors that contribute to making it difficult to obtain a good vacuum: the components of the vacuum pumping line impede the flow of gas molecules from the vacuum chamber to the vacuum pump; gases behave very differently when they flow through tubes in the high vacuum range than they do in the rough vacuum range; and, as the vacuum becomes better, gas molecules can enter the vacuum chamber by various means as fast as the pumps are able to remove them. An understanding of these factors, and the way in which they become involved in determining how long it takes for a vacuum system to pump down to a given operating pressure, and the ultimate level of vacuum attainable, is essential in obtaining the best performance from all vacuum systems. The topics discussed in this chapter are important for these and many other reasons. I hope you have found my presentation of them interesting, easy to understand, and useful.

3 Vacuum gauges

Vacuum gauges are critically important components of all vacuum systems. They are essential for determining if a vacuum system is performing as specified. They are needed during a pumpdown process to determine when it is appropriate to change from the rough pumping to the high vacuum mode of operation, and, ultimately, they are needed to show when the required operating vacuum has been attained so that the apparatus can be put into use properly. They are essential components of all automatic valve operating systems, and they can be highly useful in diagnosing the source of a problem in a vacuum system. In selecting, operating, and maintaining vacuum instruments it is extremely helpful to have an understanding of the physical principles that underlie the functioning of these gauges, and to have an appreciation of their general performance characteristics. The theory, construction, and performance of many of the different types of gauges that have been developed for measuring gas pressures in vacuum work are discussed in considerable detail by Hablanian (Chapter 10), O'Hanlon (Chapter 6), and Van Atta (Chapter 3). This chapter is mostly devoted to describing thermal conductivity and ionization gauges, because these are the gauges most widely used on instruments in electron microscopy laboratories. However, partial pressure gauges, or residual gas analysers, and capacitance manometers are also discussed briefly. Partial pressure gauges are increasingly being used for measuring levels of contamination and for finding leaks in vacuum systems on high-resolution and analytical electron microscopes, while capacitance manometers are appearing on the recently-introduced environmental scanning electron microscopes.

The references cited in this chapter are listed in Appendix 1. The manufacturers and suppliers of vacuum gauges and related equipment referred to here are listed in Appendices 2 and 3.

3.1 Thermal conductivity gauges

Thermal conductivity gauges make use of the fact that the rate at which thermal energy is transferred from a heated wire to a surrounding tube or envelope by the movement of gas molecules between the wire and the envelope varies with the pressure of the gas in the tube. This variation in the rate of heat conduction changes the temperature of the wire, and so measurement of the temperature of the wire, or of some other property dependent on it, can be used as an indirect measure of the gas pressure in the tube.

3.1.1 The thermal conductivity of gases

The ability of gas molecules to conduct heat from a body at a high temperature T_{hi} to an adjacent body at a lower temperature T_{lo} arises because the gas molecules that strike solid surfaces exchange energy with the atoms in the surfaces. Gas molecules leaving a hot surface have higher levels of vibrational, rotational, and kinetic energy than those leaving a cool surface. If these 'hot' molecules subsequently strike the cool surface, they give up some of their energy to the atoms in the cool surface. If gas molecules from the cool surface strike the hot one they will, on the average, acquire energy from it. The net result is a transfer of thermal energy from the hot surface to the cool one, in keeping with the general principles of thermodynamics. This heat transfer process is characterized by a thermal conductivity coefficient, which depends on the molecular structure of the gas and the density of gas molecules between the surfaces.

In the range of viscous flow (Section 2.2, p. 29), where the Knudsen number K_n has a value less than 0.01, the density of gas molecules is so high that molecules leaving the hot surface cannot reach the cooler one without colliding with a very large number of other gas molecules and exchanging energy with them. Therefore heat transfer in this pressure range is a complex process which, because it involves these molecular interactions in such an important way, is closely related to the gas viscosity. The capacity of gases to transfer energy in this pressure range is numerically described by a viscous thermal conductivity coefficient K, which is defined as the quantity of heat transferred per unit time from a unit area of surface under the influence of a unit thermal gradient, and is now usually given in units of $[(J/s)/m^2]/[K/m]$. For the reasons just mentioned, K is proportional to the gas viscosity η. It also depends on the molecular mass and structure of the gas molecules. Whereas monoatomic gases such as He and Ar can only absorb energy by increasing their kinetic or translational energy, polyatomic gases, such as N_2, CO_2, and NH_3, can also absorb energy by increasing the rate at which they rotate and the extent to which the atoms within them vibrate. At a given temperature, molecules with a small molecular mass travel faster than those of larger mass (Section 1.6, p. 19), and therefore make more trips carrying energy between the hot and cold surfaces in a given time. Approximate values of the viscous thermal conductivity coefficient for some common gases are: H_2, 0.18; He, 0.15; air, 0.024; CO, 0.022; CO_2, 0.015; CH_4, 0.031; and C_2H_6, 0.017 [(J/s)/m²]/[K/m]. Using this thermal conductivity coefficient, the rate of heat transfer H (J/s) between two parallel flat surfaces separated by a distance x (m) is given by:

$$H = AK\left(\frac{T_{hi} - T_{lo}}{x}\right) \tag{3.1}$$

in which A is the macroscopic area of the surfaces (m²), and $(T_{hi}-T_{lo})/x$ is the thermal gradient (kelvins/m) between them. This equation gives a molecular heat transfer rate for air of about 2.4×10^{-4} J/s per mm² for surfaces with a temperature difference of 100 kelvins and a separation of 10 mm. Since gas viscosity does not vary with pressure (Section 1.11, p. 25), the molecular thermal conductivity of a gas is also independent of pressure in the pressure range of viscous flow. The total rate of heat transfer may increase slowly as pressure increases in this range, however, due to additional heat conduction by convection. Convective heat transfer is a mass flow phenomenon in which the heated gas flows away from the neighbourhood of the hot surface, carrying thermal energy with it, as a result of localized density gradients.

As the pressure decreases into the range where K_n becomes greater than 0.01, and gas flow becomes transitional in character (Section 2.4, p. 38), the density of gas molecules decreases enough so that they have a significant probability of travelling between the hot and cold surfaces without colliding with one another. Here the rate of molecular heat transfer no longer depends on the viscosity of the gas, but is basically proportional to γ, the number of molecules that strike the surfaces per second. Since γ is proportional to the pressure of the gas (Section 1.8, p. 22), the molecular thermal conductivity now varies linearly with pressure. This process is characterized by the free molecular thermal conductivity coefficient Λ, which is usually given in units of [(J/s)/m^2]/[K·Pa], and which, like K, is also dependent on molecular mass and structure. Approximate values for some common gases are: H_2, 4.6; He, 2.2; H_2O, 2.0; CO_2, 1.3; and air, 1.2 [(J/s)/m^2]/[K·Pa]. Because the gas molecules collide with surfaces so infrequently in this pressure range, the extent to which they reach thermal equilibrium with the atoms in a surface when they do collide with it also has a strong effect on the heat transfer process. This is accounted for by a molecular accommodation factor α, which depends on the molecular structure of the gas molecules and on the composition and physical characteristics of the surface. The value for α is about 1.0 when good thermal equilibration is attained, but may be less than 0.01 under conditions of poor equilibration. Taking this factor into consideration, the rate of molecular heat transfer H (J/s) between two flat surfaces of area A (m^2) with temperatures T_{hi} and T_{lo} (K) at a pressure P (Pa) in the range where flow is transitional in character is given by:

$$H = A\alpha\Lambda(T_{hi}-T_{lo})P \tag{3.2}$$

On this basis, the rate of free molecular heat transfer by air between surfaces with a temperature difference of 100 kelvins and an accommodation coefficient of 0.5 is only about 3×10^{-6} J/s per mm^2 at a pressure of 5×10^{-2} Pa (5×10^{-4} Torr).

The useful working range for the commonly encountered types of thermal conductivity gauges is basically determined by the characteristics of these molecular heat transfer coefficients. At pressures above about 100 Pa (1 Torr), where the conductivity coefficient is constant because of the dependence of K on gas viscosity, there is essentially no response to changes in pressure. Response is best in the range of transitional flow, from about 100 Pa down to about 1 Pa (1 to 0.01 Torr) where heat conduction varies with pressure, as determined by the free molecular conductivity coefficient Λ. As the pressure decreases further the molecular conductivity also decreases, and ultimately becomes negligible compared to other effects. This usually occurs when K_n becomes greater than one, and pressures this low are frequently described as providing an 'insulating vacuum'. Here radiative heat transfer, which is proportional to the difference in the fourth power of the temperatures of the two surfaces (i.e. $H \propto T_{hi}^4 - T_{lo}^4$), is usually greater than that due to molecular conduction.

Thus, the practical lower limit of response for a thermal conductivity gauge occurs when gas molecules strike its sensing element so infrequently that heat loss by radiation, and by conduction through the metal posts that support the sensing element, becomes greater than heat loss by gas conduction. This limit depends strongly on the design of the gauge, but is usually between 10^{-1} and 10^{-2} Pa (10^{-3} and 10^{-4} Torr). These relationships are discussed in detail by Roth (Sections 2.7.3 and 6.6).

The actual response of thermal conductivity vacuum gauges to changes in pressure is further complicated by the way in which the gauges are constructed and operated. In these gauges heat transfer usually occurs between a very fine (e.g. 0.025 mm diam.) heated wire and a surrounding cylindrical metal envelope which is at room temperature. The simple equations given above must then be modified to take this geometrical arrangement into consideration. In addition, the wire is usually heated by maintaining the electric current through it constant, rather than by maintaining its temperature constant, and so its temperature can vary considerably as pressure varies. In some gauges, for example, the temperature of the wire is below 50°C at pressures in the range of viscous flow where thermal conductivity is highest, but rises above 400°C as the pressure drops into the high vacuum range. This makes it difficult to make anything more than rather general statements about the way these gauges respond to changes in pressure.

3.1.2 Thermocouple gauges

In a thermocouple gauge, a very fine thermocouple is used to measure the temperature of a heated wire inside a tube which is connected to the vacuum system. A thermocouple is an electrical device that is widely used for measuring temperature. It consists of two wires of different chemical composition that are welded together to form a thermocouple junction. The dissimilar characteristics of the wires cause a small voltage to be generated at this junction, and the magnitude of this voltage varies as the temperature of the junction changes. In solid-state physics this is known as the Seebeck effect. Fig. 3.1 shows schematically the construction of a thermocouple gauge tube, and a battery-powered circuit for operating it. In the usual mode of operation the current in the heated wire W is maintained constant by adjusting the variable resistor R to keep the reading of the milliammeter mA at a specified value, and the output of the thermocouple junction T is

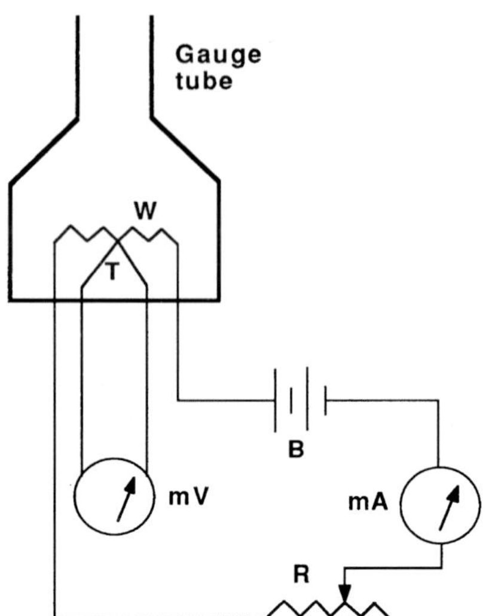

Fig. 3.1 A battery-powered control circuit for a thermocouple vacuum gauge. W, heated wire; T, thermocouple junction; B, battery; R, variable resistor; mA and mV, meters.

read on a millivolt meter mV (or a low resistance microammeter). Since the number of gas molecules that strike the wire per second to conduct heat away from it decreases as the pressure decreases, the temperature of the wire increases as the pressure of the gas decreases, causing the output of the thermocouple to vary in the manner shown in Fig. 3.2. Specifications for commercially-available tubes of this traditional design typically give 0.1 Pa (10^{-3} Torr) to 300 Pa (2 Torr) for their operational pressure range. However, the range of best sensitivity (i.e. the range over which the thermocouple output changes most rapidly with a unit change in pressure) is from about 1 Pa (10^{-2} Torr) up to about 100 Pa (1 Torr), as shown by Fig. 3.2. While there may be a measurable response above and below this range, sensitivity there is generally poor. Thermocouple gauges also can show different responses for different gases, because different gases have different molecular thermal conductivities, as noted in the previous section. Taking the response for air as equal to 1.0, rough relative values for some other gases are: Ar, 0.6; CO_2, 0.9; H_2O, 1.1; He, 1.2; and H_2, 2.5.

The circuit diagram of Fig. 3.1 is intended only to indicate the general nature and function of this type of gauge. Commercial gauge tubes may incorporate refinements in the arrangement of the components to compensate for temperature variations and to increase sensitivity. Usually these

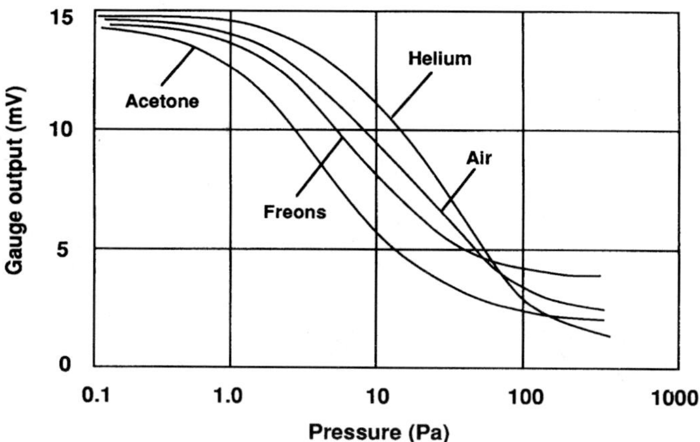

Fig. 3.2 Typical output curves for thermocouple gauges showing how response varies with pressure and illustrating the different responses produced by different gases.

gauge tubes are made of metal and are provided with a tapered pipe thread that can be conveniently sealed into a vacuum system with Teflon tape or an epoxy vacuum cement. Battery-operated circuits are seldom used now, except possibly in portable units. In units found on laboratory apparatus the heating current is usually provided by a solid-state circuit which operates from the electrical power mains. The readout circuit is invariably modified so that the meter reading increases as pressure increases, and the meter is usually calibrated directly in units of pressure. Fig. 3.3 shows thermocouple gauges and controllers of current design. One of these controllers has a set-point switch that can be used for turning off a diffusion pump or other device if the vacuum rises above a selected value. Microprocessor-based controllers, which provide digital pressure readout and which can be interfaced to a computer for automated control purposes, are also available, as well as special tubes and controllers that also utilize convective heat transfer effects to extend response to pressures near 1 atmosphere.

The signal output of thermocouple gauges is quite small, and individual tubes can differ somewhat in their output characteristics. Also, the resistance of the plugs, sockets, and cables connecting the gauge tube to the controller can have an effect on the net output received by the meter. Therefore, a thermocouple gauge controller should be recalibrated any time the gauge tube or the cable connecting it to the controller is changed. The principal step in this procedure involves installing the gauge tube in a

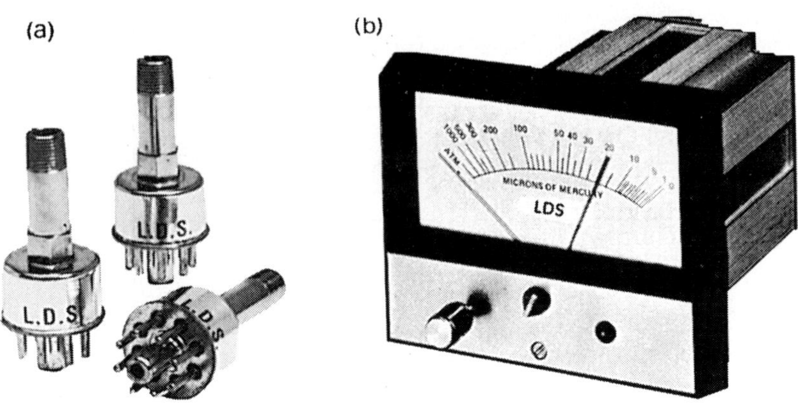

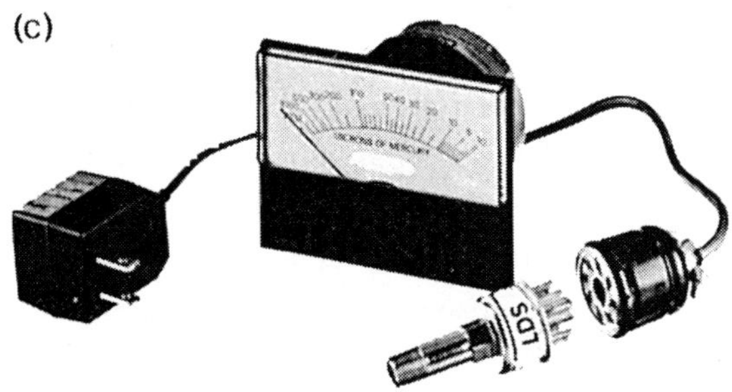

Model 200 Panel Mount

Fig. 3.3 Thermocouple gauge tubes and controllers. (a) Gauge tubes; (b) a standard controller with a single set-point; (c) a compact controller for panel mounting. (Courtesy of LDS Vacuum Products, Inc.)

system that allows it to be pumped down to a pressure well into the 10^{-2} Pa (10^{-4} Torr) range, and then adjusting a variable resistor in the controller to cause the meter to register at the bottom of its pressure range. Some controllers also have a variable resistor that is first adjusted to make the meter read full scale at atmospheric pressure.

Thermocouple gauges are rugged and relatively inexpensive, and are not damaged by exposure to air at atmospheric pressure while in operation. They therefore are very convenient for use in monitoring the pressure in rough vacuum lines. It must be recognized, however, that they do not provide a high degree of accuracy, reliability, or reproducibility. Output can change significantly with time, and can be influenced by contamination of the gauge tube with oil from a vacuum pump. Furthermore, the output of these tubes is a very small direct current (DC) voltage (typically less than 15 mV), and so the signal detected by the gauge controller can be strongly influenced by resistance effects in the electrical leads connecting the controller to the tube. Unplugging and then reconnecting one of the plugs in this line can change the total resistance of the line enough to make a major change in the reading produced by the controller. This can be particularly bothersome if the thermocouple gauge controller is serving as a sensor for a safety or control system. On one occasion I saved a friend the trouble of undertaking a major overhaul of a vacuum system by merely wiggling the plugs in the thermocouple gauge circuit to re-establish good electrical contact through them. Obviously, considerable attention should be given to the quality and condition of the components of the electrical lines in thermocouple gauge circuits. In particular, the gauge and its controller should be calibrated with the same electrical lines and connectors as are going to be used in practice. The meter used to register the output of a thermocouple gauge must have a very low resistance (typically from 50 to 70 ohms), and the mechanism supporting its needle must be very delicate, in order to register the few millivolts output produced by the thermocouple junction. Interestingly, this can also lead to some rather unexpected problems. A short time ago the thermocouple gauge on a sputter-coating apparatus in our laboratories indicated that the vacuum was not dropping below about 100 Pa (1 Torr), regardless of how long the system was evacuated. This suggested that there was either a large leak in the system or that the vacuum pump was not performing properly. However, completely overhauling the vacuum pump, cleaning all components of the vacuum line, and replacing the thermocouple gauge tube did not cure the problem; the gauge meter still consistently registered a vacuum only slightly below 100 Pa (1 Torr). Finally, George Brooks, our laboratory supervisor, happened to wipe the dust off the face of the meter and noticed that this caused a large, erratic movement of the needle. With great insight he investigated this behaviour further and finally determined that the interior surface of the plastic face plate on the meter had, in some unknown manner, acquired a large static electric charge that prevented the needle from moving normally. Once diagnosed, this problem was easily cured by cleaning the face plate and coating it with anti-static solution (available from electronic supply stores — a thin layer of evaporated carbon would probably have worked as well). Thermocouple gauges are very useful devices, but they must be handled with care if they are to give useful vacuum readings.

3.1.3 *Pirani gauges*

The Pirani gauge is the type of thermal conductivity gauge most widely used on electron microscopes. In this gauge a change in pressure causes the temperature of a heated wire to change in the same manner as described above for the thermocouple gauge. This change in temperature causes the resistance of the wire to change, and it is this change in the resistance of the wire, rather than the change in its temperature, that is measured as an indicator of the change in pressure. The gauge tube consists of a metal or glass envelope that surrounds a filament consisting of a long piece of very fine wire. Tungsten is commonly used for the filament because its resistance changes rapidly as its temperature changes. Fig. 3.4 shows a traditional battery-powered Pirani gauge circuit. Here, the active gauge tube and a sealed compensating tube (having resistance values R_a and R_s, respectively), and two resistors, R_1 and R_2, form the four arms of a Wheatstone bridge, a classical circuit for sensitive measurement of resistance. The voltage provided by the battery B causes current to flow through the two parallel branches of the bridge formed by R_a+R_s and R_1+R_2. This current heats the filaments in both tubes. When the ratio of the resistances of the two tubes equals the ratio of the resistances of the two bridge resistors (i.e. when $R_a/R_s = R_1/R_2$) no current flows through the milliammeter mA. This fact is used to calibrate the device by evacuating the active gauge tube to a pressure below its normal response range and adjusting resistor R_1 to make the reading of the milliammeter mA zero. When put into use at higher pressures, the temperature of the filament in the active tube and its resistance both decrease, while the sealed reference tube remains unchanged. This unbalances the bridge, causing the meter reading to increase. The amount of unbalance varies with the pressure of the gas in the tube, giving corresponding variations in the meter reading. The function of the sealed gauge tube is to compensate for changes in ambient temperature. This tube, which is matched to the active tube, is usually evacuated to below 10^{-4} Pa (10^{-6} Torr) and thoroughly baked out before

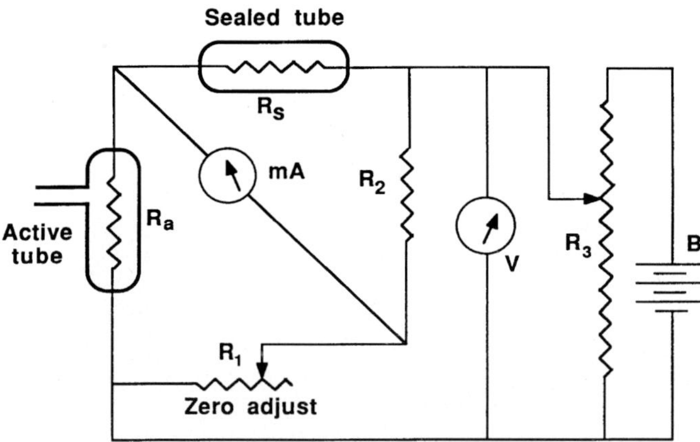

Fig. 3.4 A battery-powered control circuit for a Pirani vacuum gauge. B, battery; R_1, R_2, R_3, R_a, and R_s, resistance elements; V, voltmeter; mA, milliameter.

being sealed off, and is commonly mounted beside the active gauge tube inside a thermally-insulated enclosure so that it experiences the same external thermal effects as the active tube.

Current commercial models operate on basically the same principal, but incorporate many modifications to improve performance characteristics. To reduce cost, the sealed compensating gauge tube is usually replaced by a temperature compensating circuit, and the zeroing operation is commonly carried out when the tube is at atmospheric pressure. As with the thermocouple gauge, commercial units are usually provided with stabilized solid-state power supplies that operate from the electrical power mains, and the circuit output, whether indicated on a meter or displayed digitally, is calibrated directly in pressure units. Set-point switches are also commonly available. Fig. 3.5 is a photograph of a Pirani gauge tube and controller. This controller has two set-point switches, provisions for calibrating the tube at both atmospheric and low-pressure settings, and a high-sensitivity setting that is useful in testing for leaks (Section 10.7, p. 438).

Traditionally, Pirani gauge controllers have used the *constant voltage* mode of operation, where the voltage across the Wheatstone bridge circuit is maintained constant, because of its simplicity in both construction and operation. In the circuit shown in Fig. 3.4 this is done by adjusting resistor R_3 to maintain the reading of the voltmeter V constant. Systems using this mode typically are rated as covering the range from about 10^{-1} Pa

Fig 3.5 A Pirani gauge tube and controller. (Courtesy of HPS Division of MKS Instruments.)

(10^{-3} Torr) up to nearly 10^3 Pa (10 Torr). Many systems with solid-state electronics now employ the *constant temperature* or *constant resistance* mode of operation by incorporating into the power supply a feedback amplifier that automatically keeps the bridge circuit always in balance. The variation in voltage needed to do this is then used as the indicator of pressure variation. This mode of operation provides faster response, greater sensitivity, and allows the response range to be extended above 10^4 Pa (10^2 Torr). For all Pirani gauges, however, the range of greatest sensitivity is dictated by the physical phenomena involved in the heat transfer process, and is about the same as that given above for thermocouple gauges (i.e. from 1 to 100 Pa). Like thermocouple gauges, Pirani gauges are rugged and are not damaged by exposure to air at atmospheric pressure while in operation. They are moderately more expensive than thermocouple gauges, but are generally considered to be more reliable, more sensitive, more stable, and more accurate than thermocouple gauges.

3.1.4 Convection gauges

Several companies now manufacturer vacuum gauges that are designed to combine the mechanism of convection with that of molecular conduction to achieve a wider range of response and a somewhat more stable performance than is afforded by the standard thermocouple and Pirani gauges described above. Granville–Phillips, for example, advertise that their Convectron gauge provides good sensitivity from atmospheric pressure down to 0.1 Pa (10^{-3} Torr) with an accuracy of about one percent, a response time of less than 0.1 s, and significantly less drift than standard thermal conductivity gauges. These gauges usually make use of a bridge circuit which is operated in the constant temperature mode, somewhat as described above for Pirani gauges but with refinements to improve stability and sensitivity. In addition, the gauge tubes may contain baffles arranged to enhance the cooling of the sensing element by convective processes in the pressure range above 100 Pa (1 Torr), and so some gauge tubes must be mounted in a particular orientation to utilize these convective effects optimally. Convection gauges are not appreciably more expensive than standard Pirani gauges.

3.2 Ionization gauges

High energy electrons flowing through a gas collide with the gas molecules and produce gas ions. In general, the number of gas ions produced per unit of time is proportional to the magnitude of the electron current, and to the concentration of gas molecules which is, in turn, proportional to the pressure of the gas. By collecting these gas ions under standardized conditions the ion current can be used as a measure of the gas pressure. Numerous devices based on this principle have been designed for measuring gas pressures in the high and ultra-high vacuum ranges, and several of these are described by Weston (Chapter 4). Only hot cathode and cold cathode ionization gauges are commonly used on instruments in electron microscopy laboratories, and only these are discussed here.

3.2.1 Hot cathode ionization gauges

Some of the earliest gauges of this type were made by simply sawing the end off an ordinary triode radio tube and sealing it into the vacuum system with wax. It is therefore appropriate to explain the functioning of these gauges by reference to the battery-powered triode tube circuit shown in Fig. 3.6. As the name implies, the filament F of the gauge tube is heated to incandescence by current supplied by battery B_1. The electrons emitted from the hot filament are drawn to the grid G, which is basically a piece of wire mesh that is maintained at a positive potential of about 150 V relative to the filament by battery B_2. In travelling from the filament to the grid the electrons gain sufficient kinetic energy to ionize the gas molecules present in the tube, producing positive gas ions. These ions are collected by the plate P of the tube, which is maintained at a negative potential of 25–50 V relative to the filament by battery B_3. The electron current I_e is measured by the milliammeter mA in the grid circuit, and is maintained at a standard value by adjusting the variable resistor R to change the temperature of the filament. The ion current I_i is measured with the microammeter µA in the plate circuit. At pressures below 10^{-1} Pa (10^{-3} Torr) there is little likelihood that an electron will collide with more than one gas molecule in travelling from the filament to the grid, and so the ratio of the ion current to the electron current becomes linearly proportional to the pressure; that is, $I_i/I_e = SP$, where S is a sensitivity factor that depends on tube design and operating conditions. S is also different for different species of gas molecules. In general, S is proportional to the ionization cross section of the gas molecules (i.e. to the probability that they will be struck by the electrons and become ionized). Gases with small atoms and molecules generally have lower values of S than those with larger molecules. Taking $S = 1.0$ for air, approximate relative values for other common gases are: He, 0.2; H_2, 0.5; H_2O, 0.9; O_2, 0.9; N_2, 1.1; CO, 1.1; CO_2, 1.4; acetone, 5.0; and, hydrocarbons, 2–10. This makes it possible to use a hot cathode gauge for detecting leaks, as described in Section 10.6 (p. 437).

Even a relatively crude device of this type, such as the adapted triode radio tube described above, can easily measure pressures in the 10^{-3} Pa (10^{-5} Torr) range, while triode vacuum gauge tubes that are manufactured especially for the purpose can reach into the 10^{-5} Pa (10^{-7} Torr) range. At lower pressures these simple triode tubes lose sensitivity for a variety of reasons. However, the Bayard–Alpert vacuum gauge tube shown in

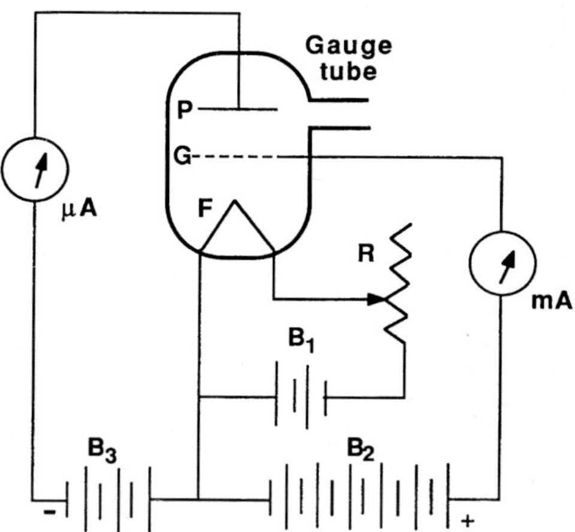

Fig. 3.6 A battery-powered circuit for a triode hot cathode ionization gauge. B_1, B_2, and B_3, batteries; R, variable resistor; mA and μA meters; F, filament; G, grid; P, plate.

Fig. 3.7, a sophisticated modification of the triode tube, exhibits useful sensitivity at pressures below 10^{-8} Pa (10^{-10} Torr). These are the most reliable and the most accurate vacuum gauges conveniently available for measuring pressures below the rough vacuum range, because manufacturing techniques developed for the production of radio tubes make it possible to produce them with highly reproducible characteristics, and the phenomena that underlie their operation are well understood. Most suppliers of vacuum equipment offer Bayard–Alpert gauge tubes and controllers with stable, reliable, solid-state power supplies and sensitive amplifiers for operating them. Fig. 3.8 shows a basic controller that will operate either a triode or a Bayard–Alpert hot cathode gauge and two thermocouple gauges, with set-point switches for all three gauges. More elaborate controllers that display the pressure reading digitally and have computer interfaces for automated process control are widely available. Commonly, Bayard–Alpert gauges are rated for use in the range from 10^{-8} Pa (10^{-10} Torr) up to 10^{-1} Pa (10^{-3} Torr). However, Terranova Scientific have recently developed a new type of gauge controller that extends the range of the Bayard–Alpert gauge to 100 Pa (1 Torr) by using a very low filament heating current in combination with a modulated grid voltage at pressures above 10^{-1} Pa (10^{-3} Torr). This approach also drastically reduces problems with filament burnout and gives exceptionally long filament life.

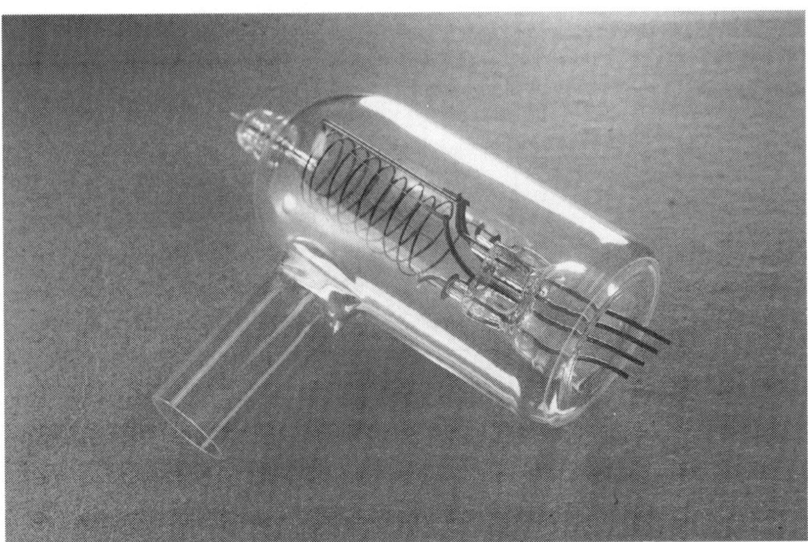

Fig. 3.7 A Bayard–Alpert hot cathode ionization gauge tube.

In the past, hot cathode gauges had two characteristics that made them somewhat unattractive for use on instruments such as electron microscopes. First, since the gauge tubes most commonly used have tungsten filaments, care must be exercised not to turn them on unless the vacuum is below about 10^{-1} Pa (10^{-3} Torr). At higher pressures the filament may burn out, or tungsten oxide may form on it and be evaporated onto the envelope and other components of the tube, altering response characteristics. Most gauge controllers have a safety circuit that will turn the filament off if the pressure rises too high while the gauge is in operation. Unfortunately, these circuits do not provide complete protection against filament burnout, because they cannot prevent someone from trying persistently to turn the gauge on before the pressure is low enough for safe operation. Inexperienced operators must therefore always be warned about the possibility of damaging the gauge tube in this way. These problems can be reduced by using tubes with iridium filaments, which are considerably more resistant to oxidation and burnout than conventional tungsten filaments.

The second unattractive characteristic was that most gauge controllers used to require an adjustment of the filament heating current to give a specified value of the electron current to the grid I_e so that the value of the

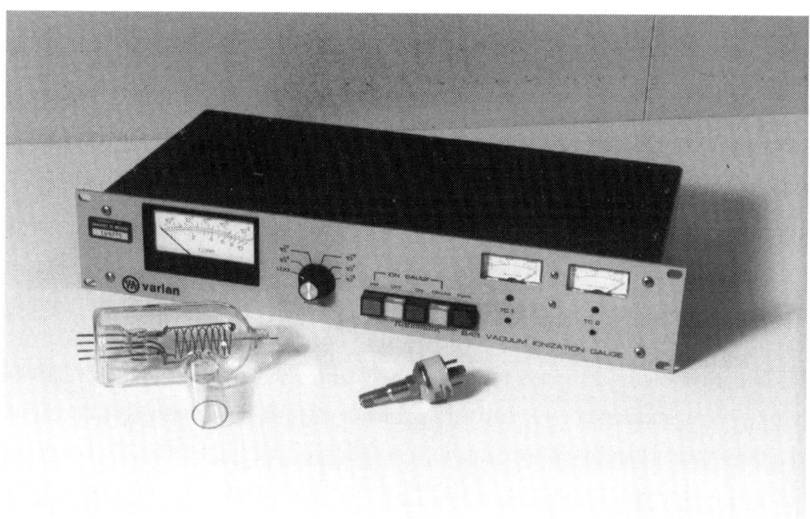

Fig. 3.8 A gauge controller manufactured by Varian that will simultaneously operate a hot cathode ionization gauge and two thermocouple gauges.

ion current I_i could be read out directly in units of pressure on a precalibrated meter. This calibration procedure, which had to be repeated more or less frequently, depending on operating conditions and the characteristics of the gauge controller, was considered annoying by many users. Varian eliminated this procedure completely by a patented process that monitors the I_i/I_e current ratio directly, while most other manufacturers have virtually eliminated it by other means, so this annoyance is absent or greatly reduced with most modern units.

When a gauge tube is exposed to atmospheric pressure a considerable quantity of gas can adsorb onto the grid wire and the internal surfaces of the tube. When the gauge is subsequently put into use the desorption of this gas by heat from the filament can contribute significantly to the concentration of gas molecules inside the tube for long periods of time, causing the pressure registered by the gauge to be higher than that actually existing in the vacuum system itself. This effect becomes particularly significant at pressures below 10^{-5} Pa (10^{-7} Torr) (i.e. in the ultra-high vacuum range). It is therefore desirable to *outgas* or *degas* a gauge tube to ensure that accurate readings are attained as rapidly as possible. This can, of course, be accomplished by heating the tube with an external heater ('torching' tubes was once a common practice), but most gauge controllers now include a circuit for heating the grid wire to incandescence for this purpose. This immediately drives adsorbed gas off the grid, and gradually heats up the walls of the tube and outgasses them.

When the grid heating current is first turned on to undertake an outgassing treatment the pressure in the gauge tube may rise above the safe operating limit and cause the safety circuit to turn the power supply off automatically. If this occurs it is necessary to wait 20 or 30 s for the vacuum in the tube to recover, then the power can be turned on and the outgassing attempted again. Alternatively, the heating current to the grid can be turned off before the cut-off pressure is reached, and the tube allowed to recover. It may be necessary to repeat this process several times before the rate of gas desorption becomes low enough to allow the tube to remain on while the grid is heated, especially if the tube has just been pumped down from atmospheric pressure where it was exposed to a high level of humidity.

The matter of deciding how long to outgas a tube is a bit complex, because it is so dependent on the operating pressure range and the conditions to which the tube has recently been exposed. Without adequate outgassing the pressure indicated by the tube will usually be significantly greater than that existing in the vacuum system, while in the extreme, excessive outgassing can reduce the concentration of adsorbed molecules on the surfaces inside the tube to the point where these surfaces will act as a pump after the outgassing treatment is stopped, causing the pressure inside the tube to be lower than that in the system until equilibrium is established. One method of judging the time needed for outgassing is to observe how the pressure registered by the gauge varies during the outgassing process. Usually, the pressure will rise suddenly as soon as the outgassing treatment is started, reach a maximum value after a period of time that depends on the level of adsorbed gas inside the tube, and then slowly decrease as outgassing progresses. Outgassing can usually be considered complete when the pressure reading begins to level out. Only a few minutes of outgassing is warranted at operating pressures above 10^{-3} Pa (10^{-5} Torr), and extended outgassing is not needed even at pressures in the upper end of the 10^{-5} Pa (10^{-7} Torr) range, because the concentration of gas molecules at these pressures is so high that surfaces inside the tube become saturated with gas molecules again almost immediately after outgassing is stopped (Section 1.9, p. 23). Here, outgassing serves mainly to reduce the time needed for a tube to approach an equilibrium reading. In

Fig. 3.9 An unprotected glass vacuum gauge tube dangerously exposed to potential breakage.

ultra-high vacuum systems that operate continuously at pressures below 10^{-5} Pa (10^{-7} Torr), an initial outgassing of 1 or 2 h is usually recommended, followed by shorter treatments of 10–30 min on a daily basis if the tube is operated continuously, or otherwise whenever it is turned on. When the operating pressure is below 10^{-7} Pa (10^{-9} Torr), an initial outgassing of 12–24 h may be beneficial, followed periodically by shorter treatments to maintain the tube in good condition. As a practical guide, the length of an outgassing period is probably about right if the pressures registered after two or three successive outgassing treatments are substantially the same. The heating current to the grid is, of course, usually turned off when pressure measurements are being made.

There is one important precaution that should be observed when installing a glass Bayard–Alpert gauge on a vacuum apparatus; namely, it should be physically protected so it cannot be struck and broken accidentally. These gauge tubes are usually made of borosilicate glass, and so they are relatively sturdy and can withstand bake-out temperatures well above 250°C. However, if a wrench is dropped on them, or if they are accidentally bumped with a metal object, they can be expected to shatter like most glass objects will. If the system is under vacuum when such an event occurs, there will be an immediate inrush of a large amount of air, which

Fig. 3.10 A nude Bayard–Alpert ionization gauge.

can have very damaging effects on most vacuum systems. As a minimum, a glass gauge tube should be installed in a remote, protected location; better yet, it should be surrounded by a sturdy protective screen. Fig. 3.9 is a photograph of a disaster waiting to happen. The unprotected Bayard–Alpert gauge on this apparatus is mounted adjacent to the main control panel of the apparatus and is near an aisle that is a main laboratory thoroughfare. It is only a matter of time until someone working on the apparatus or walking by it strikes the gauge tube and breaks it. When this happens, the diffusion pump on the apparatus will be flooded with air and serious problems will follow.

These problems of breakage, outgassing, and pressure equilibration associated with using a standard Bayard–Alpert gauge tube with a glass envelope can be virtually eliminated by using a 'nude gauge' of the type shown in Fig 3.10. In this design, the grid, filament, and collector wires are attached to sturdy electrical feedthroughs on a metal high vacuum flange, without a surrounding glass envelope. In use, the flange is installed on a matching port on the vacuum system so that the gauge head protrudes into the vacuum chamber and is exposed directly to the vacuum existing there.

With this arrangement outgassing involves only heating the grid and filament for a few minutes, and the gauge is fully protected from most causes of breakage. Nude gauge heads of this kind are used on most ultra-high vacuum systems, because they can withstand the bakeout treatments needed for such systems, and because they give pressure readings that are more representative of the level of vacuum actually existing in the system than can be obtained with the standard glass gauge tubes. Interestingly, however, Hablanian (p. 339) reports that the pressure reading obtained with a nude ionization gauge will often be about a factor of 10 higher than that produced by a standard glass gauge tube, because adsorption of gas molecules on the surface of the glass tubing that connects the gauge tube to the system reduces the concentration of gas molecules inside the tube. Nude gauges also eliminate most of the potential problems that can arise from breakage of glass gauge tubes. Unfortunately, it is usually difficult to find an appropriate site for installing a nude gauge on most electron microscopes.

3.2.2 Cold cathode ionization gauges

The cold cathode ionization gauge is also often referred to as the *Penning Gauge* or the *Philips Gauge* in recognition of the scientist and the company that pioneered in its development. Early versions of this gauge had a glass envelope that contained a pair of parallel plate cathodes with an anode ring of heavy tungsten wire located midway between them, as indicated in Fig. 3.11. In this design a DC potential of from 2000 to 4000 volts is applied between the anode ring and the cathode plates. This causes electrons to be emitted from the cathode plates and accelerated toward the anode ring, and these electrons strike gas molecules and produce gas ions on the way. This ionization process produces additional free electrons, which are also accelerated as they move toward the anode ring and so contribute to the production of gas ions. A strong magnetic field perpendicular to the plane of the cathode plates, supplied by an external permanent magnet, causes the electrons to spiral as they travel toward the anode ring. This gives them very long paths and greatly enhances the probability that they will strike gas molecules and produce gas ions. The gas ions are drawn to the cathode plates producing a current which varies with the pressure of the gas in the tube. This current is large enough to be registered directly by a milliammeter mA. The current-limiting resistor R prevents damage to the meter and power supply in the event that an electrical short occurs

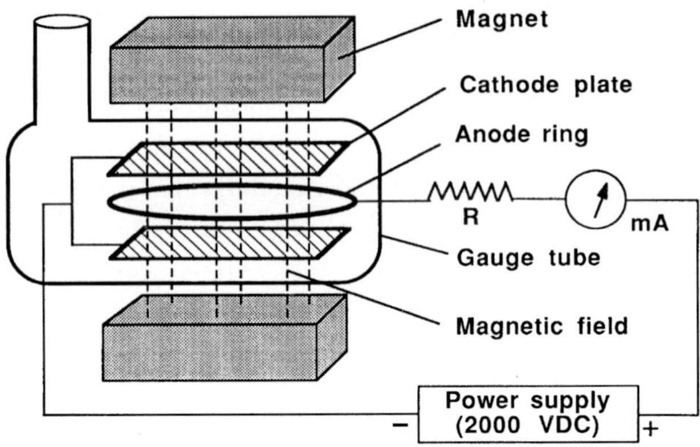

Fig. 3.11 A basic cold cathode ionization gauge circuit. R, current-limiting resistor; mA, milliammeter.

across the high voltage insulator or between the anode and a cathode plate in the tube. Since the ion current can be registered without amplification, only a relatively simple electronic circuit is needed to operate cold cathode gauges. The range of useful response for gauges of this design usually extends from about 1 Pa (10^{-2} Torr) down to about 10^{-4} Pa (10^{-6} Torr), although the sensitivity at the ends of this range may not be good. At pressures below 10^{-3} Pa (10^{-5} Torr) it becomes difficult to maintain the gas discharge. At pressures above about 1 Pa (10^{-2} Torr) the pressure sensitivity becomes poor, and ultimately the gas discharge is quenched by the high concentration of gas molecules.

The construction of current commercial versions of the cold cathode ionization gauge differ considerably from the arrangement shown in Fig. 3.11, with refinements to achieve better sensitivity and an extended range of operation. Usually the envelope is constructed of metal, and its walls, which are maintained at earth potential, serve as the cathodes. Such a gauge tube is shown assembled and disassembled in Fig. 3.12. In this design the anode loop is elongated to increase the area for gas ionization, the magnet is cylindrical, and surrounds the tube to reduce stray magnetic fields, and the magnetic field is concentrated in the centre of the anode loop by iron pole pieces that are welded into the tube wall. One modification of basically different design that is now widely used is known as the 'magnetron gauge'. In this type of gauge the cathode is a rod with circular end plates that is mounted along the axis of the tube, as shown in Fig. 3.13. The metal

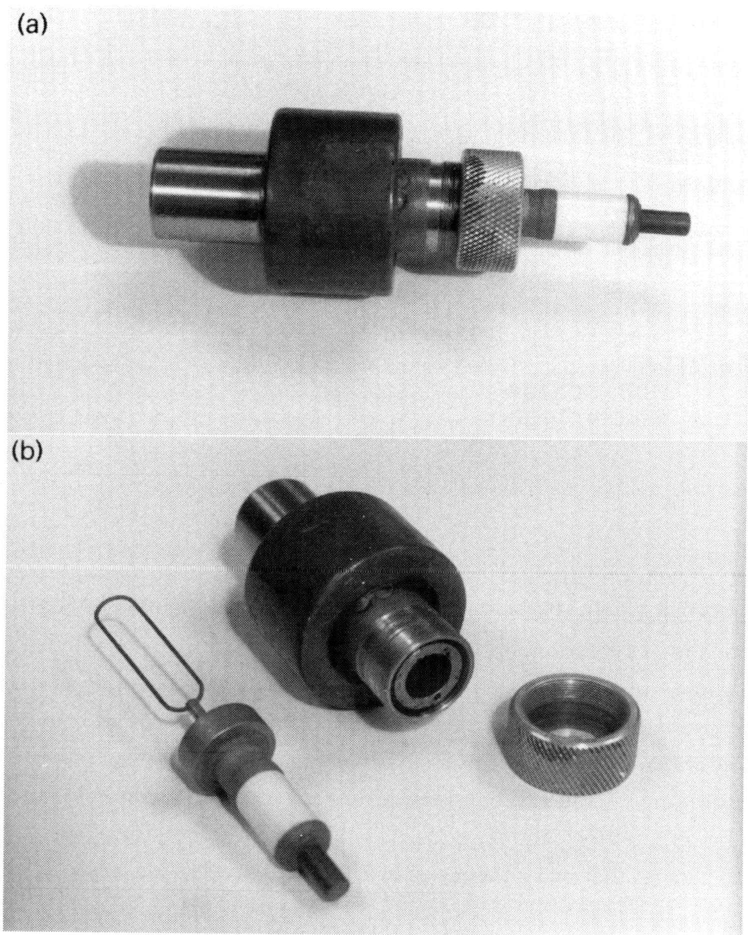

Fig. 3.12 A typical cold cathode ionization gauge. (a) Assembled; (b) disassembled.

shell of the tube then forms a cylindrical anode around the cathode, and a cylindrical magnet surrounding the shell produces a magnetic field parallel to the cathode rod. These tubes are usually specified for operation in the range from 1 Pa (10^{-2} Torr) down to 10^{-5} Pa (10^{-7} Torr). More refined designs are capable of maintaining the gas discharge down to pressures below 10^{-8} Pa (10^{-10} Torr), but are not commonly used on equipment found in electron microscopy laboratories. Gauge controllers now use solid-state circuitry, and are available with digital displays, computer interfaces, set-point switches, and similar refinements.

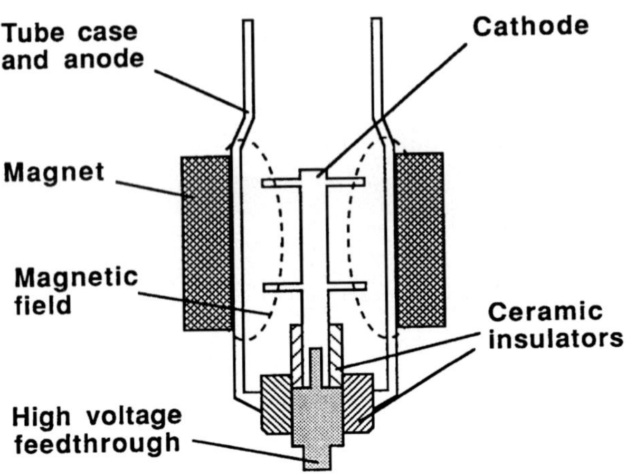

Fig. 3.13 A magnetron cold cathode ionization gauge tube.

Cold cathode discharge gauges have several advantages. Unlike the hot cathode gauges, they are not damaged if exposed to air at atmospheric pressure while in operation, because they have no heated filament to burn out. The gauge tubes are rugged and not easily damaged. The associated circuitry is relatively simple, there is usually no need for a circuit calibration procedure, and the range of sensitivity extends to the bottom of the high vacuum range. Cold cathode ionization gauges are well suited for use on vacuum evaporators, and are widely used there. However, they have several disadvantages that make them a poor choice for an electron microscope. It is highly undesirable to have a strong permanent magnet in close proximity to the column of an electron microscope. As indicated above, the gas discharge may become difficult to maintain at pressures below about 10^{-3} Pa (10^{-5} Torr), and so a gauge may cease to perform well just as the pressure enters the most critical range. Even at higher pressures the discharge tends to become erratic, particularly after a gauge tube has been used for an extended time. If a cold cathode gauge is used in a system that is pumped with a rotary-vane pump or an oil diffusion pump, organic molecules from the pump fluids eventually enter the tube and are broken down by the gas discharge to form carbonaceous deposits at various places inside the tube. This can further contribute to instabilities and may change response and sensitivity. Because of these characteristics, cold cathode ionization gauges usually cannot be relied upon to give accurate pressure readings.

In addition to the difficulty that may be involved in maintaining a discharge while operating at low pressures, it may become difficult to start a cold cathode gauge at low pressures. When the high voltage is first turned on, the discharge will not be established until an initial ionizing event, which is usually produced by a cosmic ray, occurs within the tube. At low pressures the concentration of gas molecules is so low that such events become less probable, and it may take several minutes for the discharge to strike. The delay seems to increase if the tube is old or dirty. Usually it is a good idea to turn on a cold cathode discharge gauge before the pressure drops below 10^{-1} Pa (10^{-3} Torr) so that the discharge becomes well established before low pressures are reached and these problems arise.

There are several very interesting phenomena that occur in cold cathode discharge gauges. The gas ions, moving under the influence of the high potential in the tube, acquire sufficient kinetic energy to dislodge metal atoms from the cathode. These 'sputtered' atoms deposit on the walls of the tube, burying gas molecules or reacting with them. Some of the high energy gas ions also become buried in the cathode. The net result of these processes is a pumping action that can cause the pressure inside the tube to become significantly less than that in the chamber to which it is connected. In fact, it was this pumping action that led to the development of sputter-ion vacuum pumps (Section 7.1, p. 276). Cold cathode gauges usually have large connecting tubes to promote pressure equilibration between the gauge and the system, thereby minimizing the effects of this pumping action. At higher pressures the sputtering action can become severe and has been observed to sever the cathode structure in magnetron versions of the gauge. It is therefore advisable to turn a cold cathode gauge off at pressures above 10^{-1} Pa (10^{-3} Torr) unless it is specifically designed for operation at such high pressures.

Cold cathode gauge tubes should always be installed in a location where there is little opportunity for magnetic particles to get into them, and care should be exercised to prevent magnetic objects from coming near the entrance to these tubes. Such items can be attracted into the tubes by the magnets, whereupon they can cause an electrical short when the tubes are turned on. The response of cold cathode discharge gauges differs for different gases, and so they can be used for leak detection in the same manner as other types of gauges (Section 10.6, p. 437).

3.2.3 Using ion pumps to measure pressure

Ion pumps (Section 7.1, p. 276) capitalize on the phenomena described above that produce the pumping action of cold cathode ionization gauges, and, in a sense, were developed by drastically modifying the design of cold cathode discharge gauges to enhance this pumping action. As a consequence, the relationship between the ion current and gas pressure in these pumps is essentially the same as for cold cathode ionization gauges. Most pump power control units have a meter that shows this ion current, and this meter is usually provided with a scale that indicates the total pressure of gas in the pump, based on the ion current value. This is a valuable feature of ion pumps, particularly on electron microscopes which do not otherwise have a numerical indicator of the level of the vacuum. The major shortcoming is that the pump control unit is usually located some distance from the microscope console, and so it is not easy to watch the meter on it while operating the microscope. In one laboratory, however, this problem has been solved with a closed circuit television system.

3.3 The use of thermal conductivity and ionization gauges

Thermal conductivity gauges are well suited to, and widely used for, monitoring the vacuum in pumping lines running from rough vacuum pumps to other parts of vacuum systems. When used in this way to monitor the pressure in the backing lines to oil diffusion and turbomolecular pumps, a controller with a set-point switch is highly advantageous because, as described in Section 5.8.5 (p. 220), this switch can be used to activate a protective circuit that will turn the pump off, thereby preventing damage to it, if the backing pressure rises above a safe level. Thermal conductivity gauges can also be used as leak detectors, as discussed in Section 10.7 (p. 438)

Pirani gauges are commonly used as the primary sensing elements for the automatic control systems that operate the vacuum valves of most electron microscopes. Typically, one gauge is used to monitor the pressure in the rough vacuum part of the system, and another to monitor the pressure in the high vacuum manifold leading to the gun and specimen chambers. Usually, the circuit for the gauge in the rough vacuum line produces an output signal that causes the automatic valve operating system to open and close the valves as needed to change the system from the rough pumping configuration to the high vacuum configuration when the pressure in this line drops to a value near 10 Pa (0.1 Torr). Pirani gauges are well suited for performing this function, because their response characteristics are good over the pressure range of transitional flow (Section 2.4, p. 38). Then, as the high vacuum pump takes over to bring the pressure in the instrument down into the high vacuum range, the automatic control system monitors the output of the Pirani gauge in the high vacuum manifold. As the pressure there approaches the lower end of the response range of this gauge, the control circuit activates a relay or an electronic switch that turns on a 'ready light' to signal that an 'operating vacuum' has been achieved, and then allows the high voltage and filament power supplies to be turned on. Pirani gauges are only marginally suited for this purpose. As discussed above, their sensitivity decreases markedly at pressures below the middle of the 10^{-1} Pa (10^{-3} Torr) range, and so it is likely that this operation is performed at a pressure well above the value at which it is advisable to turn on the high voltage and heat up the filament in an electron microscope. Sometimes a time delay circuit is added to give an additional period of evacuation for safety. Generally, however, no numerical pressure readout is provided, and so, if no other vacuum gauge is available, it can only be assumed that the instrument is marginally ready to be put into operation when the ready light comes on. Therefore, it is always advisable to wait an additional period of several minutes before turning the filament on, to permit the vacuum to reach a more acceptable level. Better yet, every good electron microscope should be equipped with a hot cathode ionization gauge, which can be turned on to obtain a more accurate indication of the vacuum after the ready light is activated by the Pirani gauge.

Most manufacturers of electron microscopes do not incorporate a hot cathode ionization vacuum gauge into the vacuum systems of their instruments, probably because these gauges are not suitable as sensors for the automatic valve operating systems, which have to perform critical functions during the transition from the rough vacuum to the high vacuum range where the pressure is above the safe operating range of hot cathode gauges. However, a hot cathode gauge is the best type of gauge currently available for monitoring the vacuum in an electron microscope at the lower end of its operating range. Anyone who purchases an electron microscope, especially if it uses a lanthanum hexaboride filament or is designed for analytical functions, should insist that the manufacturer install a Bayard–Alpert gauge on the instrument to provide a reliable, numerical indication of the level of the high vacuum during operation. Operators capable of working on a highly complex analytical electron microscope certainly should be expected to be able to master the minor complexities of managing a hot cathode ionization gauge with a modern controller. Some models of electron microscopes are equipped with cold cathode ionization gauges for monitoring the pressure in the high vacuum region. As we will see shortly, however, these gauges have much poorer accuracy, reproducibility, and stability than hot cathode ionization gauges. Their principal advantage is that they are less likely to burn out if misused.

3.3.1 *The accuracy of vacuum gauges*

It must be appreciated that it is very difficult to obtain an accurate measurement of the true value of the gas pressure in a vacuum system. All of the types of gauges described above are ‘indirect gauges’ in the sense that they do not measure directly the force exerted by the collisions of gas molecules with a surface, but instead measure some gas property that varies in proportion to the number of gas molecules present per unit volume. Characteristically, therefore, the response of all of these types of gauges is different for different species of gas molecules, as was indicated when they were described above. This is not a major factor for most instruments found in electron microscopy laboratories, because most gauges are calibrated for air, and this is the principal gas found in these instruments. However, this can be important in many industrial processes where complex gas mixtures are involved. The response of all gauges will show normal manufacturing variability, and manufacturers usually do not guarantee their gauges to be accurate or reproducible to much better than about ±10 percent. Cold cathode discharge gauges are generally considered to be less accurate by as much as 5–10 times this amount. In fact, one manufacturer has advertised that they do not sell these gauges because of their poor accuracy, reproducibility, and stability. The response of a given gauge system can vary with age, and can be altered by such factors as misuse, contamination, erosion of electrodes, and sagging of filaments. Periodic replacement of gauge tubes can help minimize these effects. If accurate pressure readings are indeed needed, however, it is usually necessary to calibrate the particular gauge unit being used under the actual conditions of use. Procedures for doing this are described by Hablanian (Section 10.5.3), Harris (p. 53), O'Hanlon (p. 70), and Roth (p. 324).

It must also be recognized that significant pressure gradients can exist in a vacuum system. As discussed above, the pressure in a gauge tube, and therefore the pressure registered by the gauge, can be significantly different from that existing in the vacuum system due to gas evolution and pumping effects that can occur inside the tube. For example, when one vacuum evaporator I encountered was pumped down from atmospheric pressure in the normal manner the pressure indicated by the cold cathode discharge gauge would usually drop quickly into the upper part of the 10^{-3} Pa (10^{-5} Torr) range, and then gradually fall to about 3×10^{-3} Pa (2×10^{-5} Torr). If the valve between the diffusion pump and the bell jar was then closed the pressure in the bell jar would slowly increase to the level where the gauge would automatically turn off. If the valve was then opened again the gauge reading would immediately drop into the upper end of the 10^{-4} Pa (10^{-6} Torr) range and then slowly settle down to a reading of about 5×10^{-4} Pa (4×10^{-6} Torr). Although I did not have an opportunity to investigate this behaviour fully, it appears that the reading produced by the gauge during the initial pumpdown was too high because of gas evolution inside the gauge tube. When the pressure was then allowed to rise to the top of the operating range for the gauge the intensity of the gas discharge inside the tube probably increased enough to produce significant outgassing of the tube, and so a much lower reading was obtained when the system was again evacuated. As suggested in Section 3.2.1 (p. 98), the best way to avoid such effects is to install a nude hot cathode ionization gauge inside the part of the system where the pressure is to be monitored. This approach is commonly used in apparatus, such as surface analysis instruments, which operate in the ultra-high vacuum range where these effects can be most pronounced, but is hardly feasible for most electron microscopes. We therefore have to be content with inserting a gauge tube into the system as close to the critical component, usually the specimen chamber, as other constructional constraints permit. The tube connecting the gauge to the vacuum system should be as large in diameter and as short in length as possible to promote pressure equilibration between the gauge and the system. It is also important to recognize that the pressure registered by a gauge installed in the pumping manifold can be considerably different from the pressure that actually exists in the gun or specimen chamber of an electron microscope during the early stages of an evacuation process, particularly if the gauge is located in the end of the manifold near the inlet to the high vacuum pump. When such an arrangement is encountered it is always advisable to wait an extra few minutes after the gauge registers an adequate vacuum before turning on the filament of an electron microscope. This practice will usually result in a noticeable increase in filament life.

Even though pressure readings of high absolute accuracy may not be obtained, it is very advantageous to have a good set of vacuum gauges that provide numerical readouts on an electron microscope, because the comparison of pressure readings taken over time can provide such a valuable means of detecting changes in the performance of the vacuum system. A basic gauge controller which can simultaneously operate a Bayard–Alpert gauge and two thermocouple or Pirani gauges, such as the one shown in Fig. 3.8, could form the basis for a very acceptable gauge system. Such controllers are available from most manufacturers and suppliers of vacuum equipment. The thermocouple or Pirani gauge tubes should then be installed in the pumping lines leading to any roughing or backing pumps used in the vacuum system, while the Bayard–Alpert gauge tube might be installed in the high vacuum manifold as close to the specimen chamber as possible.

3.3.2 *Installing vacuum gauges on electron microscopes*

The problem of installing a system of vacuum gauges on an electron microscope may be a non-trivial one. As mentioned above, manufacturers usually do not provide such a system of gauges, nor do they usually provide for numerical readout from such gauges as may be present. On some instruments, however, they provide vacuum couplings for vacuum gauges even though they normally do not install gauges in them. Fig. 3.14 shows a Bayard–Alpert gauge and a thermocouple gauge installed in such couplings that are thoughtfully included as standard features of the vacuum system of the Hitachi S-520 scanning electron microscope. One of the Pirani gauges used by the automatic valve operating system is also visible at the top of this photograph. If such couplings are not a standard part of the instrument it is usually possible to persuade the manufacturer to provide them as special accessories, if the matter is pursued with sufficient vigour at the time the purchase of the instrument is being negotiated. Fig. 3.15 shows a Bayard–Alpert gauge in a coupling that was installed in

Fig. 3.14 Bayard-Alpert (B) and thermocouple (T) gauges installed on a Hitachi S-520 scanning electron microscope. The Pirani gauge (P) for the automatic valve operating system is also visible.

the vacuum line leading into the specimen chamber of a scanning transmission electron microscope in response to such a special request made prior to purchase. Such arrangements should be fully explored and an acceptable solution incorporated into the instrument's specifications and the purchase agreement before a purchase order is issued.

Once an instrument has been purchased the manufacturer will probably be very reluctant to assist in installing a system of gauges on it, so you must be prepared to assume full responsibility for the installation and its consequences if you decide to pursue such a course of action. It is, of course, highly desirable to obtain as much input as possible from the manufacturer before starting the undertaking. Some degree of co-operation can usually be obtained by presenting well-informed arguments for the desirability of having such gauges to monitor the instrument's condition, thereby reducing misuse and serious service problems, perhaps accompanied by mild complaints about negligence on the part of the manufacturer in not having provided the gauges in the first place. It is also very helpful to get input from other owners of similar instruments. Many people have

Fig. 3.15 A Bayard–Alpert gauge installed on a JEOL JEM-100CX scanning transmission electron microscope.

installed gauges on their instruments in recent years due to increasing general concern with vacuum matters. Manufacturers' service engineers, who are familiar with instruments in a number of laboratories, can be very helpful in this regard. Occasionally, however, a service organization will offer objections to the installation of a set of gauges on an instrument, with accompanying suggestions that the warranty and service contract coverage may be reduced, on the basis that modifications are being made that will impair the performance of the instrument. Unfortunately, there are advantages from the standpoint of the service organization for the user of an instrument to be somewhat ignorant of exactly how well, or poorly, it is performing, and so this makes the situation rather difficult to deal with. The only recourse here is to discuss your plans fully with your service representatives and convince them that there will be no serious effects, taking the matter to the national service manager, or ultimately to an engineer at the factory, if necessary. However, no modifications should be made to an instrument as delicate as a modern electron microscope without great care and careful planning. Indeed, it may well be a major undertaking to install a coupling for a Bayard–Alpert gauge in the high vacuum part of an electron microscope. Sometimes there is a spare port available on the specimen chamber that can be used for this purpose, and sometimes it is possible to find a suitable location in the vacuum manifold that can be used with minimum effort. Fig. 3.16 shows a Bayard–Alpert gauge that was installed in the top end of the high vacuum manifold of an analytical microscope by brazing a coupling into an existing blank-off plate.

Ideally, the gauge controller should be located where it can be easily seen by the person using the electron microscope, so that the vacuum can

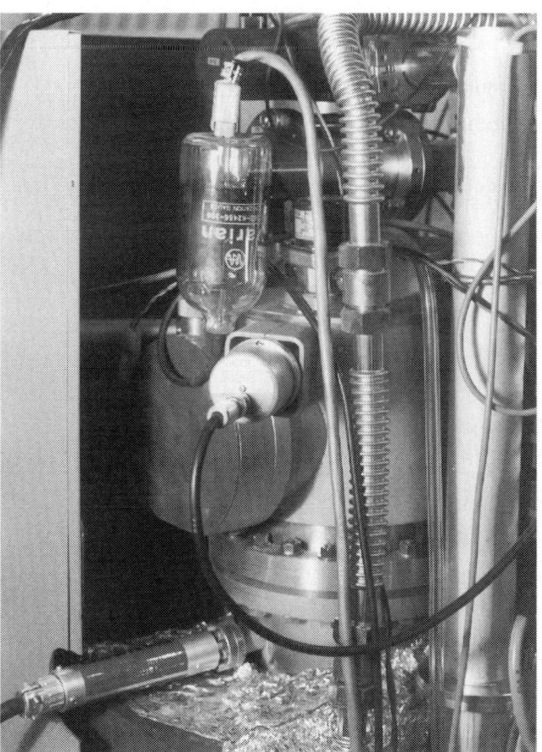

Fig. 3.16 A Bayard–Alpert gauge installed in the high vacuum manifold of an analytical transmission electron microscope.

be monitored while the instrument is in use, and particularly during specimen exchange operations. However, some problems may be encountered in doing this, because some gauge controllers may generate magnetic and electrical fields, or feed electronic noise into power mains, sufficient to degrade the performance of an electron microscope. For example, one gauge controller was found to cause severe image distortion when it was placed on top of the image display console of a scanning electron microscope. The majority of the solid-state controllers being produced now are very unlikely to cause serious problems of this kind. Nonetheless, the performance of a microscope after installation of a system of gauges should always be critically compared with its performance before the installation, just to be sure. Even if it is necessary to turn the controller off, or to unplug it from the power mains, during critical use of an electron microscope, it is still extremely valuable to have it available for periodically checking the performance of the vacuum system.

3.3.3 *Monitoring the vacuum in electron microscopes*

Once a good system of vacuum gauges is available, there are several ways in which it can be utilized advantageously. At the time an electron microscope is first installed, when it is presumably fully meeting all performance specifications, it is advisable to record the pressures registered by the gauges in the roughing lines and high vacuum manifold at the various stages of the pumpdown process. In addition, a record should be made of such characteristics as: the ultimate vacuum attainable after uninterrupted pumping overnight and over a weekend; the amount of pressure rise experienced when a specimen is inserted and when photographic film is changed, and the time subsequently required to pump down to a good operating vacuum; and, the rate at which the pressure rises when all valves leading from the pumps to the column are closed and the column is isolated from the pumps (which is often called the 'leak-up rate', but more accurately should be called the *rate of pressure rise*) as described in Section 10.2 (p. 423). The values so obtained should correspond closely to the manufacturer's specifications and to good vacuum practice. If they do not, the instrument should not be accepted until significant differences are corrected by the engineers performing the installation or are fully justified by the engineers at the factory where the vacuum system was designed. Similar measurements should then be made under identical conditions at bi-weekly or monthly intervals, depending on the level of instrument use, throughout the life of the instrument, and recorded in a log maintained for the purpose. Significant deviations from the original values are indicative of the onset of problems that should be taken care of before the system degrades to the point where specimen contamination or other serious difficulties arise.

The vacuum gauges should normally be turned on while an instrument is in use so that the operator can constantly monitor the vacuum, and the vacuum should be recorded at the beginning and end of each operating session as part of the routine operating log. Operating procedures should be based on numerical readings of the vacuum gauges, rather than on the instrument's 'ready-lights'. For example, the vacuum should be well into the 10^{-3} Pa (10^{-5} Torr) range before an instrument with a tungsten filament is turned on, and in the mid or low 10^{-5} Pa (10^{-7} Torr) range for one with an LaB_6 emitter. The readings of the gauges should also be observed while exchanging specimens to be sure the mechanism involved is being used properly and that it is not leaking. As was mentioned in the discussions of the different gauges above, a vacuum gauge can also be very useful in locating a leak in a vacuum system. This topic will be discussed in more detail in Chapter 10.

3.4 Partial pressure gauges

The types of gauges described above are broadly classified as *total pressure gauges*, because their response depends basically on the total concentration of gas molecules in the gauge tube, but not on the species of molecules present or their relative abundance. *Partial pressure gauges*, which are also often called *residual gas analysers* (RGAs), are devices that do respond to these variables. These devices, which are essentially specialized mass spectrometers, are widely used to monitor many industrial vacuum processes and can be extremely valuable for analysing sources of contamination and diagnosing leak problems on electron microscopes. O'Hanlon (Chapter 4) and Weston (Chapter 5) describe the operating principles of the considerable number of different types of partial pressure gauge that have been developed, and discuss their performance characteristics and use in considerable detail. The following discussion is more descriptive and practically orientated, and deals only with magnetic sector spectrometers, which are used in most helium leak detectors, and with quadrupole spectrometers, which are used in RGAs of the type that are most likely to be installed on an electron microscope.

3.4.1 Magnetic sector devices

As their name implies, these devices use the action of a strong magnetic field on a beam of moving gas ions to separate the gas ions according to their mass-to-charge ratio m/q. This principle was used in some of the first mass spectrometers developed prior to 1920, and is still widely used in the design of mass spectrometers for accurate chemical analyses and in helium leak detectors (Section 10.9, p. 440), because it can provide high resolution and good sensitivity for ions of low mass. The arrangement of the essential components of a spectrometer of this type is shown schematically in Fig. 3.17. Gas molecules from the vacuum system enter the ionization chamber of the spectrometer and are bombarded by electrons emitted from a heated filament. The gas ions produced in this way are drawn from the ionization chamber through the source slit of the spectrometer and are collimated and focused by a negative accelerating voltage V that is applied to one or more accelerator electrodes. The ion beam formed in this manner is directed between the poles of a strong magnet with its field perpendicular to the path of the ion beam. This magnetic field exerts a force on the moving ions that is perpendicular to both the direction of the field and the direction in which the ions are moving. This force causes the ions to follow circular paths while they are inside the magnetic field. Outside the field the ions travel in straight lines. The poles of the magnet have the shape of a sector of a circle, as shown, which usually subtends an angle of 60°, 90°, or 180°. If the magnet is carefully designed, and if the slits are properly located, the magnet acts as a primitive lens and focuses the slightly divergent beam of ions from the source slit into the detector slit for measurement by an ion detector.

Although we are not primarily concerned with the design of mass spectrometers, it is nonetheless interesting and useful to have a general understanding of the way in which individual ion species are separated in such devices. This can be obtained by making use of basic electrostatic theory, which tells us that an ion having a mass m and a charge q gains kinetic energy $KE = mv^2/2 = qV$ when it is accelerated by a voltage V in the manner described above. Solving this equation for the velocity gives $v = \sqrt{2qV/m}$.

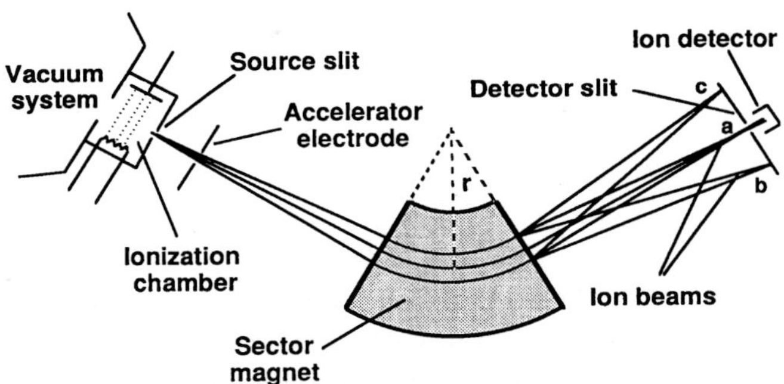

Fig. 3.17 A schematic diagram of a magnetic sector residual gas analyser.

The magnitude of the radial force exerted on the ion by a magnetic field of flux density B is equal to Bqv, and this must be equal to the centrifugal force mv^2/r that also acts on the ion; that is, $Bqv = mv^2/r$. The radius r of the circular path the ion follows in this field can now be found by solving for r and making use of the above equation for the velocity, as follows:

$$r = \frac{mv}{Bq} = \frac{m}{Bq}\sqrt{\frac{2qV}{m}} = \frac{1}{B}\sqrt{\frac{2mV}{q}} \qquad (3.3a)$$

If the mass of the ion is expressed in atomic mass units (1 amu = $1/N_A$ = 1.66×10^{-24} g) and its charge is given as ze, where z is the number of electrons removed from it in the ionization process and e is the charge per electron (1.6×10^{-19} coulomb), and if the strength of the magnetic field B is given in units of teslas (1 T = 10^4 gauss) and the ion accelerating voltage V is given in volts, this expression becomes:

$$r = 0.144\ \frac{\sqrt{V}}{B}\sqrt{\frac{m}{z}}\ \text{mm} \qquad (3.3b)$$

This equation shows that the radius of the path of ions of a given m/z ratio will increase if the accelerating voltage V is increased or if the magnetic field strength B is decreased. For fixed values of V and B, the path radius increases as the mass-to-charge ratio of the ions m/z increases, and so different ionic species follow different paths as they emerge from the magnetic field.

For example, in a spectrometer operating with an ion accelerating potential of about 270 V and a magnetic field strength of about 0.1 T, the path radius for singly-ionized water molecules (H_2O^+, $m = 18$, $z = 1$) with $m/z = 18$ is about 100 mm, while for singly-ionized methane molecules (CH_4^+, $m = 16$, $z = 1$) with $m/z = 16$ it is about 94 mm, and for doubly-ionized carbon dioxide molecules (CO_2^{2+}, $m = 44$, $z = 2$) with $m/z = 22$ it is 110 mm. If the spectrometer magnet has a radius r of 100 mm, and if the source and detector slits are placed at appropriate locations, water ions in the ion beam will be focused by the magnet into the opening of the detector slit, as indicated by ray bundle **a** in Fig. 3.17, producing an output signal from the detector. Under these same conditions methane ions, which follow paths of smaller radius in the magnetic field, are deflected above the detector slit, as indicated by ray bundle **c**. These ions will not be registered by the detector. Neither will carbon dioxide ions, because their larger path radius causes them to be deflected below the detector slit, as indicated by ray bundle **b**. An accelerating potential of about 300 V would be required to bring the methane ions into focus in the detector slit opening; the corresponding potential for the carbon dioxide ions would be about 220 V. By varying the ion accelerating potential slowly and continuously from 200 to 325 V, carbon dioxide, water, and methane ions can be detected successively, and the output signal from the detector can be registered by a chart recorder to produce a mass spectrum. A similar result can, of course, be obtained by holding the ion accelerating voltage constant and varying the strength of the magnetic field over a corresponding range.

Magnetic sector spectrometers were used in early RGAs. They are still used in most helium leak detectors because they can be designed to give very high sensitivity and resolution for He ions. However, they are usually rather heavy and bulky, and are not conveniently portable, and so they are being replaced by the smaller, lighter quadrupole devices for routine residual gas analysis applications. Their strong magnets also make them particularly unsuitable for use on electron microscopes. They have been discussed here in some detail to illustrate in a general way how mass spectrometers work.

3.4.2 *Quadrupole devices*

Quadrupole mass spectrometers were developed in the 1950s, and over the intervening years have become the most popular type for use in RGAs. As indicated schematically in Fig. 3.18, these devices usually have an ion source, circular source and detector apertures, and an ion detector, which perform functions similar to the corresponding components described above for the magnetic sector instruments. The 'quadrupole', from which these devices derive their name, consists of a set of four parallel cylindrical rods (only three of which are shown in Fig. 3.18) arranged symmetrically around the axis defined by the two apertures. A positive DC potential is applied to one opposing pair of these rods, and a negative DC potential to the other pair. Alternating radio frequency (RF) potentials are superimposed on these DC potentials for each pair of electrodes. Gas ions produced in the ionization chamber are directed along the axis of the quadrupole by the combined action of the electric potentials applied to the source and accelerator apertures, and travel in oscillatory paths under the influence of these complex potentials. The electrodynamics of this system are quite complicated but the result is that a given, appropriately-chosen, combination of the DC and RF voltages allows only ions of a single m/z ratio to reach the detector aperture. By varying the DC and RF voltages appropriately, ions of different m/z ratios can be selected, while continuous variation of these voltages produces mass spectra. It is possible to mimic this quadrupole behaviour by applying a combination of DC and RF potentials to a single rod located in the vee of a right-angle plate. This 'monopole' arrangement is used in some instruments, because it is simpler and less expensive to construct than the quadrupole, and can be very competitive in performance. A considerable variety of quadrupole RGAs, which vary widely in cost and performance characteristics, is available from manufacturers and suppliers of vacuum apparatus.

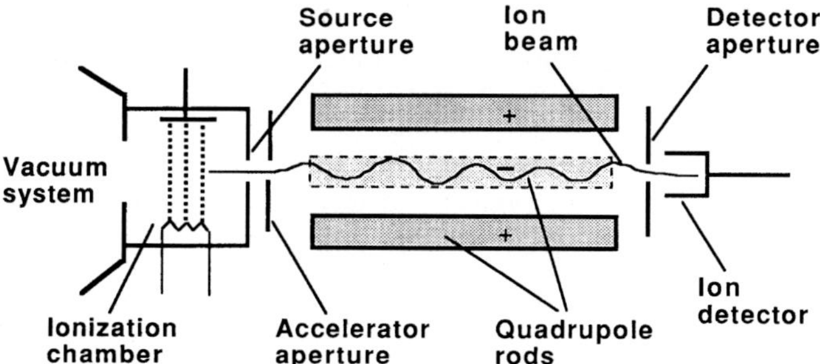

Fig. 3.18 A schematic diagram of a quadrupole residual gas analyser.

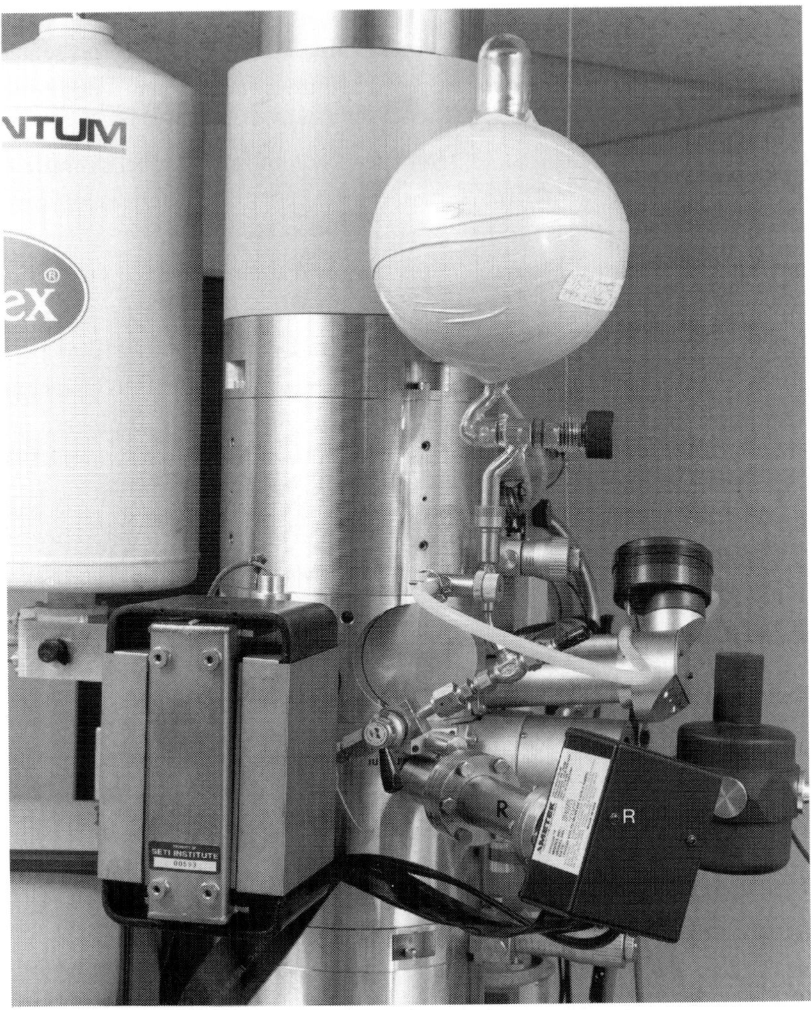

Fig. 3.19 A residual gas analyser (R) installed on a port in the specimen chamber of a scanning transmission electron microscope. (Courtesy of David Blake, NASA Ames Research Center.)

Quadrupole RGAs are preferred over the magnetic sector instruments for many applications, because they are lighter and more compact. Typically the spectrometer head, which includes the detector and first stage amplifier, is only 30–40 mm in diameter and from 100–400 mm in length, and weighs less than 1 kg. Fig. 3.19 shows one of these units installed on the specimen chamber of an analytical scanning transmission electron microscope. Control modules range from basic manually-operated units that provide only simple spectrum display and analysis, to computer-based units with sophisticated software for spectrum analysis and libraries of reference spectra. These latter instruments are designed for, and are highly advantageous in, industrial applications where it is necessary to obtain rapid analyses of complex gas mixtures for process analysis and control. The simpler and less expensive manual instruments are usually adequate for most applications in electron microscopy. Most spectrometers can be set to scan over a mass range from 1 to 50 or 1 to 100 for high resolution analyses or from 1 to 200 or 300 for survey purposes, or can be used to monitor a single mass peak for leak detection or process control. Minimum detectable partial pressures range from 10^{-8} Pa (10^{-10} Torr) for spectrometers that are equipped with a simple Faraday cup detector to below 10^{-10} Pa (10^{-12} Torr) for those that have electron multiplier detectors. The less expensive and more rugged Faraday cup detectors are usually adequate for electron microscopy. In addition, most analysers can be adjusted to measure the total ion current from the ionization chamber, thereby providing a means of measuring total pressures in the range from 10^{-2} Pa (10^{-4} Torr) to below 10^{-7} Pa (10^{-9} Torr). The maximum pressure at which most spectrometers can be operated is about 10^{-2} Pa (10^{-4} Torr), making it unacceptable to install an RGA directly on a system that normally operates at higher pressures. However, most manufacturers offer independent mobile pumping stations for their spectrometers that can be used to keep the spectrometer tube at an acceptable operating pressure while gas is sampled from a system which is at a higher pressure by means of a needle valve.

3.4.3 *Molecular fragmentation*

There is one very important phenomenon that occurs in the ionization chambers of RGAs that must be understood when using these instruments; namely, gas molecules undergo fragmentation as well as ionization. Each type of molecule has a unique 'cracking pattern'; that is, it breaks down into a group of ions having a characteristic set of m/z ratios. The relative abundance of ions with various m/z ratios produced in this fragmentation process depends on the geometry of the ionization chamber and the value of the electron accelerating voltage used in it, while the relative intensities of the corresponding peaks that appear in a mass spectrum are also influenced by the characteristics of the spectrometer and the efficiency of its detector for ions of different *m/z*. Table 3.1 lists these *m/z* ratios and approximate relative intensities for spectra from several common compounds as recorded by a current RGA. Table 3.2 contains similar data for four diffusion pump oils. In these tables the intensity for the most abundant ion species is arbitrarily set equal to 100, and peaks with lower intensities are given values relative to this. Peaks having relative intensities less than one, and peaks with *m/z* greater than 100, are not listed in this table. Spectra recorded for these compounds with other instruments may be slightly different. Most manufacturers provide a set of reference spectra for use in carrying out qualitative and quantitative analyses of unknown spectra with their instruments.

The data presented in these Tables illustrate some of the more important characteristics of the mass spectra of different types of compounds. There is usually a peak representing the singly ionized parent molecule. These are the peaks of greatest intensity in spectra of compounds such as oxygen, nitrogen, water, carbon monoxide, and carbon dioxide, because these simple molecules are relatively resistant to fragmentation. More complex molecules tend to be more susceptible to fragmentation, however, and so their most abundant ionic species may be a molecular fragment rather than the singly ionized parent molecule. For example, the most abundant ionic species for acetone (CH_3-CO-CH_3) is the singly-ionized molecular fragment $(CH_3\text{-}C{=}O)^+$, with $m/z = 43$, which is formed by removal of a methyl group $(CH_3)^+$ from the parent molecule. These methyl groups then produce the second strongest peak at $m/z = 15$, and, after further fragmentation to methylene groups $(CH_2)^+$, the weaker peak at $m/z = 14$. Hydrocarbon pump oils typically break up into several fragments of the general type $H(CH_2)_n^+$, each of which contains an integral number n of methylene groups. These primary fragments undergo further ionization and fragmentation producing groups of ions that cluster around *m/z* ratios of 15, 29, 43, 57, 71, for n = 1, 2, 3, etc. Peaks at *m/z* values of 41, 43, 55, 57, 67, 69 and 71 are usually the most prominent, as shown for hydrocarbon mechanical pump oil in Table 3.2. The peak with $m/z = 43$, for instance, corresponds to the ion $H(CH_2)_3^+$. The presence of several of these peaks in a spectrum is usually indicative of the presence of hydrocarbons in the system. The polyphenyl ether diffusion pump fluids usually produce spectra with strong peaks at *m/z* values of 39, 51, and 77, while

Table 3.1 Approximate relative intensities of the major peaks in residual gas analyser (RGA) spectra of some common gases

m/z	H_2	He	H_2O	N_2	CO	MeOH	O_2	Ar	CO_2	Acet
1	3		3							
2	100					15				1
4		100								
12					10	3			5	4
13						6				4
14				8		11				15
15						50				67
16			3		6		10		10	1
17			25			4				
18			100							
20								6		
26										9
27										13
28				100	100	16			10	17
29						75				8
30						10				
31						100				1
32						67	100			
37										2
38										3
39										6
40								100		1
41										3
42										10
43										100
44									100	2
58										15
M	2	4	18	28	28	32	32	40	44	58

Based on data taken with a UTI 100C RGA using: electron energy = 70 eV, ion energy = 15 eV, focus voltage = –20 V, filament emission = 2 mA, electron multiplier gain = 1000, and resolution $m/\Delta m = 2$. *m/z* = mass to charge ratio, M = molar mass (g/mol), MeOH = methyl alcohol, Acet = acetone. Peaks with *m/z* greater than 100 or relative intensities less than 1 percent are not listed here. (Courtesy of J. Blessing and E. Browning, UTI Instruments Co.)

Table 3.2 Approximate relative intensities of the major peaks in residual gas analyser (RGA) spectra of some common pump oils.

m/z	Mechanical pump oil	Silicone 705 fluid	Convalex 10 fluid	Fomblin Y-25
12				6
16				15
20				28
31				8
39	20	75	65	
41	90	15		
42	15			
43	100	60		
44	8			
47				8
51			100	
53	7			
54	5			
55	65	12		
56	20			
57	75			
58	5			
63			25	
64			30	
65			22	
67	10			
69	15			100
70	10			
71	20			
77		85	90	
78		100		
97				7
M	400	546	454	3400

Based on data taken with a UTI 100C RGA using: electron energy = 70 eV, ion energy = 15 eV, focus voltage = –20 V, filament emission = 2 mA, electron multiplier gain = 1000, and resolution $m/\Delta m = 2$. *m/z* = mass to charge ratio, M = molar mass (g/mol). Peaks with *m/z* greater than 100 or relative intensities less than one percent are not listed here. (Courtesy of J. Blessing and E. Browning, UTI Instruments Co.)

the silicone 705 fluids have strong peaks at m/z values of 43, 77, and 78. The peak at $m/z = 77$, which is prominent in these spectra, is produced by phenyl group fragments $(C_6H_5)^+$. The perfluorinated vacuum oils and greases usually produce spectra with a characteristic strong peak at $m/z = 69$, which is produced by $(CF_3)^+$ ions. O'Hanlon (Chapter 5 and Appendix E) and Harris (Chapter 4) discuss cracking patterns in considerable detail, and cite data for a number of the compounds commonly used in industrial vacuum systems. Since $z = 1$ for most ions formed in RGAs, it is common (although somewhat incorrect) practice to use only mass values in referring to the peaks in RGA spectra (e.g. "water produces peaks at mass values of 16, 17 and 18").

3.4.4 *The use of partial pressure gauges*

In vacuum systems that must handle complex gas mixtures, such as those used in processing semiconductor devices in the electronics industry, RGAs are invaluable for monitoring the composition of the gas mixtures during the course of process runs, and can even be adapted to control the composition by operating automated inlet valve systems. In such applications the RGA may also serve as a diagnostic tool to determine the cause of variations in product quality. For these purposes it is usually necessary to make a quantitative determination of the concentrations of the individual components of the process gas mixture and to analyse for traces of unwanted impurities. Detailed analyses of this kind require accurate information about the characteristics of the spectra of the various components of the mixtures, plus the use of careful calibration and analysis procedures. O'Hanlon (Chapter 5) discusses these methods in some detail. Many current instruments are operated by dedicated computers which have programs that greatly facilitate such procedures.

The uses of RGAs in electron microscopy are usually somewhat simpler and more straightforward. However, an RGA can be highly valuable for monitoring the condition of the vacuum system of an ultra-high resolution or analytical electron microscope. Since we are usually most concerned with monitoring the characteristics of the specimen environment, it is most advantageous to locate the RGA as close to the specimen chamber as possible. Installation in a port in the specimen chamber, somewhat as shown in Fig. 3.19, is probably about as good an arrangement as can normally be attained. The procedure for using an RGA on an electron microscope is basically the same as that outlined above for total pressure gauges. It is important to record a reference spectrum under well characterized operating conditions when the instrument is known, or at least thought, to be in good working condition. Ideally, this would be done soon after the instrument is installed, at which time it can also serve to check the performance of the instrument against the manufacturer's specifications prior to final acceptance. Then spectra should be recorded periodically, perhaps monthly or quarterly, under identical conditions, and kept on file as a record of the instrument's condition. Spectra should also be recorded after any significant service or maintenance work is done on the vacuum system, and any time there is any reason to suspect that changes are occurring in the performance of the vacuum system. Any significant difference between these spectra and the reference spectrum suggests the need for corrective measures.

The mass spectra obtained from the gases in an electron microscope are usually uncomplicated and relatively easy to interpret. An instrument that is substantially free of active leaks will usually yield a spectrum somewhat like the one plotted in grey bars in Fig. 3.20. As shown, the most prominent peaks are usually those produced by the water vapour that constantly desorbs from the internal surfaces of the instrument (Section 2.10.3, p. 62). These occur at mass values of 16 (O^+), 17 (OH^+), and 18 (H_2O^+), and will have individual heights corresponding roughly to the relative intensities given in Table 3.1. The absolute height of these peaks will be proportional to the partial pressure of water vapour in the instrument, and should decrease as the pumpdown period increases, or when liquid nitrogen traps or specimen anticontamination devices are put into use. Hardly any instrument is totally free from leaks, and both oxygen and nitrogen also desorb from internal surfaces, although usually in far smaller amounts than water. Therefore, a peak due to nitrogen at $m/z = 28$ (N_2^+) will usually be present, but should be much smaller than the peaks due to water, and one due to oxygen at $m/z = 32$ (O_2^+) may also be present. If these peaks are more than about one-tenth the height of the water peak at $m/z = 18$, or if they are larger than the water peaks as shown by the spectrum drawn with black bars in Fig. 3.20, the instrument probably has a serious leak. In these spectra the peak at $m/z = 40$ was produced by argon (Ar^+), the peak at $m/z = 14$ by nitrogen (N^+ and N_2^{2+}), and those at m/z values of 1 and 2 by hydrogen (H^+, H_2^{2+}, H_2^+). Very few instruments are totally free of carbon-containing organic compounds. Such materials react

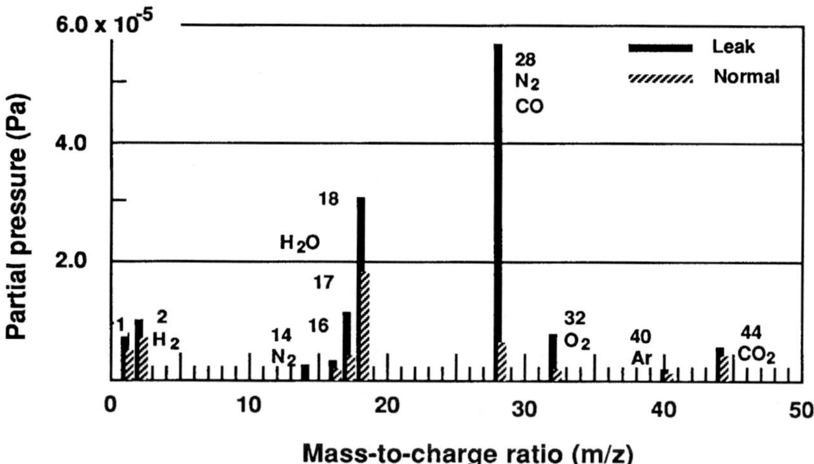

Fig. 3.20 Residual gas analyser spectra showing evidence for a small leak in an analytical electron microscope. (Replotted from data provided by L.F. Allard, Oak Ridge National Laboratory.)

with the hot tungsten filaments in vacuum gauges, in the RGA itself, and in the electron gun, forming tungsten carbide, which then reacts with oxygen and water to produce both carbon dioxide and carbon monoxide, giving rise to peaks at *m/z* values of 28 (CO^+) and 44 (CO_2^+). The relative heights of these peaks varies, depending on the pumping speed and operating conditions for the RGA.

If hydrocarbon pump oils are present peaks will usually appear at *m/z* values of 41, 43, 55, 57 and 69, as indicated in Table 3.2. The magnitude of these peaks should be carefully noted when an instrument is first installed. If they are detectable at the highest sensitivity range of the RGA the microscope has already developed a detectable level of hydrocarbon contamination during assembly and testing at the factory. This is very undesirable because it indicates that the design of the vacuum system is such that it allows contamination to develop in a relatively short time. Since the level of contamination can only be expected to increase with continued use, the instrument is poorly designed and probably should not be accepted. The appearance of these peaks after an instrument has been in use for some time is a clear indication that a problem with hydrocarbon contamination is developing. Once hydrocarbon contamination is present, it is unlikely that it can ever be completely eliminated again; however, measures can be taken to locate its source and to bring that under control. Obviously,

RGA spectra can also be useful for such purposes as checking for evolution of volatile materials from specimens and for determining the effectiveness of cleaning procedures used in service and maintenance operations.

Fig. 3.21 shows three somewhat more complex spectra obtained from the electron microscope shown in Fig. 3.19, which had been modified for special analytical purposes by the addition of a sputter-ion pump, a large anticontamination device cooled with liquid nitrogen, and a bulb and manifold system for introducing gas mixtures under controlled conditions, to ports in the specimen chamber. When looking at spectra plotted in this way it is important to keep in mind the fact that the use of a logarithmic scale for the vertical axis tends to make smaller peaks appear relatively larger than they would if the scale were strictly linear. For example, the peak at $m/z = 45$ appears to be about half the height of the one at $m/z = 44$, whereas numerically its value is only about 1.5×10^{-5} compared to 2×10^{-4} or less than one-tenth as great. Likewise, the peak for water at $m/z = 18$ in the spectrum that is plotted with the diagonally hatched bars is numerically nearly 100 times smaller than the corresponding peak in the spectrum plotted in black bars. It has been remarked that the years of our lives seem to pass by on a logarithmic time scale — those before age 20 seem interminably long, while later years are unbelievably short.

The spectrum drawn in solid black bars was obtained when the instrument was evacuated with only the oil diffusion pump that was originally supplied by the manufacturer. As usual, the most prominent peaks in this spectrum are those of water at m/z values of 16, 17, and 18. There were some problems involved in interpreting the remainder of this spectrum, however, until it was learned that some surfaces of the specimen chamber and anticontamination device had been coated with carbon paint to reduce the scattering of electrons inside the specimen chamber. Then it was realized that the peaks at mass values of 27, 29, 43, and 45 were produced by isopropyl alcohol (CH_3-CHOH-CH_3), a solvent commonly used in carbon paints. The strongest peaks for this compound are:

m/z	19	27	29	31	43	45
Intensity	7	16	10	6	17	100

and most of these are present in this spectrum. The peak at $m/z = 43$ is larger, relative to the one at $m/z = 45$, than would be expected for isopropyl alcohol alone, but undoubtedly contains a substantial contribution from

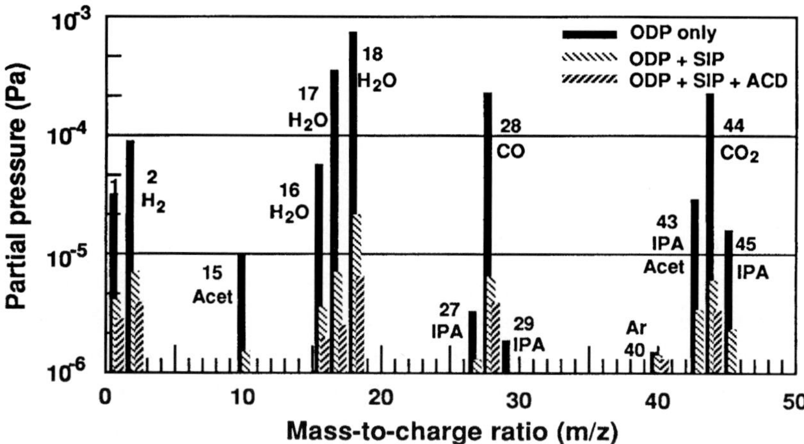

Fig. 3.21 Residual gas analyser spectra from the electron microscope shown in Fig 3.19. ODP, oil diffusion pump; SIP, sputter-ion pump; ACD, anticontamination device; IPA, isopropyl alcohol; Acet, acetone.

acetone (CH_3-CO-CH_3), which is a solvent that is also commonly used in carbon paint. This is supported by the appearance of the peak at $m/z = 15$, corresponding to the second strongest peak of the acetone spectrum (see Table 3.1). The peaks from CO and CO_2, at $m/z = 28$ and 44, are unusually large, undoubtedly because of the considerable amount of acetone and isopropyl alcohol in the system, both of which are carbon-containing compounds that can react with hot filaments to produce these gases. The peaks for hydrogen at $m/z = 1$ and 2, and particularly the peak at $m/z = 1$, are also unusually large, undoubtedly because large numbers of H^+ and H_2^+ ions were formed by fragmentation of water, acetone, and isopropyl alcohol molecules. The spectrum plotted in grey bars shows that all molecular species decreased markedly in concentration when the sputter-ion pump was turned on, because it markedly increased the speed of evacuation for the specimen chamber. The spectrum plotted in diagonally hatched bars shows a further decrease in the partial pressure of water and the disappearance of isopropyl alcohol and acetone, accompanied by a marked decrease in the partial pressures of H_2, CO, and CO_2, when the anticontamination device was also cooled with liquid nitrogen. Collectively, these spectra show the effectiveness of adding supplemental pumps, and of using traps and anticontamination devices cooled with liquid nitrogen, in improving the level and the quality of the vacuum in an electron microscope. They also show the general utility of an RGA in determining the kind and amount of gases present inside an instrument, and in revealing the contributions of various practices, such as applying carbon paint inside an instrument, to the characteristics of the vacuum.

3.5 Capacitance manometers

Pressure is the force per unit area produced by the collisions gas molecules make with solid surfaces. The thermal conductivity and ionization gauges, which are the types of gauges most widely used on vacuum apparatus found in electron microscopy laboratories, do not measure pressure directly. Instead, they determine it indirectly by measuring phenomena which depend on the number of gas molecules per unit volume, a property that is related to the pressure. Because of this, these types of gauges are often classified as 'indirect' pressure gauges. Several types of pressure gauges have been developed which do respond directly to the force of the molecular collisions, however, and these are called 'direct' pressure gauges. The sensing element in these devices is often simply a sealed, thin metal bellows which elongates and contracts slightly in response to changes in atmospheric pressure. This movement of the bellows is registered by a sensitive mechanism and displayed on a calibrated dial. Simple gauges of this kind are not widely used in vacuum work because they usually are not capable of measuring pressures below about 100 Pa (1 Torr) and because they have generally poor response characteristics and limited accuracy. Capacitance manometers, which are highly refined diaphragm gauges, are an exception to this generalization. These devices are now being used to measure the pressure in the specimen chambers of the recently-developed environmental scanning electron microscopes (Section 9.6, p. 416), and so they will be described here briefly.

The way a capacitance manometer works can be explained by referring to the diagram in Fig. 3.22. A thin metal diaphragm divides the manometer housing into two chambers, one of which is sealed under a high vacuum while the other has an open tubulation for connecting the device to the vacuum system. Each chamber contains a stationary metal electrode that is mounted parallel to the diaphragm in such a way that each electrode, together with the diaphragm, forms a parallel-plate capacitor.

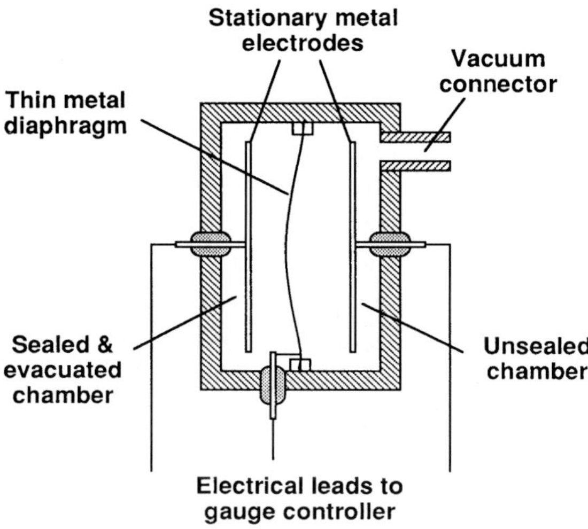

Fig. 3.22 A schematic diagram of a capacitance manometer.

These two capacitors in turn form two elements in a capacitance bridge, a sensitive electrical circuit for measuring capacitance values that is somewhat analogous to the Wheatstone bridge circuit shown in Fig. 3.4, which is part of the circuitry in the manometer controller. Pressure changes in the open chamber of the manometer cause the diaphragm to deflect slightly, changing the spacings between the two electrodes and the diaphragm. This produces changes in the capacitance values of the two capacitors which are then registered by the sensitive bridge circuit. With modern electronics and gauge head design, deflections in the diaphragm which are of the order of 10^{-8} mm can be detected in this way. With proper calibration, this makes it possible to measure very small changes in pressure with a high degree of accuracy and a response time of less than 25 ms.

Capacitance manometers are available from most major manufacturers and suppliers of vacuum equipment. Although most of these commercial models have quite different arrangements of components in the gauge head from that shown in Fig. 3.22, the general principle of their operation is basically the same as described above. The big advantage of these devices is that over their operating pressure ranges they provide the most convenient currently-available means of making reliable, accurate, sensitive, continuous pressure measurements. The gauges currently being used on the environmental scanning electron microscopes typically are capable of measuring pressures over the range from 1 to 10^4 Pa (10^{-2} to 10^2 Torr) with an accuracy of the order of 0.5 percent. This is more than 10 times the accuracy that can be obtained with most other types of gauges that operate in this pressure range, and special manometers with even better accuracy are available at extra cost. Capacitance manometers also provide exceptionally good stability and reproducibility. The calibration of a manometer may change if dust or dirt particles, or drops of a liquid find their way into the gauge head and become attached to the surface of the diaphragm, or if the temperature of the gauge head varies appreciably. Filters and thermostatically controlled heaters are available to help avoid these disturbances. Manometers can also be damaged by a sudden influx of gas at high pressure. Otherwise, they are quite rugged and highly reliable. Unlike the thermal conductivity and ionization gauges, the response of capacitance manometers depends only on the pressure of the gas in the measurement chamber, and is not a function of the type or composition of the gas.

The pressure required to produce the maximum tolerable deflection of the diaphragm depends on the stiffness of the diaphragm, and so it is possible to make gauges with different full scale pressure values by using diaphragms of different thickness. Manufacturers currently offer gauges with full scale pressure capabilities of 1, 10, 100, 1000, 5000, etc., up to 50 000 mbar (1–50 000 Torr). A given gauge can typically make measurements over a total range that extends three or four orders of magnitude below the full scale maximum value, and so overall capacitance manometers can be used to measure pressures from about 10^{-3} mbar (10^{-3} Torr) up to about 50 atmospheres. Controllers for capacitance manometers now use solid-state electronics, and so are compact, convenient to use, and highly reliable, and are available with a variety of display options and with computer interface circuits.

3.6 Concluding remarks

Any vacuum system can be operated more effectively, more intelligently, and more confidently if it is equipped with an appropriate set of vacuum gauges which provide a numerical display of the pressure in strategic parts of the system. This is true for nearly all kinds of vacuum instruments, but is particularly true for modern, high-resolution and analytical electron microscopes, because of the critical role the vacuum system plays in determining the performance of these instruments. Users should include a good system of gauges in the specifications for all vacuum instruments at the time their purchase is being negotiated, and should refuse to purchase instruments that do not have such gauges. The added cost should be a negligible fraction of the total purchase price of most instruments, while the benefits will be a significant improvement in the use, management, and performance of the instruments throughout their life. The current trend by many manufacturers of substituting a couple of blinking green lights, or an obscure code number in the 'back page' of a computer display system, for a good numerical display of pressure should not be tolerated. An RGA is a particularly useful device to have on analytical and ultra-high resolution microscopes because of the detailed information it can provide about the compounds present in the vacuum of the specimen chamber. As will be discussed in Chapter 10, vacuum gauges and RGAs are also very helpful in diagnosing vacuum problems and in locating leaks in vacuum systems.

4 Rough vacuum pumps

Pumps with ultimate pressures above about 10^{-2} Pa (10^{-4} Torr) are called 'rough vacuum pumps'. These pumps perform several important functions in vacuum systems. They are, in fact, necessary components of all vacuum systems. At one time or another every vacuum system is opened to the atmosphere. Since most pumps that produce the high and ultra-high vacuums needed in electron microscopes and similar apparatus cannot operate at pressures above about 1 Pa (10^{-2} Torr), separate rough vacuum pumps must be used to reduce the pressure to this level, establishing a rough vacuum, before the high vacuum pumps can be put into operation to produce the final working vacuum. This operation of removing gas at atmospheric pressure from a vacuum system is often referred to as 'roughing out' the system, and since most rough vacuum pumps function effectively in performing this operation, they are also commonly called 'roughing pumps'. Oil diffusion pumps and turbomolecular pumps, two of the most widely used types of high vacuum pumps, are unable to develop output pressures greater than about 1 Pa (10^{-2} Torr), and so rough vacuum pumps must be used to back up these pumps. When used in this way they are commonly called *backing pumps*. In addition, a rough vacuum pump may be the only type of pump needed on apparatus that do not require pressures below the rough vacuum range.

At the present time nearly all rough pumping is done with oil-sealed rotary-vane pumps, and so most of this chapter is devoted to discussing these pumps. In recent years, however, many vacuum processes have arisen in which the presence of hydrocarbons has a deleterious effect. The almost inevitable contamination of these processes by oil from these oil-sealed mechanical pumps has led to a rising interest in the development of oil-free pumping systems. Several of the types of rough vacuum pumps being used for this purpose are described briefly in Section 4.2 (p. 155).

The references cited in this chapter are listed in Appendix 1. The manufacturers and suppliers of the vacuum equipment referred to here are listed in Appendices 2 and 3.

4.1 Oil-sealed rotary-vane pumps

At least three different types of oil-sealed rough vacuum pumps are presently in use. These include rotary piston pumps, vane pumps with eccentric rotors, and vane pumps with concentric rotors. These pumps are particularly useful because of their ability to discharge gas from a vacuum system directly into the surrounding atmosphere. They are so widely used in such a variety of roles that, overall, they probably outnumber all other types of vacuum pumps. Only the vane pumps with concentric rotors, which are usually referred to simply as 'rotary-vane pumps', will be discussed here, because they are used on virtually all electron microscopes, and on most other types of research apparatus that have vacuum systems. In fact, this is almost the only type of rough vacuum pump currently found in research laboratories. Hablanian (Chapter 5) and Harris (Chapter 5) also discuss the characteristics and uses of pumps of this type. The other types of rotary pumps, which are more widely used for industrial applications requiring very large pumps, are described by Van Atta (Chapter 5) and O'Hanlon (Chapter 7).

4.1.1 Construction and function

The way these common rotary-vane pumps are constructed and their function can be understood by reference to the schematic diagram of Fig. 4.1. As shown, these pumps have a cylindrical rotor which is mounted off-centre in a cylindrical or slightly elliptical chamber of larger diameter. The rotor nearly touches the inside wall of this chamber along a line between the openings to the inlet and outlet ports. The clearance between the rotor and the chamber is typically only about 0.025 mm in this region, and in some models the chamber is machined to match the curvature of the rotor. The crescent-shaped space between the rotor and the chamber is divided into two parts by two vanes, which are held in slots in the rotor and pressed against the chamber wall by springs. Fig. 4.2 is a photograph of a rotary-vane pump with one end-plate removed to show the rotor and sliding vane assembly inside the pumping chamber. When the rotor is in the position shown in Fig. 4.1, gas molecules can diffuse from the vacuum system into the space between the rotor and the chamber via the inlet port. Remember, no pump exerts a force that attracts or draws gas molecules into it. Gas molecules enter a pump under the influence of their constant,

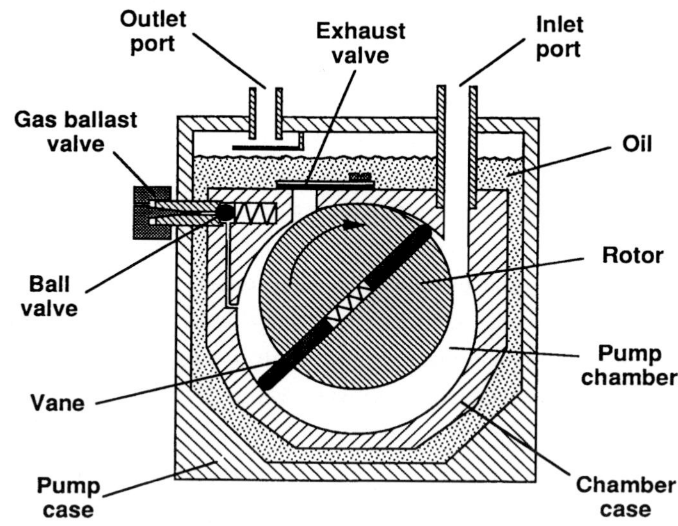

Fig 4.1 A schematic diagram of an oil-sealed rotary-vane rough vacuum pump

Fig 4.2 The pumping chamber of an oil-sealed rotary-vane vacuum pump

random, thermally-activated motion (Section 2.5.1, p. 45), because the concentration of molecules in the pump is lower than in the vacuum system to which the pump is connected. As the rotor turns, the upper vane moves past the inlet opening and the gas that has entered the pump becomes trapped in front of the advancing vane. The space defined by the vane decreases as the rotor continues to turn, and the trapped gas is compressed until its pressure ultimately becomes great enough for it to be forced out of the pump through the spring-loaded exhaust valve. Rotary-vane pumps are available in either single-stage or two-stage models. As the name implies, single-stage pumps have only a single pumping chamber. Two-stage pumps have two chambers mounted side by side in a single pump body, with their rotors sharing a common drive shaft, and with the outlet port of the first stage connected via an internal duct to the inlet port of the second stage. One important advantage of the rotary-vane pumps, compared with the other types of mechanical pumps mentioned above, is that they have lower levels of mechanical vibration because the drive shaft is concentric with the rotor.

For many years all rotary mechanical pumps were 'belt-driven'. That is, the pump was mounted on a base plate with a separate electric motor that drove the pump by a belt-and-pulley arrangement as shown in Fig. 4.3. Half of the belt guard was removed from the pump when this photograph was taken to show the design more clearly. However, a belt guard is a required safety device that should always be fully and properly installed when a belt-driven pump is in use. Usually the pulley on the pump is considerably larger than the one on the motor and so the pump turns at a speed which is considerably less than that of the motor, typically at about 500 rpm. Few manufacturers currently produce belt-driven pumps; most now favour *direct-drive* pumps, such as the one shown in Fig. 4.4, which have the motor connected directly to the shaft of the pump by a flexible coupling. With this arrangement, the rotor of the pump turns at the same speed as the motor, which is usually about 1800 rpm for 60 Hz induction motors and about 1500 rpm for 50 Hz motors. This makes it possible to achieve the same free-air displacement with much smaller rotors and pump chambers, and so direct-drive pumps can be considerably smaller, lighter, and easier to handle than belt-driven pumps, for a given pumping capacity. The absence of the drive belt and pulleys is a safety advantage of direct-drive pumps, and they are usually slightly less expensive than belt-driven

Fig 4.3 A belt-driven rotary-vane vacuum pump manufactured by Welch Vacuum Technologies

Fig 4.4 A direct-drive rotary-vane vacuum pump manufactured by Leybold Vacuum Products

pumps of comparable pumping speed. Consistent with this trend, most manufacturers of electron microscopes, vacuum evaporators, and similar apparatus now use direct-drive pumps. It is not entirely clear that this is an advantage to the purchasers of these instruments, however. When the service and maintenance procedures recommended by the manufacturer are followed, a good belt-driven pump, with its slow speed of rotation, can be expected to perform satisfactorily for more than 10 years of continuous operation. Experience to date indicates that the lifetimes of direct-drive pumps may be considerably shorter because their higher speed of rotation leads to more rapid mechanical wear.

The oil that leads to the classification of these pumps as 'oil-sealed' performs several important functions. It serves as a lubricant to minimize wear of the moving parts, it acts as a heat transfer agent to aid in cooling the pumping mechanism, it provides corrosion protection, it seals the exhaust valve, and it is the ultimate sealing agent in preventing the high-pressure gas on the outlet side of the pump from leaking past the rotor and the vanes back into the inlet. Commonly this oil is a high quality mineral oil which has been chemically treated to remove unsaturated and aromatic compounds and then vacuum distilled to remove volatile components. Doubly- and triply-distilled grades, which give longer service, lower pressures, and lower backstreaming rates, are also available. In chemical and industrial applications synthetic diester fluids, which are more resistant to cracking and sludge formation than the hydrocarbon oils, may be used to pump hot gases, while perfluoropolyether fluids, sold under the Fomblin and Krytox trade names by Ausimont and Du Pont, respectively, are used to pump oxygen and corrosive gases. Specially designed pumps may be required for these synthetic fluids.

In earlier pumps, the entire pumping mechanism was immersed in the oil, somewhat as shown in Fig. 4.1. There are shortcomings associated with this arrangement, however, because it is difficult to optimize the amount of oil fed into the pump for lubrication, and because large amounts of oil can be sucked out of the pump into the vacuum line if the pump is turned off while under vacuum. Most currently manufactured pumps have more sophisticated designs, which include anti-suckback features and oil distribution systems that deliver oil from a reservoir to the critical parts of the pump in the correct amount for optimum performance.

4.1.2 Pumping speed

The usual practice is to rate all types of vacuum pumps on the basis of their pumping speeds and the ultimate pressures they can attain (Section 2.5, p. 39). When rating rotary-vane pumps, manufacturers usually quote the free air displacement for the rated pumping speed. This is the volume of air that can be pumped per unit time when the inlet and outlet ports are both at atmospheric pressure. Speed ratings can be calculated from the volume of the space between the rotor and the chamber and the number of revolutions the rotor makes per minute. Measured values are generally considered to be more accurate, however. Traditionally the speeds of smaller pumps have been given in units of litres per minute (l/min), while units of cubic feet per minute (for which the unusual abbreviation of 'cfm' is commonly used; 1 cfm = 28.3 l/min) were used for larger industrial pumps (see Table 2.1 in Section 2.1 p. 28). Recently some manufacturers have begun to use units of cubic metres per hour (1 m^3/h = 0.589 cfm = 16.7 l/min). The pumps used on most research apparatus typically have speeds in the range from 20 to 500 l/min, while those in industrial applications range up to 1000 cfm. Pumps on electron microscopes and vacuum evaporators usually have speeds in the range from 100 to 200 l/min.

Rotary-vane pumps exhibit maximum pumping speed when the inlet is at atmospheric pressure. As the inlet pressure decreases so does the pumping speed. Although there is some variation depending on pump design, it typically drops to about two-thirds the rated speed when the inlet pressure decreases to about 10 Pa (10^{-1} Torr) for a single-stage pump, and to about 1 Pa (10^{-2} Torr) for a two-stage pump, and then falls rapidly until it becomes zero at the ultimate pressure of the pump. Performance curves typical of those obtained from single- and two-stage pumps with rated speeds of about 150 l/min are shown in Fig. 4.5. Because of their high pumping speeds in the range from atmospheric pressure down to about 10 Pa (10^{-1} Torr), rotary-vane pumps are commonly used for the initial stages of evacuation of electron microscopes, vacuum evaporators, and a host of other apparatus. Because of their unique ability to discharge directly into the atmosphere they are also used as backing pumps for high vacuum pumps, such as oil diffusion and turbomolecular pumps, which do not develop discharge pressures great enough to do this.

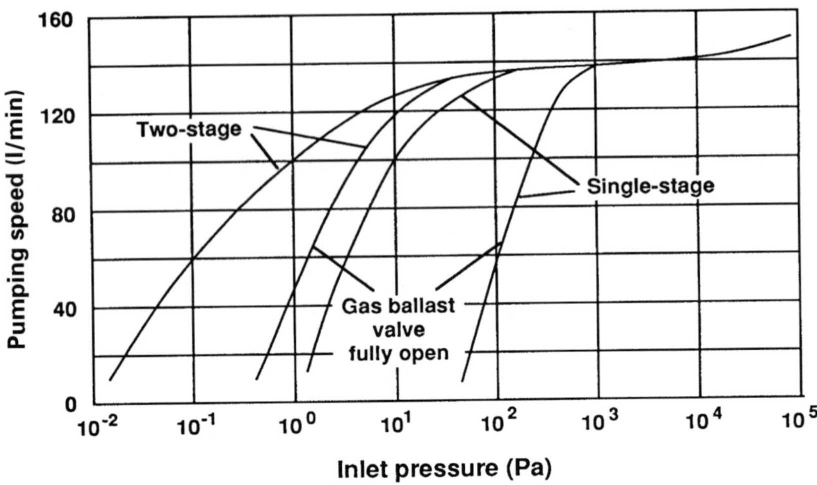

Fig 4.5 The variation of pumping speed with inlet pressure typical of single- and two-stage rotary-vane pumps with rated speeds of 150 l/min when operated with the gas ballast valve closed or fully open.

4.1.3 *Ultimate pressure*

The ultimate pressure used in rating mechanical rough vacuum pumps is the 'blank-off pressure'. This is the lowest pressure a new pump filled with fresh, clean, dehydrated oil can attain when a vacuum gauge is sealed directly into its inlet port. This pressure limit is determined by the vapour pressure of the pump oil, the amount of air that dissolves in the oil, and the extent to which air leaks past the vanes and rotor from the exhaust side to the inlet side of the pump. Interestingly enough, it also depends on the type of gauge used to measure the pressure. If a McLeod gauge, a specialized mercury manometer, is used a partial pressure value is obtained, because the gas is placed under pressure during measurement and so the pump oil and other condensable gases do not contribute to the gauge reading. If a Pirani gauge or a capacitance manometer is used, all gases contribute to the reading and a total pressure value is obtained which is usually about a factor of 10 greater than the partial pressure value. Some manufacturers report both values, although the partial pressure value looks most favourable and is the one most often used in advertisements.

Typically, single-stage pumps have ultimate pressure ratings of about 1 Pa (10^{-2} Torr) as measured by a McLeod gauge, and 10 Pa (10^{-1} Torr) as measured with a total pressure gauge. Two-stage pumps are able to achieve lower ultimate pressures than single-stage pumps because the second pumping stage keeps the pressure at the outlet port of the first stage well below atmospheric pressure. This greatly decreases the pressure drop across the first stage, giving a correspondingly lower rate of gas leakage past the vanes and rotor into the intake port of the pump. Typically the ultimate pressure ratings for two-stage pumps are about 10^{-2} Pa (10^{-4} Torr) on the partial pressure basis, and 10^{-1} Pa (10^{-3} Torr) on the total pressure basis. Rotary-vane pumps seldom produce pressures in practical vacuum systems that approach their rated ultimate pressures for a variety of reasons, including: increased leakage of gas past the vanes and rotor resulting from mechanical wear; degradation and contamination of the pump oil; and gas influx into the system being evacuated.

4.1.4 *Contamination of the pump oil*

The performance of rotary-vane pumps can be severely degraded, and the pumps may even be seriously damaged, if they are used to pump corrosive or condensable vapours. This should be an unlikely occurrence for pumps used on electron microscopes; however, it is a great temptation for uninformed personnel to use a vacuum evaporator to 'dry something out'. Extreme examples include the students who, on separate occasions over the years, used the vacuum evaporator in our laboratory to dry out a wet pair of expensive tennis shoes, to attempt to dehydrate a jar of instant coffee that had water accidentally spilled into it, and to remove excess toluene from a polymer solution. Admittedly, these are rare and unusual events that were performed by people of unusual 'ingenuity'; however, rotary-vane pumps are routinely used for similar, although perhaps less bizarre, applications in many laboratories. Most electron microscopists, for example, are accustomed to using them for pumping large amounts of water in the process of desiccating photographic film.

The problems associated with using rotary-vane pumps in this way arise because the pressure of the gas in a rotary-vane pump must exceed one atmosphere by a significant amount in order for it to be forced out of the pump against the spring-loaded outlet valve. Under such pressures, the vapours of water and most common organic solvents, such as acetone, alcohol, Freon, and toluene, condense to liquids and either dissolve in the pump oil or become dispersed in it as droplets. As the oil circulates through the pump these materials are carried over to the inlet side where they again vaporize in the reduced pressure existing there, diminishing the capacity of the pump to accept gas molecules from the inlet port. As a result, the pumping speed of the pump decreases markedly and the ultimate pressure it can attain rises toward the vapour pressure of the contaminating material at the operating temperature of the pump. Some contaminants may even corrode the pump or impair the lubricating action of the oil. If a pump becomes badly contaminated, the only solution is to drain out the contaminated oil, flush the pump thoroughly (use only fresh pump oil for this purpose) (Section 4.1.7, p. 150), and refill it with fresh oil.

4.1.4a *The gas ballast valve*

Notwithstanding these potential problems, it is often very convenient to use a rotary-vane pump to pump modest amounts of condensable vapours. In such applications severe contamination of the pump oil can often be prevented by using a gas ballast valve (also called an 'air ballast' valve or 'vented exhaust valve'). As shown schematically in Fig. 4.1, this device functions like a needle valve in series with a small spring-loaded ball valve located on the exhaust side of the pump. When the needle valve is open, atmospheric pressure can force a small amount of air through the ball valve into the pump chamber before the pressure of the gas there becomes high enough to cause the condensable material to liquefy. Dilution of the gas mixture in the pump chamber with air in this manner prevents the partial pressure of the condensable material from exceeding its vapour pressure, and so it is swept out through the exhaust valve without condensing to a liquid. As shown in Fig. 4.5, the use of the gas ballast valve degrades both the pumping speed and ultimate pressure of a pump, because the admitted air increases the pressure on the outlet side of the pump, and this increases the rate at which gas leaks past the vanes and rotor and back into the inlet port. When the gas ballast valve is fully open the ultimate pressure can rise as high as 1 Pa (10^{-2} Torr) for a two-stage pump, and 100 Pa (1 Torr) for a single-stage pump. Therefore, the gas ballast valve should be adjusted to admit the minimum amount of air needed for a particular situation, and should be fully closed when the gas ballast action is not needed and best pump performance is desired. In addition to helping prevent serious contamination of a pump, a gas ballast valve can often be remarkably effective in cleaning up a moderately contaminated pump if the inlet of the pump is blanked off and the pump is run for several hours with the gas ballast valve operating, preferably admitting dry air or dry nitrogen. Harris (p. 78) and Van Atta (p. 180) discuss the theory and use of the gas ballast valve in considerable detail.

4.1.4b *The liquid nitrogen trap*

In applications that absolutely require the pumping of large amounts of condensable vapours, the best procedure is to prevent the vapours from reaching the pump by installing a liquid nitrogen trap in the inlet line to the pump. A trap that is relatively inexpensive to build and which works well for this purpose is shown in Fig. 4.6. In this design, the liquid nitrogen is contained in a glass tube, which provides good thermal insulation and minimizes the rate of evaporation of the liquid nitrogen. This tube can easily be made from a standard fitting of the type used to form an O-ring joint between two pieces of borosilicate glass pipe in chemical piping systems. In the United States such fittings are supplied with diameters of 40, 50, and 75 mm. The part of the fitting that is designed to hold the O-ring is thickened, and so it has good strength, and it is flared out, and so will seal against the O-ring that is held in the shallow groove in the tapered surface

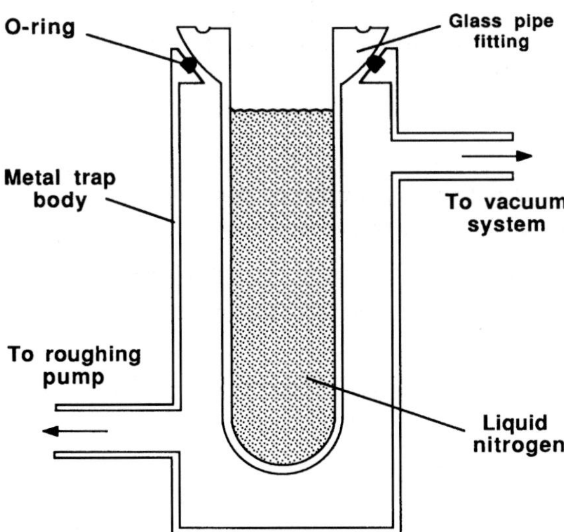

Fig 4.6 A liquid nitrogen trap for preventing a rotary-vane pump from becoming contaminated with condensable materials from the vacuum system.

of the metal ring on the top of the trap body. The other end of the fitting is sealed off to form the bottom of the liquid nitrogen reservoir. A reservoir 75 mm in diameter and 200 mm long will hold three-quarters of a litre of liquid nitrogen, which is enough to last a convenient time between fillings. This reservoir should be made of high-silica or borosilicate glass for strength and thermal shock resistance, and should be fully annealed to minimize the potential for accidental breakage. Since no screws or clamps are needed to hold it in place, the reservoir can be lifted out of the trap body easily and quickly for cleaning as soon as the pumping line is let up to atmospheric pressure. Stainless steel traps that also can be taken apart easily for cleaning are available from some suppliers of vacuum equipment (e.g. Lesker).

During use, it is important to keep such a trap constantly filled to a level near the top of the liquid nitrogen reservoir, particularly if large amounts of condensable materials are being pumped; otherwise, vapours will be released from the trap as the level of liquid nitrogen lowers, and may enter the pump if the liquid nitrogen level gets too low or if the lower part of the trap becomes covered with a very heavy coating of condensed material. Once a trap has accumulated condensed material it *must not be allowed to warm up* while the system is under vacuum or the condensate will evaporate and contaminate the pump. Instead, the trap must be kept cold until the pump is turned off and air is admitted to it. Then, the trap *must be removed* from the pumping line *immediately* and cleaned before it is used again, so that the condensate is removed from the system.

4.1.5 *Contamination of the vacuum system by pump oil*

All oil-sealed mechanical pumps have one characteristic that is causing them to fall out of favour for many vacuum applications; that is, they can be very serious sources of oil contamination in vacuum systems. This arises because the oil in these pumps is somewhat volatile, and because the oil molecules are broken down by the mechanical action of the rotor and vanes rubbing against the wall of the pump chamber into low molecular weight fragments which have even greater volatility. Molecules of the pump oil, and particularly these more volatile molecular fragments, gradually diffuse out through the inlet port of the pump, and eventually move into the vacuum system by vapour diffusion and by creeping along the walls of the pumping line. An impressive amount of research has been devoted to this problem, and manufacturers have tried a great many variations in the design and construction of these pumps, without eliminating it. At present it appears that this is just an inherent feature of oil-sealed rotary mechanical pumps. This is very unfortunate, because these pumps have so many useful characteristics, and because no other pumps perform quite as well in rough pumping vacuum systems from atmospheric pressure.

It is possible to reduce the rate at which oil molecules backstream from a rotary-vane pump into the high vacuum part of the system significantly, however. The most common approach is to keep the pressure in the line that leads from the rotary-vane pump into the system in the range of viscous flow whenever the line is open to the system, so that the oil molecules are physically prevented from entering the system by the viscous flow of air molecules toward the pump (Section 2.2, p. 29). For pumps that are only used for roughing purposes this can be accomplished by closing a valve in the inlet line to the pump before the pressure drops below the range of viscous flow, and then admitting air to the pump and shutting it off immediately. Likewise, when a rotary-vane pump is used to rough pump an instrument such as an electron microscope, the rough pumping operation can be stopped, by closing the valve between the roughing line and the high vacuum part of the system, before the pressure drops into the viscous flow range. Fig. 2.1 (Section 2.2) shows that these methods limit the attainable rough vacuum to about 10 Pa (10^{-1} Torr). For pumps that are used to back-up turbomolecular or oil diffusion pumps, the desired effect can be accomplished by installing an adjustable leak valve in the inlet line and setting it to admit air at such a rate that the pressure in this line does not fall below 10 Pa (0.1 Torr). This is not a good solution, because the ultimate pressure attainable by these types of pumps depends strongly on the pressure in their backing lines (Section 6.1.3, p. 234). However, the XEI Scientific company have recently introduced a device (SEM CLEAN) that makes use of this principle by automatically admitting dry nitrogen into pumping lines and vacuum systems at a pressure in the range of viscous flow to remove contaminating oil molecules. This device is discussed in more detail in Section 5.5.5 (p. 199).

It is also important to prevent oil-sealed mechanical pumps from overheating during use. An increase of five or ten degrees in the temperature of the oil in a pump can increase the rate at which it undergoes thermal degradation enough to double or triple the rate of backstreaming and contamination. Pumps that are installed inside the cabinets of the currently-fashionable, compact desktop instruments are particularly likely to overheat unless very good forced ventilation is provided. In applications where the control of the backstreaming rate is critical, an effort should be made to keep the temperature of a pump from rising above 20°C during operation.

4.1.5a Foreline traps

Another approach to controlling oil contamination from a rotary-vane pump is to install a trap in the line that leads from the pump to the vacuum system. Traps designed for this purpose are usually referred to as foreline traps (rather than backing line traps) in manufacturers' literature. Usually it is impractical to use a liquid nitrogen trap in this application, because it is bothersome and expensive to keep filling it, especially for vacuum systems like those on electron microscopes where the rotary-vane pumps may run continuously for months at a time. However, several other types of traps are available. Traps filled with molecular sieve materials are widely used, and are considered by many to be the most effective. Molecular sieves are alkali aluminosilicate compounds (synthetic zeolites) that are very active adsorbents because they have very high surface areas. In fact, they are the adsorbent most commonly used in sorption pumps (see Section 4.2.1, p. 155). There are two problems associated with the use of these materials, however. First, because they contain numerous pores with dimensions of about one nanometer, they adsorb such large quantities of water, oxygen, and nitrogen when exposed to the atmosphere that the evolution of these gases when they are subsequently evacuated can delay the attainment of a good vacuum unless the molecular sieve material is baked out at a temperature near 200°C under vacuum. For this reason it is desirable to install valves on both sides of a molecular sieve trap so that it can be isolated whenever the pumping line is brought up to atmospheric pressure. Second, molecular sieves are usually produced in the form of sintered pellets, and these have a tendency to degrade to dust particles that can be very damaging if they get into a mechanical vacuum pump. After a period of use, the molecular sieve material becomes loaded with adsorbed gases and must be regenerated by heating for several hours at about 200°C, preferably under vacuum.

Several other types of filler are available which do not have these problems. The Micromaze traps produced by the Lesker company have a proprietary filler that is in the form of interleaved sheets of a ceramic material which has a very high surface area and which adsorbs oil molecules nearly as well as molecular sieves. This material does not readily degrade to dust particles, and it does not adsorb water and atmospheric gases as tenaciously as molecular sieves, because its pores are too small (about 5 nm) to accept the molecules of these gases. As with the molecular sieves, the Micromaze filler is regenerated by heating to about 200°C for several hours under vacuum, using heaters available from the manufacturer. Ed Birko, of the Hitachi company, has suggested that it is advantageous to heat these traps continuously at 70–80 percent of full power (by using a variable-voltage autotransformer to control the power input) while they are in use. It is postulated that doing so causes oil molecules that reach the Micromaze elements to be broken down into volatile fragments which are easily pumped away. Somewhat similarly, the Model URB trap produced by Balzers contains a catalyst which, when heated to 250°C, converts hydrocarbons to carbon dioxide and water. Spheres of activated alumina (Al_2O_3) are recommended as fillers for foreline traps by the Edwards company, and are claimed to be effective in collecting more than 95 percent of the oil vapours emanating from rotary-vane pumps. These spheres do not produce dust particles, nor do they adsorb significant quantities of water or atmospheric gases, and they can be regenerated by heating under vacuum at about 200°C for several hours (or, perhaps more simply, by putting them in an ordinary furnace for an hour at about 550°C, which will burn off hydrocarbons quickly). Hydrocarbon molecules readily adsorb onto machined copper surfaces, and so copper wool is often used as a filler for foreline traps. Stainless steel wool, glass fibres, and carbon granules are also used when reactive gases are being pumped. It is not easy to regenerate these materials reliably, and so they are usually replaced. Fig. 4.7 is a photograph of a trap with copper wool filler.

It is extremely *important* to recognize that a foreline trap can be effective only if it is properly cleaned, and if the filler material is replaced or regenerated frequently enough to prevent oil molecules from penetrating through it into the part of the pumping line that is connected into the vacuum system. Otherwise, it will become saturated with oil, whereupon it will degenerate into a storage site for oil and become a source of

Fig 4.7 A foreline trap designed to be easily disassembled and cleaned. The filler is copper wool.

contamination itself. The frequency with which such service should be performed will, of course, depend on the conditions under which a trap is used. Manufacturers usually recommend doing so at least as often as every 6 months. In addition, a trap must be installed before a pumping line is first put into use, because it cannot be effective if oil molecules have already been allowed to reach the high vacuum end of the pumping line.

Foreline traps are available in three basically different designs, and maintenance procedures are quite different for each. Some traps are designed so they cannot be opened, and so that the filler material in them cannot be replaced. Maintenance of these traps is relatively simple, although it may be somewhat expensive; it involves merely discarding the old trap and installing a new one. If replacement is performed frequently enough, traps of this kind are likely to give the best overall protection of a vacuum system with the least physical effort. Some traps, such as the one shown in Fig. 4.7, are designed so that they can easily be removed from the pumping line, opened, disassembled, and cleaned. Satisfactory cleaning can usually be accomplished by wiping all internal surfaces of the trap several times with cloth pads moistened with a 50/50 mixture of toluene and acetone, rinsing the trap several times with this mixture, preferably in an ultrasonic cleaner, then rinsing it thoroughly with reagent grade isopropyl alcohol, and finally drying it with a hair dryer or heat gun (Fisher Scientific). As a general rule, the safest practice is to replace the filler material in these traps. Again, very dependable results can be obtained if these maintenance procedures are performed carefully and frequently enough. Finally, many traps, particularly those that have a molecular sieve material as the filler, are made so that they cannot be opened or taken apart, but are supplied with heaters for use in regenerating the filler. Regeneration usually involves heating the trap to about 200°C for several hours while pumping on it with the mechanical roughing pump. For this treatment to be fully effective, however, it is necessary to install a bakeable valve adjacent to the trap on the high vacuum side of the line, and to heat this valve to a slightly higher temperature than the trap during the regeneration treatment. Otherwise, hydrocarbon molecules driven out of the trap will simply condense in the valve or be driven into the vacuum system, and the intended function of the trap will be totally defeated. Traps of this kind require considerably more careful management than the other two types. Finally, it might be noted that oil molecules can also reach the high vacuum part of a vacuum system by migrating along the walls of the pumping line. A well designed trap that is properly maintained can also serve as an effective barrier to oil contamination by this process.

All of these approaches to controlling hydrocarbon contamination have serious drawbacks. The viscous flow method requires accurate pressure measurement and careful attention to operating procedures, and is readily subject to error. The use of a liquid nitrogen trap is bothersome and expensive; furthermore, it becomes ineffective if the trap warms up while under vacuum. The other types of trap are very likely to be neglected, whereupon they become saturated with oil and lose their effectiveness. Therefore, contamination of vacuum systems by oil from rotary-vane pumps remains a serious problem which, from a practical standpoint, remains unsolved.

4.1.6 *Contamination of the surroundings by oil mist*

The sealing oil in mechanical pumps can cause problems in another way. When a pump is operated at or near atmospheric pressure a very considerable amount of gas flows through it. As it is ejected through the oil that seals the outlet valve, this gas produces tiny drops of the oil and carries them through the outlet port of the pump and into the surrounding atmosphere. The quantity of oil mist produced in this manner can be quite large, particularly from large pumps. If the pump is installed inside the cabinet of an instrument, everything inside the cabinet quickly becomes coated with oil. If the pump is standing in an open room, the floor and nearby walls and furniture soon become covered with an oil film. In addition, it is clearly unhealthy to breathe this oil mist. The best solution to this problem is to install a mist filter on the outlet of the pump. Most manufacturers sell mist filters for their pumps that meet current clean-air standards. Some of these are exorbitantly expensive, however, so it pays to shop around. Even with a mist filter in use, a pump will eventually become covered with oil which will then drip onto the floor. It is therefore a good idea to set a mechanical pump in a shallow pan to prevent the oil from soiling the floor and spreading around the laboratory. Incidentally, the clay products commonly sold for filling cat litter boxes usually absorb oil readily. It is handy to keep a bag of such material in the laboratory for cleaning up spilled oil. Sprinkle it over the area where the oil is spilled, brush it around occasionally over a period of a few days, and then sweep it up.

4.1.7 *Service and maintenance of rotary-vane pumps*

Rotary-vane pumps should be carefully serviced on a regular basis to ensure that they remain in good operating condition. It is of primary importance to be sure that an operating pump always contains the correct amount of oil. If the oil level becomes too low both the pumping speed and the ultimate pressure will degrade, and the pump may suffer mechanical damage. There are at least two ways in which a rotary-vane pump can lose oil. First, oil is continually ejected as a vapour through the outlet port of a pump whenever it is in operation, and as a fine mist during times when gas throughput is high. Second, after a period of use the seal around the drive shaft tends to develop a leak and allow oil to escape from the pump. Therefore, the level of oil in an operating pump should faithfully be checked *at regular intervals* and replenished as needed. As related in Section 10.3 (p. 430), failure to do this can lead to embarrassing service problems. At the same time the pump should be inspected for signs of a leak around the drive shaft. A gradual increase in the frequency with which the oil needs to be replenished, accompanied by the appearance of a steadily growing pool of oil underneath the pump, usually indicates that the shaft seal needs to be replaced. This should be done as soon as it is convenient to do so. It is a minor problem if taken care of promptly, but can lead to serious damage to the pump if neglected too long. Most manufacturers will supply the parts and instructions for replacing the shaft seals on their pumps, or will perform the replacement for a modest fee.

Over a period of time the oil in an operating rotary-vane pump degrades from mechanical wear and becomes contaminated with water and other material that enters the pump during use, and so it is necessary to replace the oil periodically. The frequency with which this needs to be done will depend on the conditions under which the pump is used. A pump that serves only as a backing pump for an oil diffusion pump or a turbomolecular pump, and is not frequently cycled to atmospheric pressure, experiences rather mild operating conditions and needs to have the oil replaced only after the maximum interval recommended by the manufacturer, perhaps once a year. Pumps that are cycled to atmospheric pressure frequently, such as those on scanning electron microscopes and vacuum evaporators, should have the oil replaced somewhat more frequently, while pumps that have to handle corrosive gases or large amounts of water may have to have the oil replaced every week or two. In all situations the oil should be replaced at least as frequently as recommended by the manufacturer of the pump. Furthermore, when the oil is replaced, only the grade of oil recommended by the manufacturer should be used. Following the manufacturer's recommendations in this way will minimize the opportunity for the pump to become mechanically damaged, and will maintain it in the best possible operating condition over the longest period of time. Proper replacement of the oil is particularly important for the direct-drive pumps, because the rotors on these pumps turn at very high speeds. The oils used in them must contain special additives to prevent excessive wear under these severe operating conditions, and the oil must be replaced regularly to maintain the necessary concentration of these additives.

Manufacturers always prescribe the procedure to be followed when changing the oil in their rotary-vane pumps, and it is wise to follow this procedure faithfully. Usually this involves blanking the inlet of the pump off and running it for half-an-hour so that it is thoroughly warmed up, turning the pump off, admitting air to it, and then draining the oil off through the drain plug provided in the pump case for this purpose. The pump is then flushed by adding about 250 ml of fresh oil, running it for 30 s, and draining the oil again. Flushing is repeated until the oil is clean and odour-free, whereupon the pump is filled to the proper level with clean fresh oil. Although special 'flushing oils', which are less expensive than the regular operating oils, are available, I prefer to never put anything in a pump other than the grade of oil recommended for use during operation.

In particular, it is not a good practice to use a solvent of any kind to clean out a rotary-vane pump. If this is done it will be almost impossible to remove the solvent completely from the pump, and only a trace needs to remain in the pump to seriously degrade both its pumping speed and its ultimate pressure. I remember one instance when a technician informed me that he was setting up a practice to service rotary-vane pumps. A bit of discussion revealed that the key step in his method involved 'really cleaning the pumps out' by flushing them with kerosene. I don't know whether or not I was ever successful in convincing him that this was not a good practice, but I never took the risk of sending one of my pumps to him to be serviced.

The belts on belt-driven pumps should be inspected for cracks once or twice a year, and should be replaced if the faintest sign of a crack appears. Even in the absence of visible cracks, it is a good idea to replace these belts yearly as a relatively inexpensive safety measure against the more serious problems that can arise if a belt breaks. The sparing use of a belt dressing compound of the type sometimes used on automobile fan belts will help prevent the belts on pumps from slipping and may prolong their life somewhat. Finally, most manufacturers have programmes for exchanging and rebuilding pumps if a problem develops that is too difficult to be taken care of in the laboratory.

4.1.8 *The operation of systems with rotary-vane pumps*

Vacuum ovens, prepump chambers for photographic film, sputter coaters, plasma ashers, and similar apparatus that operate in the rough vacuum range usually have vacuum systems with only a rotary-vane vacuum pump. Fig. 4.8 is a schematic diagram of a system of this kind with a vacuum valve and a liquid nitrogen trap in the pumping line. When such a system is to be started up, the air inlet valve to the pump and the valve in the pumping line are closed, the pump is turned on, and the trap is immediately

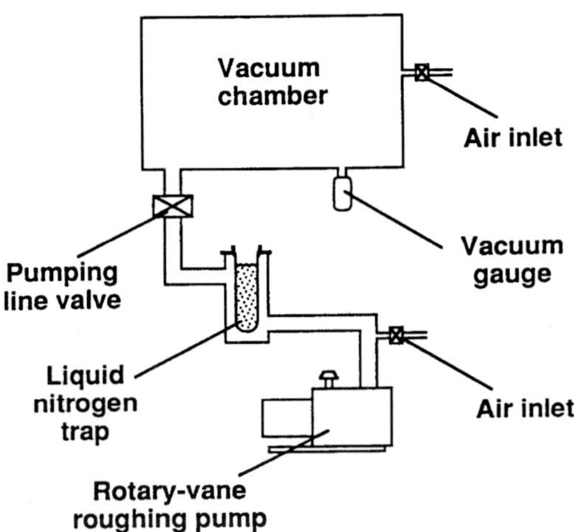

Fig 4.8 A vacuum system with a rotary-vane rough vacuum pump.

filled with liquid nitrogen. The material to be processed is then placed in the vacuum chamber, the chamber door and the air inlet valve on it are closed, and the valve in the pumping line is slowly opened to evacuate the chamber. The initial flow of air from the chamber through the trap will probably boil off a considerable amount of liquid nitrogen, making it necessary to refill the trap at this time. The trap should, of course, be checked periodically and filled as necessary throughout the evacuation process. As discussed in Section 4.1.4b (p. 143), its function is defeated if it is allowed to warm up while the system is under vacuum because then the vapours that have condensed in it will evaporate off it again and diffuse into the pump and back into the system. Throughout the evacuation process the vacuum in the system is monitored with the vacuum gauge, which may be either a thermocouple or Pirani gauge. When this process is completed, the material is removed from the vacuum chamber by closing the valve in the pumping line and slowly admitting air to the chamber through the air inlet valve. This allows the trap to be kept refrigerated and under vacuum while another load of material is placed in the chamber. The evacuation process can then be repeated. When it is finally desired to shut the system down completely the valve in the pumping line is closed, the air inlet in the pumping line is opened, and the pump is shut off immediately. The liquid nitrogen reservoir must then be removed immediately from the trap and cleaned, so that the condensable material it collected is not released into the pump or back into the system.

Somewhat simpler systems may also be encountered. If there is no concern about controlling condensable materials, the trap may be omitted. The valve in the pumping line may be retained, however, if it is planned to leave the evacuated materials standing under vacuum. This could be advantageous in preventing desiccated photographic film from absorbing water from the atmosphere again, for example. The valve in the pumping line is still needed because it is not advisable to turn an oil-sealed mechanical pump off and leave it standing under vacuum. This promotes the diffusion of oil vapours from the pump into the system and may allow the oil in the pump to be 'sucked' out of the pump into the pumping line. Most pumps are built with anti-suckback features. However, if you do not want something to happen, it usually will. It is an awful job to clean a litre of oil out of a vacuum system, while it is quite simple to eliminate the possibility of its ever getting there in the first place. The *safe procedure* for all types of systems is always to open the air inlet to a rotary-vane pump immediately before the pump is turned off. In fact, it is prudent to use a normally-open solenoid valve wired in series with the pump for this purpose. It will automatically close when the pump is turned on, and will open again when the pump is turned off or in the event of a power failure. Even simpler systems are sometimes encountered in which there is neither a trap nor a valve in the pumping line. Then, only the air inlet valve on the chamber is needed. For this configuration, this air inlet valve should be opened immediately before turning off the electrical power whenever the pump is shut down. Sputter coaters have vacuum systems of this kind, but with a needle valve added so that argon can be admitted to the vacuum chamber at a rate sufficient to maintain the gas pressure needed for efficient sputtering, usually about 10 Pa (0.1 Torr), during the coating process.

Rotary-vane mechanical pumps are very useful devices. They are available in a wide range of sizes, and are relatively inexpensive. They are simple to install, maintain, and operate. They are highly dependable, and have very long service lives. They have been, and probably will continue to be, the most widely used type of vacuum pump. Overall, however, hydrocarbon contamination is such a serious problem in high-resolution and analytical electron microscopes, and in various apparatus used in the manufacture of semiconductor devices, that considerable effort is being devoted to developing pumping systems that do not involve oil-sealed rough vacuum pumps. In fact, this is probably the most difficult task presently facing the designers of vacuum systems for electron microscopes.

4.2 Oil-free rough vacuum pumps

To date, activity to develop oil-free rough pumping systems has been greatest for applications such as food processing, medical and dental devices, surface analysis instruments, and in the semiconductor industry. However, many users and some manufacturers of electron microscopes have begun to experiment with some of the methods that have been developed in these other applications, and so new types of rough pumping systems are already beginning to appear on electron microscopes (e.g. Section 9.5, p. 410). Some of the types of pump available for this purpose are described briefly below.

4.2.1 Sorption pumps

We are all familiar with the fact that vapours adsorb readily on the fine particles of surface-active materials. This is the basis for filling certain types of gas masks with powdered charcoal and for using molecular sieves and activated alumina in the traps for oil pumps described in Section 4.1.5a (p. 146). Sorption pumps make use of the fact that this phenomenon of surface adsorption is greatly enhanced at low temperatures. As shown schematically in Fig. 4.9, a sorption pump consists simply of a canister for containing an adsorbent material with a tube and coupling welded to it for attaching it to a vacuum system. Typically, the canister is only 100–150 mm in diameter and 250–500 mm in length. A large insulated container is also needed to hold the liquid nitrogen used to cool the pump during use, because the sorptive capacity of most adsorbents is from 10^3 to 10^5 times greater at the temperature of liquid nitrogen (–196°C, 77 K) than at room temperature. Sorption pumps must also have a pressure release valve as a safety feature to prevent them from exploding if they warm up while charged with gas and not open to the atmosphere, because much of the sorbed gas is released from the adsorbent when a pump is allowed to warm up after a pumping operation is completed. *Great care* should be exercised to ensure that this safety valve is always in good operating condition.

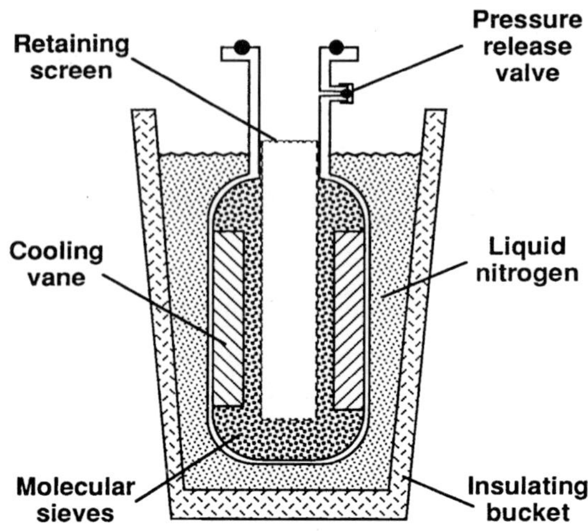

Fig 4.9 A schematic diagram of a sorption pump

The materials most frequently used as the adsorbent are synthetic zeolites that are commonly called *molecular sieves*. These are alkali aluminosilicate compounds that have an open crystalline structure with pores of the order of 1 nm in diameter. Because of their chemical composition and this unusual structure, these materials have very active surfaces and more than 500 m^2 of surface area per gram. This gives them tremendous cryosorption capacities for oxygen, nitrogen, and water vapour. Typically, a pump containing 2 kg of a molecular sieve adsorbent can pump of the order of 10^7 Pa–l (10^5 Torr–l) of these gases when cooled with liquid nitrogen. Such a pump can reduce the pressure in a 100 l vacuum chamber from atmospheric pressure to near 2 Pa (10^{-2} Torr) in 10–20 min, using only 15 or 20 l of liquid nitrogen in the process. Physically, the speed of a sorption pump depends on how well the incoming gas can gain access to the adsorbent, and on how rapidly the heat released by the condensing gas is removed by the liquid nitrogen. Manufacturers use various arrangements of internal screens, tubes, and vanes to ensure that the gas has good access to the adsorbent and to provide good thermal conductivity for heating and cooling the adsorbent.

The ultimate pressure attainable with sorption pumps depends somewhat on the strategy employed when using them. The sorption capacity of molecular sieves for hydrogen is less than for oxygen, nitrogen, and water by a factor of at least 10^2, while for the inert gases it is less by a factor of about 10^6, and so for all practical purposes these gases are not pumped. Therefore, the lowest pressure practically attainable with a single sorption pump is about 2 Pa (10^{-2} Torr), which is about the sum of the partial pressures of neon, helium, and hydrogen in the atmosphere. Lower pressures can be attained by using two pumps in sequence, however. To do this, the first pump is used to evacuate the system to the bottom of the range of viscous flow. As the pressure drops below about 50 Pa (0.5 Torr) a valve leading to this pump is quickly closed so that the hydrogen and inert gases that were swept into it by the viscous flow of oxygen and nitrogen are retained inside it and are not allowed to diffuse back into the vacuum system. The second pump is then used to complete the evacuation process. Since the first pump decreased the partial pressure of the inert gases in the system considerably, the second pump can usually achieve a final pressure near 10^{-1} Pa (10^{-3} Torr).

The effectiveness of this two-stage procedure depends critically on closing the valve to the first pump at the proper time to prevent the inert gases from diffusing back into the system. The same result can be obtained more reliably and more easily by carrying out this initial stage of evacuation with a type of pump that is effective in pumping hydrogen and the inert gases so that there is no inherent problem with back-diffusion. If an oil-sealed pump, such as a common rotary-vane pump, is used for this purpose there is always a danger that the system will become contaminated with oil, whereupon the basic advantage inherent in the use of a sorption pump will be lost. However, virtually any of the other types of oil-free rough vacuum pumps described below could be used. It would, of course, also be possible to remove most of the hydrogen, helium, and argon from a vacuum system by thoroughly flushing it with pure dry nitrogen before starting the pumpdown process.

Sorption pumps are known as 'entrapment pumps' because the gases they pump are trapped inside them during the pumping process. Although their sorptive capacity is great, it is still finite. For sorption pumps to be effective it is generally recommended that they provide a minimum of 1 kg of molecular sieve adsorbent for every 100 l of system volume. During the pumping process the molecular sieve adsorbent becomes loaded with adsorbed gases, causing both the capacity and pumping speed to decrease. When a pump is allowed to warm up to room temperature after use, oxygen, nitrogen, and other gases with non-polar molecules desorb and escape from the pump into the atmosphere via the pressure release valve. This will restore most of the pumping capacity for subsequent use. However, water molecules, which are very polar, adsorb so strongly to molecular sieves that they are not released at room temperature, and so a pump will eventually become loaded with water. This condition will be indicated by a decrease in pumping speed and an increase in the ultimate pressure a pump produces. At this point the pump must be *regenerated* to restore its performance. This is done by heating it to 250°C for about 5 h to drive off the water contained in it. Manufacturers provide heaters for this purpose. It is not a good idea to attempt to speed up a regeneration treatment by pumping on the sorption pump with an oil-sealed rotary-vane pump while it is being heated, unless a liquid nitrogen trap of the type shown in Fig. 4.6 is used in a meticulous manner to prevent oil from contaminating the molecular sieve in the sorption pump. Only a very small amount of oil contamination is needed to destroy the sorptive capacity of molecular sieves. The frequency with which regeneration is required will depend on the capacity of the pump and the level and conditions of use.

The principle of sorption pumping was explored by Dewar and others in the late 1800s, and pumps filled with activated charcoal were used to some extent in the early 1900s. However, reliable high-performance pumps were first introduced by Varian in the early 1960s, after the development of molecular sieves by the Linde company provided a stable high-capacity adsorbent. Sorption pumps are reliable, simple to use, inexpensive, and have long useful lives. They have become widely used in recent years for rough pumping ultra-high vacuum systems because they are totally oil-free. They are available in sizes that contain from 300 g to 2.5 kg of molecular sieve absorbent. The principle disadvantages associated with using sorption pumps involve the cost of the liquid nitrogen needed to cool them, and the inconvenience of performing the regeneration treatment.

4.2.1a Using sorption pumps

The most common use for sorption pumps is rough pumping vacuum systems in which the operating high vacuum is produced by cryogenic or sputter-ion pumps. Figure 4.10 is a schematic diagram of such a system. While the vacuum chamber is being prepared for evacuation, insulating containers are placed around the sorption pumps and filled with liquid nitrogen to cool the pumps. During this time valves V_2 and V_3 are both

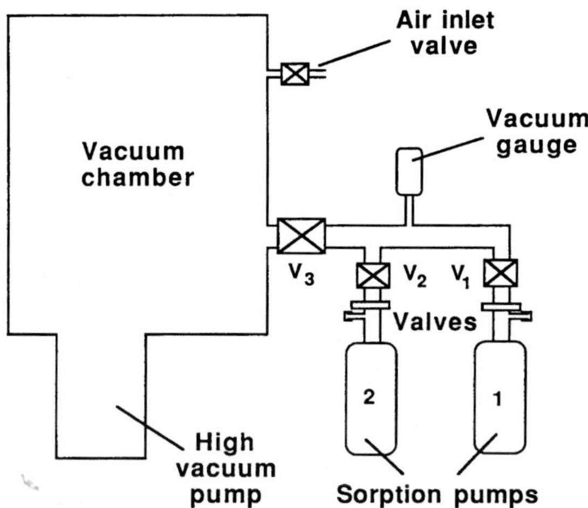

Fig 4.10 An arrangement for rough pumping a vacuum system with sorption pumps.

closed; however, valve V_1 may be opened to evacuate the manifold. When the time comes to evacuate the chamber, valves V_1 and V_3 are opened. If a rough vacuum of the order of 2 Pa (10^{-2} Torr) is acceptable the second sorption pump is not needed and can be omitted from the system; alternatively, it can be used as a backup to evacuate the system while the first pump is being regenerated. When the pressure reaches 2 Pa (10^{-2} Torr), the high vacuum pump is started up and valve V_3 is closed.

If a better rough vacuum is required, then the two pumps can be used in sequence, as described in the previous section. To do this the initial steps are the same as in the procedure just described, but after valves V_1 and V_3 have been opened the pressure in the manifold is monitored carefully with the vacuum gauge, which may be either a thermocouple or Pirani gauge. As soon as the pressure approaches the lower end of the range of viscous flow, at about 20 Pa (0.2 Torr), valve V_1 is quickly closed. The second sorption pump is then used to complete the rough pumping operation by opening valve V_2. It should be possible to reach a pressure near 10^{-1} Pa (10^{-3} Torr) in this way. The high vacuum pump is then started up and valve V_3 is closed.

An equivalent way to achieve this lower rough vacuum is to replace the second sorption pump with an oil-free carbon-vane, diaphragm, piston, or Venturi pump, as shown in Fig. 4.11. As will be seen in the following

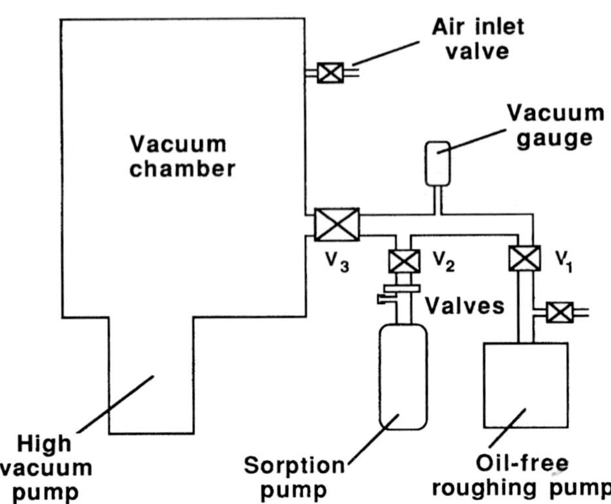

Fig 4.11 An arrangement for rough pumping a vacuum system using a sorption pump and another oil-free roughing pump.

sections, most of these pumps cannot reach pressures below about 7000 Pa (50 Torr). However, this is sufficient to remove more than 90 percent of the gas molecules from the vacuum system, greatly reducing the gas load the sorption pump must handle. In addition, it decreases the partial pressure of the inert gases in the system by a factor of 10, thereby allowing the sorption pump to produce the desired lower final rough vacuum. For this arrangement, valve V_2 is kept closed initially while valves V_1 and V_3 are opened and the system is evacuated to 10 000 Pa (75 Torr) or lower using the oil-free roughing pump. Then valve V_1 is closed, valve V_2 is opened and the sorption pump is used to complete the rough pumping operation. When an acceptably low pressure is reached the high vacuum pump is started up and valve V_3 is closed.

4.2.2 *Carbon-vane pumps*

Carbon-vane pumps operate on the same general principle as oil-sealed rotary-vane pumps except that no oil is used to provide a seal between the vanes and the pump chamber. Instead, the vanes are made of a carbon-based composite that is self-sealing and self-lubricating. The rotor may contain two, three, or four vanes. In many designs these vanes are not spring-loaded, as are the vanes in the oil-sealed rotary-vane pumps, but are held against the chamber wall only by centrifugal force. Because of the limitations inherent in such a sealing mechanism, these pumps usually do not produce pressures below about 10 000 Pa (100 Torr); however, this is sufficient to remove 90 percent of the gas molecules from a vacuum chamber, and they are totally oil-free. These pumps are widely used in industrial applications, and so their design and construction have been refined to the point where they provide long service life and require minimal maintenance. Models are available with pumping speeds ranging from 10 to 1500 l/min.

4.2.3 *Diaphragm pumps*

Diaphragm pumps have been used for years in industrial, medical, and dental applications where oil-free vacuum systems are required. The construction of this type of pump is shown schematically in Fig. 4.12. Basically, an electric motor, acting through a crank and connecting rod mechanism, drives an elastic diaphragm up and down in the pump chamber. On the downward stroke the diaphragm is pulled away from the pump head and a quantity of gas enters the space thus created through the inlet valve and port. On the upward stroke the diaphragm is forced against the pump head, compressing this gas and expelling it through the outlet valve and port. Since these pumps depend only on the flexing of the diaphragm for their pumping action, they require no lubricant in the gas path and are therefore totally oil-free. They are particularly useful

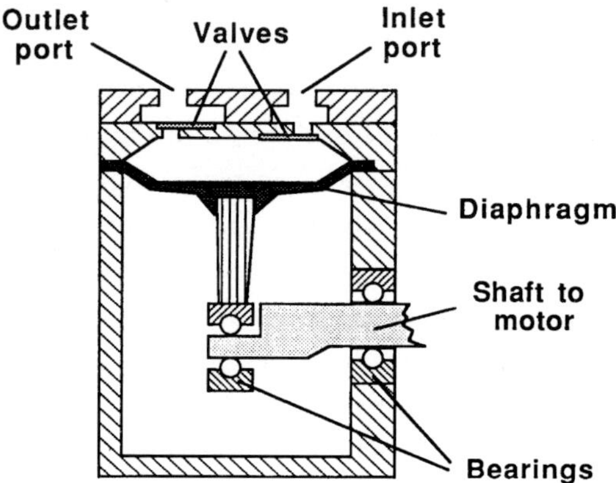

Fig 4.12 A schematic diagram of a diaphragm pump.

Fig 4.13 A compound pump consisting of four diaphragm units connected in series and operated by a single motor. (Courtesy of Elnik Systems.)

in many industrial applications because they can be used to pump explosive and corrosive gases with a high degree of safety. Diaphragms of materials such as Viton, Neoprene, and Teflon, and pump heads of various polymers, Hastelloy and stainless steel, are available for such applications.

Diaphragm pumps are available in standard models with speeds ranging up to 150 l/min. The best vacuum attainable with such single-stage pumps is about 10 000 Pa (100 Torr), while two-stage versions can reach 7000 Pa (50 Torr). However, at least two companies have recently introduced compound pumps, such as the one shown in Fig. 4.13, that utilize several standard diaphragm units working in multi-staging arrangements to achieve ultimate pressures near 200 Pa (2 Torr), with pumping speeds ranging from 40 to 110 l/min. These are the Barodyn and Elnivac pumps produced by Danielson Associates and Elink Systems, respectively. The pumping speed of these pumps remains nearly constant at the rated value from atmospheric pressure down to about 10^4 Pa (10^2 Torr), and then drops slowly to zero at the ultimate pressure. Like the carbon-vane pumps, these pumps are highly reliable, because of their wide use in industry. The diaphragms may need to be replaced at 6–10-month intervals in pumps that run continuously. This operation can usually be carried out in less than an hour with standard wrenches and screw drivers.

4.2.4 *Piston pumps*

Piston pumps have been used for a wide variety of industrial applications for many years, but are only recently beginning to find use in laboratory apparatus. The pumping mechanism of these pumps consists of a piston that moves back and forth in a cylinder, as do the pistons in an automobile engine, with inlet and outlet valves to control the direction of gas flow. The seal between the piston and the cylinder is usually provided by piston rings made of composite materials that do not require oil for lubrication, although other sealing mechanisms are used by some manufacturers. Fig. 4.14, for example, shows the construction of the ROC-R series of piston pumps produced by the Gast Manufacturing Corporation. In these pumps the seal between the piston and cylinder is made by a flexible cup-shaped gasket, which gives these pumps the combined best features of diaphragm and piston pumps. Piston pumps are completely oil-free. Their ultimate pressure rating is usually given as being about 10 000 Pa (100 Torr) for single-stage pumps, and 3000 Pa (20 Torr) for two-stage pumps. Typically, piston pumps are available in sizes ranging from 10 to 500 l/min. Although the performance of these common industrial piston pumps is modest in terms of high vacuum requirements, they can be used very advantageously with sorption pumps, in the manner described in Section 4.2.1 above, or in series with other types of rough vacuum pumps to achieve oil-free evacuation.

The technology of piston pumps has been refined considerably by the highly innovative design used in the Model DVP-500 oil-free pump manufactured by Varian Associates. In this pump, four pistons are driven from a common drive shaft by a single motor. The pistons are of an unusual, double-acting, 'stepped' design that can produce compression in both directions of a stroke. Two of the pistons act in parallel as the intake stage of the pump, while the other two work in series with them. Overall, however, four stages of compression are achieved by connecting the two sides of the fourth piston in series. Gas leakage around the pistons is minimized by making the pistons fit tightly in the cylinders, and by connecting the back sides of the double-action pistons of the first two stages to the inlet ports of the piston of the next stage, so that the pressure drop across these stages is kept small. Oil-free sealing combined with wear resistance is achieved by

Fig 4.14 A piston rough vacuum pump with part of the pumping chamber wall cut away to show the pumping mechanism. (Courtesy of Gast Manufacturing Corp.)

coating the pistons with reinforced poly(tetrafluoroethylene). A number of other highly innovative features have been employed to optimize the performance of this pump. Its pumping speed decreases only slowly from its rated value of 450 l/min at atmospheric pressure to about 350 l/min at 100 Pa (1 Torr), but then falls more rapidly to its ultimate pressure near 1 Pa (10^{-2} Torr). Both the speed and ultimate pressure of this pump are comparable to those of typical rotary-vane pumps, yet it is totally oil-free.

4.2.5 *Venturi pumps*

Venturi pumps work on the same principle as the water aspirators used to produce 'suction' in many chemistry and physics laboratories. That is, when a fluid under pressure flows through a tube with a constriction in it the velocity of the fluid increases in the region of the constriction. Although it is perhaps not intuitively obvious, basic principles of fluid dynamics show that this increase in velocity is accompanied by a decrease in pressure. In the Venturi pump, a side tube is added in the region of the constriction, as shown in Fig. 4.15. Gas molecules enter the Venturi tube

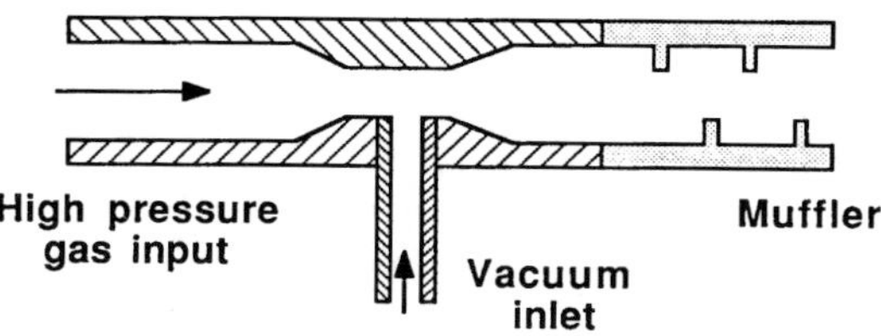

Fig 4.15 A schematic diagram of a Venturi pump.

through this side tube under the influence of the reduced pressure that exists in the constricted region, and are carried away by the fluid stream.

Venturi pumps are used in a variety of industrial applications, such as devices that lift objects by means of suction pads, where rapid vacuum response is advantageous. For such purposes, air is fed into the pump at a pressure in the range from 350 to 550 kPa (50–80 psi), and the outlet is vented to the atmosphere. A muffler is needed to reduce the noise generated by the escaping high speed air stream. The ultimate pressure attainable is about 8000 Pa (60 Torr), which again may not seem impressive by normal vacuum criteria. However, as mentioned above, this is sufficient to remove more than 90 percent of the gas molecules from a vacuum system, which drastically reduces the gas load for other types of pumps. Venturi pumps are small (typically about 30 mm in diameter and 150 mm long), inexpensive, simple to operate, highly reliable, totally oil-free, and require only a source of compressed air for operation.

4.2.6 *Mechanical blowers*

Devices called 'mechanical blowers' represent yet another approach to achieving oil-free pumping that has been used extensively in industrial applications but has not been adapted to laboratory apparatus. The most widely used of these devices is the Roots blower, shown schematically in Fig. 4.16. Here two figure-of-eight shaped impellers intermesh and rotate in opposite directions in an oval chamber. In the orientation shown, a charge of gas is trapped between the top impeller and the wall of the chamber. As this impeller turns further this gas is swept into the outlet. In the meantime, the bottom impeller similarly traps a charge of gas and ultimately sweeps that into the outlet. The impellers are driven, and their relative orientations are synchronized, by a pair of timing gears and a motor

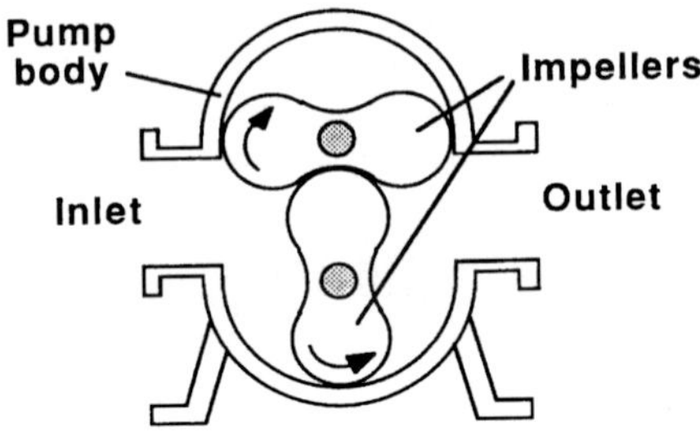

Fig 4.16 A schematic diagram of a Roots blower.

located outside the pump chamber. These gears and the bearings that support the impeller shafts must be well lubricated; however, by careful control of the clearances between the impellers themselves and between the impellers and the chamber a significant pumping action is obtained without the need for lubrication inside the pump chamber. Manufacturers use a variety of sealing devices to prevent oil and air from entering the pump around the drive shafts. Apart from possible failure of these shaft seals, these pumps are oil-free. The impellers on current models of Roots blowers range from 200 to 500 mm in length and typically rotate at 2000 to 4000 rpm. Blowers are available with pumping speeds ranging from 100 to 30 000 m^3/h (1500 to 500 000 l/min), which can produce ultimate pressures below 10^{-2} Pa (10^{-4} Torr) when backed by a two-stage oil-sealed mechanical pump of appropriate size. Van Atta (p. 185) gives a detailed discussion of the physical principles underlying the pumping action of Roots blowers, while Harris (Chapter. 6) describes their current characteristics and use.

To eliminate the need for an oil-sealed backing pump, and thereby give totally oil-free pumping, Alcatel have assembled five Roots blower stages into a single unit in their ADP-80 Dry Pump, producing a pump that can work at atmospheric pressure without a backing pump. The speed of this pump remains nearly constant at about 90 m^3/h (1550 l/min) from atmospheric pressure down to about 10 Pa (10^{-1} Torr), and then falls rapidly to zero at an ultimate pressure slightly below 1 Pa (10^{-2} Torr) .

Another type of blower, known as the 'claw blower', has also recently come into use for a similar purpose. This blower has impellers with the unusual shape shown in Fig. 4.17. Again, the impellers are synchronized and driven by external timing gears, and have sufficient clearance so that

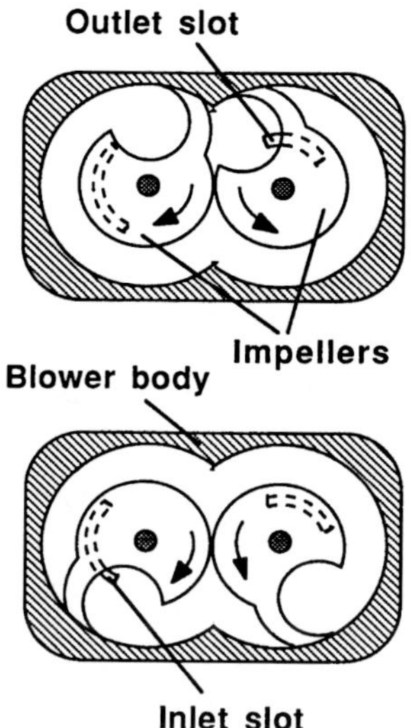

Fig 4.17 Schematic diagrams of a claw blower showing the impellers in two different positions.

no lubrication is required inside the pumping chamber. In this design the inlet and outlet ports are slots in the walls of the pumping chamber. This makes it very convenient to use several chambers in series by arranging the outlet slot from one stage so that it serves as the inlet for the next, in a common wall separating the two chambers. At least two companies are currently marketing multi-stage pumps of this kind. The Drystar pump manufactured by Edwards has a Roots stage followed by three claw stages, while the Dryvac pump produced by Leybold has four claw stages. Like the Alcatel pump described above, these pumps do not require a backing pump, but can discharge gas directly to the atmosphere. They have maximum speeds greater than 50 m^3/h (800 l/min) at inlet pressures above 10 Pa (0.1 Torr). As the inlet pressure drops below 10 Pa their speeds decrease rapidly to zero at ultimate pressures near 1 Pa (10^{-2} Torr).

At the present time, these mechanical blower pumps are being designed and produced to meet the needs of industrial processes in which large volumes of gas must be handled at intermediate pressures. They are much too large and noisy to be acceptable for use in most laboratory applications. However, they are interesting because they represent a technology that is now highly refined and that might someday be adapted for electron microscopes and similar laboratory apparatus if a sufficient demand develops.

4.2.7 *Molecular drag pumps*

Another interesting and innovative approach to achieving oil-free rough pumping is found in the Drytel and Tribodyn pumps being produced by Alcatel and Danielson, respectively. These pumps, which are described in more detail in Section 6.2 (p. 268), combine a molecular drag pump with a compound diaphragm pump, such as the Elnivac and Barodyn pumps described above, and provide oil-free pumping from atmospheric pressure to well below the 10^{-4} Pa (10^{-6} Torr) range. Unlike the compound blower pumps just described, these pumps are small enough to be very attractive for use on electron microscopes, vacuum evaporators, and similar laboratory apparatus.

4.2.8 *Scroll pumps*

Danielson Associates have recently introduced yet another type of oil-free rough vacuum pump. In these pumps, which are marketed under the Spiradyn trade name, the pumping element consists of a pair of aluminum plates which each have an Archimedes spiral machined into them, and which are mounted against one another in such a way that these spirals are interleaved. A pumping action is then achieved by holding one plate fixed while the other is moved against it in a complex circular-oscillatory pattern. Gas that enters the inlet port is trapped between the spirals and then compressed as it is forced by this motion to move along between the spirals to the outlet port. Interestingly, neither an inlet nor an outlet valve is needed with this pumping mechanism, which is totally oil-free, and which pumps all gases with equal speed.

Spiradyn pumps are currently available with rated speeds of 7.5 and 15 cfm (210 and 425 l/m). Their pumping speed remains nearly constant at the rated value from atmospheric pressure down to about 1000 Pa (10 Torr) and then decreases almost linearly to an ultimate pressure near 10 Pa (0.1 Torr). Danielson Associates also produce two-stage combination pumps with rated speeds of 50 and 70 cfm (1400 and 2000 l/min) in which the first stage of pumping is produced by a Roots blower with a helical geometry (the Helidyn design), and the second stage by a Spiradyn unit. The speed of these pumps, which are sold under the Cyclodyn trade name, remains at these rated values from atmospheric pressure down to about 10 Pa (0.1 Torr) and then drops rapidly to an ultimate value near 1 Pa (10^{-2} Torr).

4.3 Concluding remarks

The rough pumping operation, which involves the initial removal of atmospheric gases from a vacuum system, is an essential part of every evacuation process. Unfortunately, most pumps capable of reducing the pressure into the high and ultra-high vacuum ranges are not capable of performing this rough pumping function; therefore, roughing pumps are essential components of all vacuum apparatus. In the past, oil-sealed mechanical pumps were virtually the only ones available for performing this function. Although these pumps are very effective, reliable, and economical, it is extremely difficult to prevent the vacuum systems to which they are attached from becoming contaminated with the oil used in them. They are therefore falling out of favour for use on apparatus where such contamination cannot be tolerated. In fact, this problem is of such serious consequence in electron microscopy that oil-sealed rotary-vane pumps really should not be used in the vacuum systems of high-resolution and analytical electron microscopes. Although a number of other types of roughing pumps have been developed, and can be expected to appear on electron microscopes in the near future, a truly convenient, effective, reliable, and economically-attractive solution to the problem of attaining oil-free rough pumping is not yet available.

5 Oil diffusion pumps

Oil diffusion pumps have been the most widely used type of high vacuum pump for most industrial and research applications for more than 50 years. Vacuum pumps that were referred to as diffusion pumps were introduced prior to 1920. These pumps were generally made of glass and used mercury as the pump fluid. Mercury began to be replaced by oil as the working fluid for these pumps in the late 1920s. Pumps made of metal and designed specifically for the use of an oil as the working fluid began to appear in their present form in the late 1930s, and for the next 25 years were the principal devices employed to achieve pressures below the rough vacuum range. Although early diffusion pumps were rather crude devices with relatively poor performance characteristics, the need for better pumps in the space research programme and in many manufacturing processes led to extensive research and development work in the late 1950s and early 1960s that produced greatly improved pumps. Van Atta (Chapter 6) gives an interesting discussion of much of this work, while Hablanian (Chapter 6) presents considerable information on the theoretical principles involved in the design and construction of oil diffusion pumps, and, together with Harris (Chapter 7) and O'Hanlon (Chapters 8, 10, 11, and 12), describes the present status and use of these pumps. Oil diffusion pumps are very widely used, and are being replaced by other types of high vacuum pumps only in applications where an oil-free environment is a critical requirement.

The references cited in this chapter are listed in Appendix 1. The suppliers and manufacturers of vacuum equipment referred to here are listed in Appendices 2 and 3.

.1 Construction and function

The name 'diffusion pump' is perhaps not the most appropriate one that could have been applied to these pumps. Apparently this name was chosen because these were among the first pumps developed which operate primarily at pressures in the range of molecular flow. At these pressures gas molecules enter the pumps individually by virtue of their thermally-activated motion, which is also the basis of the classical gas diffusion phenomenon. This name then emphasized the contrast with rotary mechanical pumps, the major type of vacuum pump in use at the time, which operate at higher pressures where the co-operative viscous flow process dominates. More accurately, they are a form of vapour jet pump in which gas is compressed by high molecular mass oil vapour molecules moving at high speeds (300–400 m/s). Fig. 5.1 shows schematically the construction of these pumps. The 'body' or 'barrel' of the pump usually consists of a steel or stainless steel cylinder which is closed at the bottom and has a flange at the top for connecting it to the vacuum system. The oil from which these pumps derive their name is a highly refined organic liquid with a low vapour pressure which is placed in the bottom of the pump where it is heated to boiling by an electric heater. The molecules of the hot oil vapour that rise from the *boiler* are trapped inside a *jet assembly* at a pressure in the range from 100 to 300 Pa (0.75 to 2 Torr). Carefully designed jet openings in the jet assembly direct these oil molecules outward and downward into the space between the jet assembly and the barrel of the pump as high-speed vapour streams. Gas molecules that diffuse into the pump are struck by these fast-moving, heavy oil molecules and, on the average, are driven downward towards the bottom of the pump. The barrel of the pump is cooled so that the oil vapour condenses upon striking it and flows back into the boiler as a liquid. Fig. 5.2 is a photograph of an oil diffusion pump and its jet assembly.

The jet assemblies in most pumps have three or four jets. The top jet, which is located in the inlet of the pump, is usually small in diameter so that it offers minimum interference with the entry of gas molecules into the pump. This design gives a high pumping speed, but produces only a relatively small degree of compression because the jet openings are located so far from the wall of the pump that the density of oil vapour in the jet stream is low. The lower jets, which are successively closer to the barrel of the pump, provide more dense oil vapour streams and correspondingly higher degrees of compression, so that the pressure of the gas at the bottom of the pump becomes high enough for it to be pumped out into the atmosphere efficiently by a mechanical roughing pump.

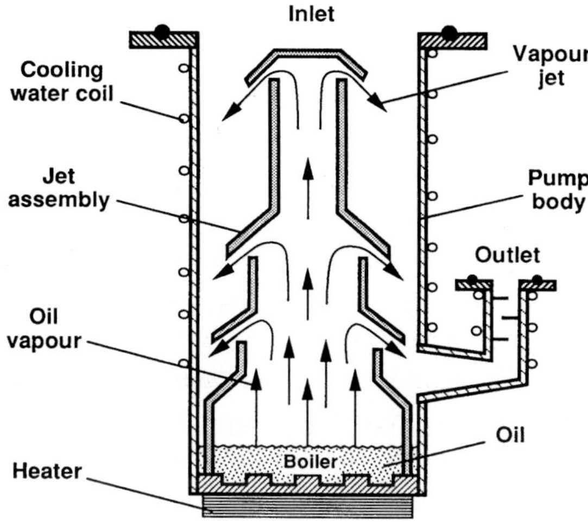

Fig. 5.1 The main features of an oil diffusion pump.

Fig. 5.2 An oil diffusion pump and its jet assembly.

5.2 Pumping speed

Oil diffusion pumps are basically suited for use at pressures below 10^{-1} Pa (10^{-3} Torr). In this pressure range gas flow is molecular in character, conductance is independent of pressure, and the pumping speed reaches a constant maximum value which is determined by the area of the pump inlet and the effectiveness of the top jet in capturing incoming gas molecules. The theoretical maximum possible pumping speed S_{tm} when pumping air at room temperature is given by eqn. 2.13b in Section 2.5.1 (p. 46), that is: $S_{tm} = 0.09D^2$ l/s for a pump with an inlet diameter of D millimetres. For example, the highest speed one should expect to be possible for a pump 200 mm in diameter is about $S_{tm} \approx 0.09 \times 40\,000 \approx 3600$ l/s, while for a pump 300 mm in diameter it is about 8100 l/s. Eqn. 2.13d ($S_{tm} \approx 0.0286\, D^2\sqrt{T/M}$) also predicts that the pumping speed for a light gas such as helium ($M = 4$ g/mol) should be significantly greater than for air ($M \approx 29$ g/mol), that is:

$$\frac{S_{He}}{S_{Air}} = \frac{\sqrt{M_{Air}}}{\sqrt{M_{He}}} = \sqrt{\frac{29}{4}} = 2.7$$

This same approach predicts a pumping speed for hydrogen (H_2, $M = 2$ g/mol) 3.7 times that for air. This trend is observed in practice; however, the magnitude of the effect is strongly dependent on the design of the pump. For pumps presently available the speed for hydrogen is usually only about 1.5 times greater than for air.

Well-designed pumps typically achieve maximum speeds that are a little less than half the theoretical value given by eqn. 2.13b. Fig. 5.3 is a plot showing typical maximum speeds for present pumps, based on data given in catalogues of four different manufacturers. The values shown for pumps 200 and 300 mm in diameter are about 1750 and 3750 l/s, which are just less than one-half the values of S_{tm} calculated above for pumps of these sizes. Pumps are normally rated on the basis of this maximum pumping speed. The smallest pumps presently produced have diameters of about 60 mm and rated speeds of about 150 l/s. Pumps of this size are found on many of the compact, desk-top evaporators and similar apparatus now on the market. Pumps having diameters greater than 1 metre and rated speeds greater than 100 000 l/s are available for industrial processes, particle

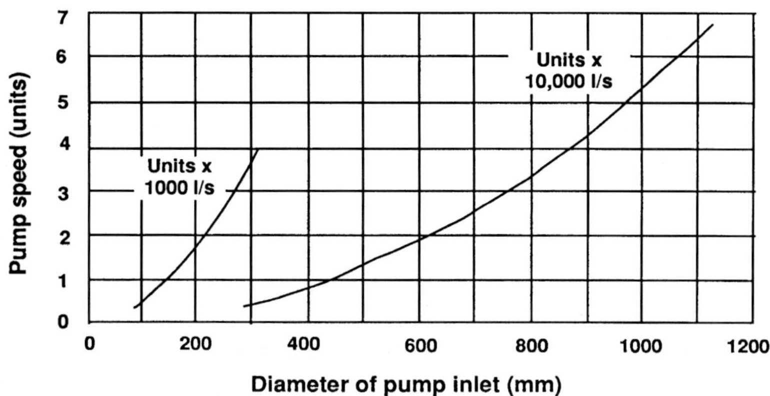

Fig. 5.3 The variation of the pumping speed of oil diffusion pumps with inlet diameter.

accelerators, and similar large-volume pumping applications. Electron microscopes, full-size vacuum evaporators, and similar laboratory apparatus usually have pumps with diameters from 150 to 300 mm and rated speeds in the range from 1000 to 3000 l/s.

Manufacturers tend to assign model designators that indicate the size of the inlet flange provided for attaching the pump to the vacuum system. This is, of course, an important characteristic because the flange on the pump must match the flange on the vacuum system to which it is to be attached. In addition, the size of the inlet flange is directly related to the inlet area, and so it is an indicator of the pumping speed via the relationships described above. European manufacturers usually use the ISO (International Standards Organization) system in which flange sizes (63-, 100-, 160-, 250-, 350-, 400- and 500-mm) correspond closely to the actual internal diameters of the flanges. Manufacturers in the United States usually base their designations on ASA (American Standards Association) flange sizes, which bear no rational relationship to any actual diameter. Common sizes are 2-, 4-, 6-, 10-, 16-, and 20-inch pumps, which have flanges with outside diameters of 6, 9, 11, 16, 23.5, and 27.5 inches, respectively, but no standard internal diameters. Consequently, pumps of a given 'size' produced by different US manufacturers often have slightly different internal diameters. Typically, speeds for 4-, 6-, and 10-inch pumps, the sizes most often encountered on laboratory apparatus, are about the same as for pumps with inlet diameters of 150, 200, and 300 mm. Some manufacturers now include an indicator of the rated pumping speed in their pump model designations. Unfortunately, this does not entirely eliminate ambiguities because two different methods are presently in use for measuring pumping speeds. The ISO method, which is most widely used in Europe, gives pumping speed values that are from 10 to 15 percent lower than the AVS (American Vacuum Society) method commonly used in the United States.

Most diffusion pumps have straight bodies, as shown in Figs. 5.1 and 5.2. However, Varian Associates manufacture 4-, 6-, and 10-inch pumps of advanced design which have bodies that are bulged outward in the region of the top jet but which are otherwise about the same size as standard pumps (see Fig. 5.7, Section 5.5, p. 191). These 'VHS' pumps have speeds that are nearly 50 percent greater than standard pumps of comparable flange size, and include many design features, which will be discussed later, that optimimize performance characteristics. Although they cost slightly more than standard pumps, their extra speed is highly advantageous in most applications.

At inlet pressures above about 10^{-1} Pa (10^{-3} Torr) the pumping speeds of oil diffusion pumps drop rapidly, as shown in Fig. 5.4. This occurs because the gas pressure exceeds the maximum value that can be tolerated by one or more of the jets, which then become overloaded. When a jet becomes overloaded its vapour stream develops a shock front between the jet assembly and the body of the pump. Beyond the shock front the motion of the oil molecules becomes random and the pumping action of the jet ceases. The pressure at which the top jet becomes overloaded is about 10^{-1} Pa (10^{-3} Torr) for most pumps, and corresponds to the pressure at which the decrease in pumping speed begins. This is known as the 'critical inlet pressure' (CIP), because it is the maximum inlet pressure for safe operation of the pump. The middle jet in a typical three-stage pump becomes overloaded at about 2 Pa (10^{-2} Torr), and the bottom jet at about 40 Pa (0.3 Torr). This last value, which represents the maximum pressure a standard three-stage pump can develop at its outlet, varies somewhat depending on heater input, the boiler temperature, the type of oil being used, and the design of the pump. The addition of an ejector jet, as shown in Figs. 5.7 and 5.10, may increase the maximum output pressure to nearly 70 Pa (0.5 Torr). These output pressures are not sufficient to discharge gas to the atmosphere, and so it is always necessary to back up a diffusion pump with a rough vacuum pump, which is then referred to as the 'backing

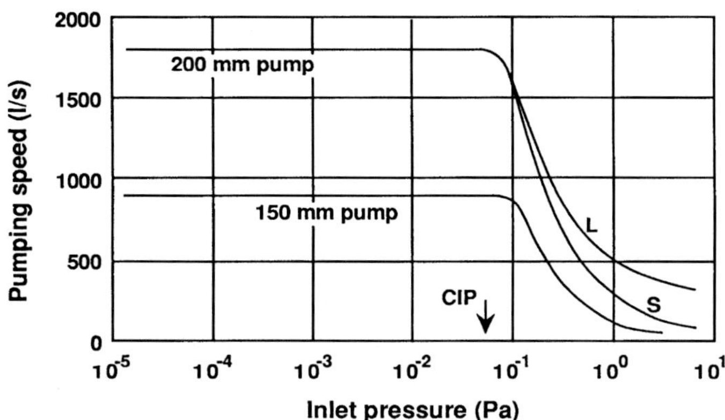

Fig. 5.4 The variation of pumping speed with inlet pressure for oil diffusion pumps with inlet diameters of 150 and 200 mm. CIP is the critical inlet pressure of these pumps. The branches L and S show the effect of large and small backing pumps on the speed of the 200 mm pump at pressures above its critical inlet pressure.

pump' (or, sometimes, the 'forepump'). At pressures above about 0.5 Pa (5×10^{-3} Torr) the pumping speed of the backing pump usually begins to exceed that of the diffusion pump, and the overall pumping speed at higher pressures becomes dependent on the size of the backing pump, as indicated schematically in Fig. 5.4. In practice it is essential to ensure that the backing pump always maintains a pressure in the outlet tube of the diffusion pump that is less than the maximum output pressure the diffusion pump can develop. Otherwise, all jets in the diffusion pump will become overloaded, it will cease all pumping action, and several deleterious phenomena, which will be described shortly, may occur. This pressure is therefore referred to as the 'critical backing pressure' (CBP) of the diffusion pump.

5.3 Throughput

Throughput is a quantity closely related to pumping speed that is also of importance in characterizing oil diffusion pumps. Using eqn. 2.9 (Section 2.5, p. 39), the throughput of a pump Q_p (Pa-l/s or Pa-l/min) can be defined as:

$$Q_p = S_p P_p \tag{5.1}$$

where S_p is the speed of the pump (l/s or l/min) and P_p is the pressure at its inlet (Pa). When discussing oil diffusion pumps and other pumps capable of producing pressures in the high vacuum range, Q_p is usually expressed in units of Pa-l/s (or Torr-l/s). The calculation in Section 1.3 showed that one pascal-litre of gas consists of 2.47×10^{17} gas molecules, and therefore the throughput of a pump actually indicates the number of gas molecules it moves from its inlet to its outlet each second. Since the process of moving gas molecules from a region of low pressure to a region of higher pressure requires the expenditure of energy, the maximum throughput of a pump is directly related to the power input to its boiler. For present pump designs, each Pa-l/s of throughput capacity requires about 5 watts of power. Thus, while the maximum speed of a pump depends on its inlet area, its maximum throughput is determined by its boiler wattage.

Fig. 5.5 shows how throughput typically varies with inlet pressure for a 150 mm pump with the speed curve of Fig. 5.4. At pressures below the CIP for the pump the throughput varies linearly with inlet pressure, since pumping speed is constant in this range. Throughput reaches its maximum value at the CIP where the diffusion pump first becomes overloaded, and remains essentially constant at higher pressures, even though the pumping speed decreases markedly in this range (see Fig. 5.4). As inlet pressure increases above about 0.5 Pa (5×10^{-3} Torr) the backing pump dominates the pumping process, and so throughput at higher pressures depends on the size of this pump.

An exploration of these throughput characteristics provides valuable insight into the relationships that exist between a diffusion pump and its backing pump. Since all gas molecules that pass through a diffusion pump must also pass through the backing pump, the throughput of the backing pump must be the same as that of the diffusion pump (assuming there are no serious leaks in the backing line). As an example, let us make some calculations for the operation of the 150 mm pump characterized by Figs. 5.4 and 5.5 at two different inlet pressures.

During operation in the high vacuum range: Assume $P = 10^{-3}$ Pa (10^{-5} Torr), then $S_p \approx 800$ l/s and $Q = S_p P_p = 800 \times 10^{-3} = 0.8$ Pa-l/s.

Assuming a backing pressure of 10 Pa (7.5×10^{-2} Torr), the backing pump would need a speed of only $S_p = Q/P = 0.8/10 = 0.08$ l/s or 4.8 l/min.

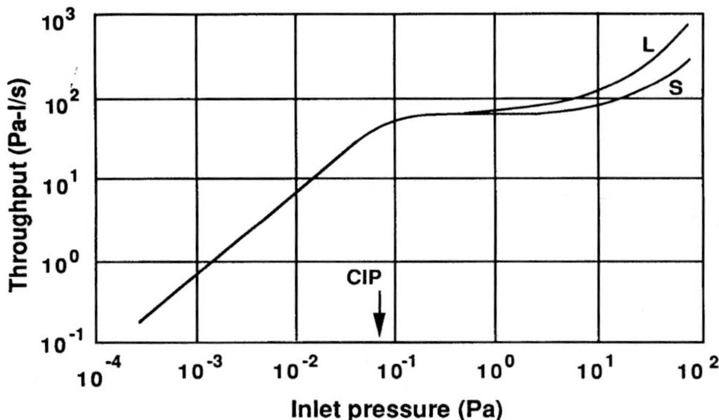

Fig. 5.5 The variation of throughput with inlet pressure for an oil diffusion pump with a rated pumping speed of 750 litres/s. CIP is the critical inlet pressure for this pump. Branches L and S show the effect of large and small backing pumps on throughput at pressures above the critical inlet pressure.

Even allowing for a substantial decrease in the speed of the backing pump due to the low operating pressure (Section 4.1.2, p. 139), a rotary-vane pump with a rated speed (at atmospheric pressure) of 20 l/min would be more than adequate to handle this gas load. Manufacturers do not make pumps with such low speeds.

During operation above the critical inlet pressure: Now, $Q = Q_{max} \approx 80$ Pa-l/s. Again assuming we try to maintain the pressure in the backing line near 10 Pa, the speed of the backing pump would have to be: $S_p = Q_p/P_p = 80/10 = 8$ l/s or 480 l/min. Again, allowance for the effects of conductance and pressure on the effective speed of the backing pump suggests that a pump with a rated speed of about 700 l/min would be required to handle this throughput. This is a much larger pump than is normally used on laboratory apparatus.

Designers of vacuum systems with oil diffusion pumps must take such relationships into consideration when selecting a backing pump. Inevitably some compromise is involved. For instruments such as electron microscopes that are used mostly in the high vacuum range, and that are not cycled to atmospheric pressure frequently, it is usually considered acceptable for the pumping system to be overloaded during the few seconds required to make the transition from the rough to the high vacuum range, and so relatively small backing pumps are chosen. If you watch the reading of the pressure gauge in the backing line of most electron microscopes, you will notice the reading rises slightly for a few seconds just after the main vacuum valve is opened due to this overload condition. For systems that are cycled to atmospheric pressure frequently, or that must handle high gas loads generated by chemical or physical processes carried out in them, relatively larger backing pumps must be specified. We are not directly concerned with designing vacuum systems; however, an awareness of these effects can lead to a better appreciation of the characteristics of the systems we have to work with.

5.4 Pump oils

The working fluid or oil used in an oil diffusion pump contributes critically to the pump's performance. For one thing, the ultimate pressure a pump can attain is determined basically by the characteristics of the oil used in it. It should be intuitively obvious that very few gas molecules from the vacuum system will enter the inlet of the pump if the pressure of oil molecules at the inlet is greater than the pressure of the gas molecules in the system. Pump oils also contribute to contamination of the vacuum system, and their characteristics can strongly influence operating procedures.

The first working fluid used in vapour jet pumps was mercury. While it has many advantages, its toxicity and relatively high vapour pressure at room temperature (0.2 Pa or 1×10^{-3} Torr) are major factors that have caused it to be largely replaced by organic fluids. Since this trend began in the early 1930s, five major categories of organic fluids have come into use. These are described briefly in the following subsections. The properties of a few typical members of each category are summarized in Table 5.1. It must be recognized that data of this kind are very difficult to measure, and so such data as are available are often incomplete and frequently contain inconsistencies. The data presented here are derived from literature recently obtained from manufacturers, but may contain some values that are somewhat inaccurate. Plots based on these data showing the variation of the vapour pressure of some of these fluids as a function of temperature are given in Fig. 5.6. The rather unusual temperature scale used in this figure arises from the fact that the variation of the vapour pressures P_v of liquids with temperature T (K) is described by the classical Clausius–Clapeyron equation of physical chemistry:

Table 5.1 Properties of some diffusion pump fluids

Fluid	M	P_v at 20°C (Pa)	T_b at 100 Pa
Hydrocarbons			
Convoil 20	400	5×10^{-5}	200
Apiezon C	479	4×10^{-7}	250
Alcatel 220	408	6×10^{-8}	260
Synthetic esters			
Dibutyl phthalate	278	2×10^{-3}	140
Octoil	391	2×10^{-5}	190
Octoil-S	427	3×10^{-6}	205
Silicone fluids			
DC-702	NA	4×10^{-5}	190
DC-704	484	3×10^{-6}	220
DC-705	546	3×10^{-8}	250
Polyphenyl ethers			
Neovac-SY	405	6×10^{-7}	215
Convalex-10	454	5×10^{-8}	285
Santovac-5	454	5×10^{-8}	290
Perfluorinated polyethers			
Fomblin Y HVAC-18/8	2800	2×10^{-6}	220
Fomblin Y HVAC-25/9	3400	2×10^{-7}	270
Krytox 1618	4300	3×10^{-6}	290
Krytox 1625	4600	3×10^{-7}	325

M, average molecular mass; P_v, vapour pressure (Pa); T_b, boiling temperature (°C)

$$\log P_v = \frac{-\Delta H_v}{2.303R}\left(\frac{1}{T}\right) + C \tag{5.2}$$

in which R is the ideal gas constant (8.314×10^{-3} kJ/mol·K), ΔH_v is the heat of vaporization of the liquid (i.e. the amount of heat needed to convert a mole of liquid to a gas, in kJ/mol), C is a constant, and $\log P_v$ is the logarithm (base 10) of the vapour pressure. With this functional dependence it is advantageous to plot the vapour pressure on a logarithmic scale versus $(1000/T)$ because doing so yields a straight line which is easy to draw and to use for extrapolation, as shown. A plot of this kind also makes

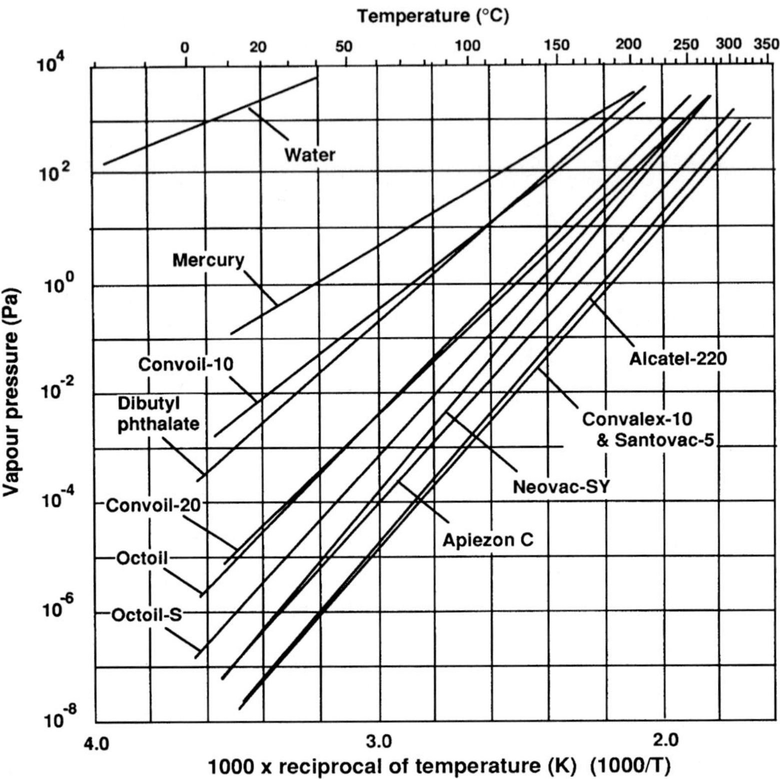

Fig. 5.6 The variation of the vapour pressure of some common diffusion pump fluids with temperature.

it possible to estimate the value of the heat of vaporization of a fluid, because ΔH_v is related to the slope of the line S, which can be calculated from the vapour pressures of the liquid P_1 and P_2 at two different temperatures T_1 and T_2, respectively, as follows:

$$S = \frac{\log P_2 - \log P_1}{\left(\frac{1000}{T_2}\right) - \left(\frac{1000}{T_1}\right)} = \frac{-\Delta H_v}{2.303 \times 10^3 R}$$

Actually, ΔH_v varies slightly with temperature for most liquids, and so the function is not truly linear. Nevertheless, Fig. 5.6 shows in an interesting way how the vapour pressures of these fluids vary with temperature.

5.4.1 *Hydrocarbon oils*

The first organic fluids to be used in diffusion pumps were refined hydrocarbon oils. Oils of this kind are presently available from many sources under trade names such as Apiezon A, B, and C (GEC Alsthom), Balzers Oil 61 and 71 (Balzers), Convoil-10 and -20 (CVC Products), and Diffoil-20, -30 and -40 (Lesker). These oils are usually natural paraffinic hydrocarbons, with molecular masses in the range from 350 to 500 g/mol, which have been highly refined and vacuum distilled to remove unstable molecular species and volatile components. Depending on the degree of refinement and distillation, their vapour pressures at 20°C may range from 10^{-7} to 10^{-4} Pa (10^{-9} to 10^{-6} Torr). Because of their relatively low cost, hydrocarbon oils are widely used in routine industrial applications where a single large diffusion pump may require a charge of 15 l or more, or where harsh operating conditions may make it necessary to change the fluid frequently. As would be expected, these natural hydrocarbon oils are good lubricants, but when used for this purpose in vacuum systems they tend to creep (i.e. to spread over surfaces) away from sites where they are most critically needed. They degrade readily when exposed to ionizing radiation, usually producing carbonaceous deposits which are slightly electrically conductive. They are more susceptible to thermal and oxidative degradation than other types of pump fluids, and should not be exposed to air at pressures above about 10 Pa (10^{-1} Torr) while the diffusion pump is hot. As will be pointed out later, this is a serious deficiency in many applications. However, a naphthalene-based synthetic hydrocarbon fluid (Alcatel 220, eicosyl naphthalene) is advertised as being nearly as stable as the silicone and polyether fluids at about one-third of their cost. Hydrocarbon fluids are not often used in electron microscopes because they are generally considered to cause higher levels of specimen contamination than other available fluids. Most hydrocarbon fluids are soluble in hydrocarbon solvents such as toluene, xylene, cetane, and petroleum ether (mineral spirits).

5.4.2 *Synthetic esters*

The first synthetic organic compounds to find wide use were esters of dicarboxylic acids, such as phthalic and sebacic acids. Dibutyl phthalate, di(2-ethyl-hexyl) phthalate (Octoil, CVC Products and Diffoil, Lesker), and di(2-ethyl-hexyl) sebacate (Octoil-S, CVC Products and Diffoil-S,

Lesker) are compounds of this type which are presently used as general purpose diffusion pump fluids. Because these are unique synthetic compounds they have molecules of a single molecular mass, rather than being a mixture of molecules of different molecular masses as are the natural hydrocarbons. Careful synthesis procedures, and refinement by vacuum distillation, make it possible to produce these compounds with a high degree of purity. This is a distinct advantage in diffusion pump fluids because it minimizes backstreaming of components with low molecular mass and gives more uniform and predictable boiling characteristics. These compounds typically have vapour pressures in the range from 10^{-6} to 10^{-3} Pa (10^{-8}–10^{-5} Torr) at 20°C. Although they are considerably more resistant to thermal and oxidative degradation than hydrocarbon oils, they should not be exposed to air at pressures above 100 Pa (1 Torr) while at the operating temperature of a diffusion pump, and it is generally recommended that they be cooled well below 100°C before being exposed to air at atmospheric pressure. These fluids cost somewhat more than the least expensive hydrocarbon oils but are still inexpensive enough to find wide use in industrial processes and research apparatus. The synthetic ester fluids are soluble in acetone and methyl-ethyl-ketone, and in 50/50 mixtures of these solvents with toluene or xylene.

5.4.3 Silicone fluids

A major improvement in stability and resistance to oxidation was achieved in the 1940s when the Dow Corning Corporation introduced a series of synthetic silicone fluids with the trade names of DC-702, DC-703, and DC-704. About 1960 a fourth fluid, DC-705, was introduced. The DC-703 fluid has not been marketed for a number of years, but the others are widely used today because of their outstanding resistance to chemical, thermal, radiation, and oxidative degradation. For example, they are used in valveless pumping systems in many manufacturing operations to achieve rapid cycling from high vacuum to atmospheric pressure by admitting air directly into the diffusion pump without allowing it to cool appreciably below operating temperature. Typically, they will withstand several thousand such cycles, and will handle high loads of air and water vapour, and some corrosive gases, without significant degradation. DC-702 is a refined mixture of phenylmethyl-dimethyl-cyclosiloxane compounds, while DC-704 is a single synthetic compound (tetraphenyl-tetramethyltrisiloxane). These fluids are comparable in cost to the synthetic esters, and are used for similar applications. The DC-704 fluid is a good choice for vacuum evaporators and similar laboratory apparatus because it will tolerate considerable misuse without degradation. The DC-705 fluid is also a single compound (pentaphenyl-trimethyl-trisiloxane). The five phenyl groups per molecule raise its molecular mass to 546 and decrease its vapour pressure to near 3×10^{-8} Pa (2×10^{-10} Torr) at 20°C, making it suitable for ultra-high vacuum applications where its higher cost can usually be justified.

The silicone fluids are rarely used in electron microscopes, however, because electron bombardment produces siliceous deposits which are non-conducting and highly insoluble. These deposits can be removed from a platinum aperture by flaming the aperture to convert the siliceous materials to silicon oxides and then immersing the aperture in concentrated hydrofluoric acid overnight to dissolve these oxides. It is difficult, however, to remove these deposits from a molybdenum aperture because the standard cleaning procedure of heating the aperture to white heat in a vacuum is not always effective. The silicone fluids themselves are also very difficult to remove from the parts of an electron microscope because they are so viscous and insoluble. Some instrument manufacturers have refused to issue service warranties on instruments if silicone oils are used in the diffusion pumps. The cleaning procedure recommended by the Dow Corning Corporation consists simply of wiping away as much of the oil as possible with dry cloth or tissue, followed by repeated wiping with cloth pads moistened with toluene, xylene, trichloroethylene, or perchloroethylene. I recommend finally wiping with a cloth pad moistened with reagent grade isopropyl alcohol.

5.4.4 *Polyphenylethers*

About 1960 the Monsanto Chemical Company developed two fluids, OS-124 and OS-138, for use in high-temperature hydraulic, heat-transfer, and lubrication applications in the space programme. Based on co-operative developmental work with the laboratories of CVC Products, highly refined relatives of OS-124 have been marketed as premium diffusion pump fluids under the trade names of Convalex-10 (CVC Products) and Santovac-5 (Monsanto) for more than 20 years. These are mixtures of five-ring polyphenylether compounds with average molecular masses of about

450 and vapour pressures in the 10^{-8} Pa (10^{-10} Torr) range at 20°C. The ether linkage (C-O-C) that forms the backbone of these compounds is very stable chemically, and so their resistance to degradation by heat, oxidation, and exposure to ionizing radiation is much better than the synthetic esters and nearly as good as the silicone fluids. They have relatively high viscosities, exhibit very low creep rates in vacuum systems, and have very low backstreaming rates. Under electron bombardment they polymerize, producing films that are slightly electrically conductive. Overall, their properties are such that they are very competitive with DC-705 as diffusion pump fluids for ultra-high vacuum applications, and are preferred for such applications when a silicone fluid cannot be tolerated. They are widely used in electron microscopes. Because they have good lubricating properties and very low creep rates, they can also be used to lubricate mechanisms in high vacuum systems. One major disadvantage of these fluids is their very high cost, which can usually be justified only for ultra-high vacuum applications. However, Varian sell a synthetic mono-*n*-alkyldiphenylether fluid under the trade name of Neovac-SY which has physical properties very competitive with these fluids but which is comparable in cost to the synthetic esters.

Because of their high viscosities and high molecular masses, these fluids are also somewhat difficult to remove from the surfaces of a vacuum instrument. The procedure usually recommended is similar to that described above for the silicone fluids: namely, removal of as much of the fluid as possible by wiping with dry cloth or tissue, then wiping repeatedly with pads moistened with acetone, methyl-ethyl-ketone (MEK), xylene, or a mixture of MEK and xylene, followed by wiping with pads moistened with isopropyl alcohol.

5.4.5 *Perfluorinated polyethers*

In the early 1970s several perfluorinated-polyether fluids that had been developed as specialized lubricants by Montecatini Edison S.p.A. were refined and adapted for use in diffusion pumps through co-operative research and development work with the Edwards company. Two of these are in use today under the Fomblin trade name (now produced by Ausimont, S.p.A.) with the designations Y HVAC-18/8 and Y HVAC-25/9 (18 and 25 indicate that the fluids have viscosities near 180 and 250 centistokes while the 8 and 9 indicate that their vapour pressures are in the

10^{-8} and 10^{-9} Torr ranges, at 25°C). Similar fluids are now available from the Du Pont Company under the trade names Krytox 1618 and Krytox 1625.

In perfluorinated organic compounds, all the hydrogens of the normal organic molecules are replaced by fluorines. Thus, the perfluorinated counterpart of diethylether, $H_3C\text{-}CH_2\text{-}O\text{-}CH_2\text{-}CH_3$, is $F_3C\text{-}CF_2\text{-}O\text{-}CF_2\text{-}CF_3$. The Fomblin and Krytox fluids are high molecular mass compounds of generally similar chemical structure. The C-F chemical bond is exceptionally stable, and the abundance of these bonds in these perfluorinated compounds gives them very unusual properties. They are immiscible with, and insoluble in, virtually all other liquids. Replacing the hydrogens (atomic mass = 1) with fluorines (atomic mass = 19) gives them much higher molecular masses (>2500) than the other types of pump fluids, yet their high molecular symmetry and the electronic configuration of the C-F bonds reduce the forces between molecules enough so that their boiling points are comparable to those of the other fluids. When subjected to bombardment by electrons and other ionizing particles, and when thermally degraded, it is the C-C and C-O bonds in the backbone chains of these compounds that rupture, rather than the stronger C-F bonds. Therefore, these compounds break down into relatively low molecular mass fragments which are volatile and easily pumped away, rather than cross-linking and forming polymerized solid products as other diffusion pump fluids do. They are therefore essentially non-contaminating in the usual sense. For this reason it was initially hoped that they would provide a simple solution to the specimen contamination problem in electron microscopes. Unfortunately, however, it was found that, while contamination problems were greatly reduced, electron microscopes in which these fluids were used eventually developed high-voltage instabilities due to micro-discharges along the ceramic insulators in the electron guns. Thoroughly cleaning the insulators relieves this problem for a time, but it eventually reappears. The problem is more severe in transmission electron microscopes, which usually operate at voltages of 100 kV and higher, than in scanning electron microscopes, which generally operate below 30 kV. Consequently, the use of the perfluorinated polyether fluids in transmission electron microscopes was very short lived. However, they have been used successfully for several years in scanning electron microscopes in some laboratories.

These fluids are used principally in applications where their exceptional thermal and chemical stabilities are advantageous. For example, they can be used to pump pure oxygen without risk of fire or explosions, and they are unaffected by such reactive gases as fluorine, chlorine, hydrogen chloride, and ammonia. Unlike the silicone fluids, they have good lubricating properties, and can even be used in rotary-vane pumps. This, combined with their low vapour pressures and their low creep rates, makes them good lubricants for mechanisms inside high vacuum apparatus. Like the silicone fluids, however, they are extremely difficult to remove from the parts of a vacuum system. They are insoluble in, and immiscible with, most classes of organic compounds, including common solvents such as acetone, toluene, xylene, and the alcohols. They are miscible with, and very slightly soluble in, ethyl ether, chloroform and carbon tetrachloride, and are modestly soluble in perfluoro-octane and Freon 113.

5.4.6 *Selecting a diffusion pump fluid*

The choice of the best fluid to use in a diffusion pump can be a complex and difficult task in many industrial applications where large apparatus are involved and matters of cost and chemical and thermal stability are critical. Fortunately, the situation is usually much simpler in most applications in electron microscopy. We are generally most concerned with having a low vapour pressure to ensure a low ultimate pressure and a low backstreaming rate to minimize specimen contamination. The silicone fluids are generally undesirable because of the siliceous deposits they produce under electron bombardment, and the perfluorinated polyethers cannot be recommended because of the high-voltage micro-discharge problem they apparently cause. This leaves the polyphenylether fluids as the best general choice. Although they are considerably more expensive than the other fluids, only relatively small quantities are required in the types of apparatus we deal with, and so the extra cost is generally a small item in the overall expense of operating an electron microscope. Incidentally, prices of pump fluids vary considerably among suppliers, and also depend strongly on the quantity purchased. For example, recent catalogue prices for a half-litre bottle of Santovac-5 ranged from \$395 to nearly \$700, while the cost of a 100 cc bottle was typically about \$160 (which is equivalent to a price of \$800 per half-litre).

Consideration of the choice of pump fluid inevitably brings up the question of whether or not it is safe to change from one type of fluid to another. This is a somewhat complex matter that depends strongly on the design of the pump. The proper functioning of an oil diffusion pump requires that the boiler delivers a sufficient number of oil vapour molecules to the jets each second to sustain the vapour streams that produce the pumping action. Physically, this means that there must be a proper balance between the heat input of the boiler and the heat of vaporization per mole ΔH_v of the pump fluid. (Technically, ΔH_v is the number of kilojoules of thermal energy needed to convert a mole of a liquid to vapour at its boiling temperature.) If the heat input is too low, or the heat of vaporization of the fluid is too high, the number of vapour molecules produced per second will fall below the normal level for which the pump was designed, causing a reduction in the critical inlet and backing pressures, and in the throughput capacity. In the extreme, the production of vapour could become so low that some of the jets would not receive sufficient vapour, whereupon the pump would not function properly. If the heat input is too high, or the heat of vaporization of the fluid is too low, excessive boiler temperatures and pressures will be produced. The resulting increased rate of boiling feeds larger quantities of vapour to the jets, causing the throughput capacity and the critical inlet and backing pressures to increase slightly. The rate of backstreaming usually increases markedly, while the pumping speed may decrease slightly, because of complex processes that disrupt the structure of the vapour stream produced by the top jet. The rates of thermal and oxidative degradation of the fluid also increase markedly. The heaters supplied on older pumps, and on pumps designed specifically for hydrocarbon fluids, may not provide sufficient heat input for such fluids as DC-705 and Santovac-5, while the heat input for pumps designed specifically for these fluids may be too high for the hydrocarbon and synthetic ester fluids. Under these circumstances it becomes necessary to change heaters when changing the type of fluid used.

Ideally, it should be permissible to replace the fluid in a given pump with another fluid having nearly the same molar heat of vaporization and boiling temperature, both evaluated at the normal boiler pressure (about 200 Pa or 1.5 Torr). It must be recognized, however, that it is very difficult to measure the vapour pressures and to calculate the thermal properties of fluids such as these accurately. Such data as are available on these properties do not inspire the highest levels of confidence. Therefore it is advisable to discuss the matter with the manufacturers of the pump and the pump fluid before implementing a change in the fluid used in a particular pump. Manufacturers usually measure the performance of their pumps and fluids under a variety of different service conditions, and can give reliable advice and guidance based on actual experience. Information presented in recent catalogues indicates that most diffusion pumps are now designed to operate satisfactorily with several different fluids. Whenever a change is made to a new type of fluid, the old fluid should be thoroughly cleaned out of the pump and all associated valves, traps, baffles, and pumping lines.

5.5 Backstreaming

'Backstreaming' is the general term used to refer collectively to the processes that lead to the movement of gases, including the pump fluid and its contaminants and decomposition products, from the pump into the vacuum system. This is a major problem associated with the use of oil diffusion pumps and the principal reason they are being replaced by other types of pumps on electron microscopes and other similar apparatus where the presence of the pump fluid in the vacuum system cannot be tolerated. Because of its overriding importance, a considerable amount of research and development work has been devoted to this problem since the early 1950s, much of which is reviewed in Van Atta (Chapter 6).

This work has led to the introduction of many innovative features into the design of diffusion pumps, some of which are shown in Fig. 5.7. A grooved boiler plate or some other device that ensures efficient heat transfer and uniform boiling is used to produce a steady high yield of oil vapour with a minimum of superheating, bumping, and thermal degradation. Internal baffles above the boiler prevent oil droplets from being carried from the boiler to the jets where they can become entrained in the vapour streams and deflect oil molecules upward toward the inlet of the pump. Fractionation tubes prevent low molecular mass oil molecules from reaching the upper jets where they are most likely to diffuse into the vacuum system. An ejector jet, which is fed vapours from the outer region of the boiler, forcibly directs low molecular mass fractions of the oil into the tube that leads from the body of the pump to the outlet port. A portion of this tube is not cooled by the cooling coils, and so it runs

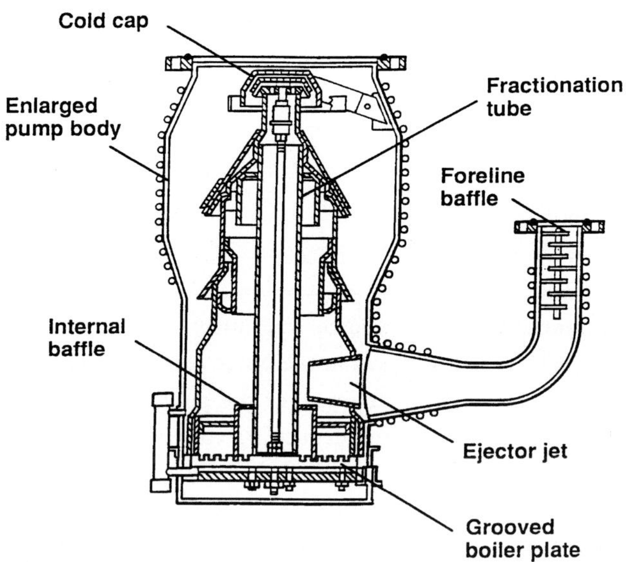

Fig. 5.7 An oil diffusion pump with advanced design features. (Courtesy of Varian Vacuum Products.)

hot enough to promote vaporization of the most volatile components from the oil, facilitating their removal by the backing pump. This tube also contains baffles which minimize loss of oil to the backing line during times when the pump is operating at maximum throughput. To minimize backstreaming, the jet assembly is designed to prevent oil vapours from leaking out in random directions, and the top jet is surrounded by a cooled cap which catches oil molecules that leave the top jet in an upward direction. The sizes of the jet orifices, and the spacings of the jets from the wall of the pump, are chosen to optimize pumping speed and the overall compression ratio. With features such as these, only the most persistent sources of backstreaming remain in pumps of advanced design.

Fundamentally, however, there must always be molecules of the pump fluid in the inlet of the pump at a concentration corresponding to the vapour pressure of the fluid at the temperature prevailing there. These molecules have random, undirected motion and are therefore free to migrate up into the vacuum system. With present pump fluids, most of which have vapour pressures well below 10^{-6} Pa (10^{-8} Torr) at room temperature, the contribution from this source should be small. However, the diffusion of oil vapours from the mechanical pump into the diffusion pump, the introduction of volatile organic materials into the diffusion pump from the vacuum system, and the thermal and oxidative degradation of the diffusion pump oil itself, all contaminate the pump oil with low molecular mass molecules and effectively raise its vapour pressure. This problem can be reduced to some extent by feeding the upper jet from the centre of the boiler through a fractionation tube as shown in Fig. 5.7. With this arrangement, molecules having higher vapour pressures evaporate out of the oil and are fed to the lower jets in the outer regions of the boiler as the oil flows into the boiler from the wall of the pump. Only molecules with low vapour pressures reach the central part of the boiler, which is enclosed by the fractionation tube, and are fed to the upper jet. The use of an ejector stage in conjunction with the fractionation tube, as shown in Fig. 5.7, provides further help by purposefully directing the low vapour pressure fractions, which evaporate from the outer region of the boiler, into the outlet line of the pump where they have a high probability of being removed by the backing pump.

5.5.1 *Cold caps*

Vapours from the top jet make the greatest contribution to backstreaming. This occurs in at least two ways; some molecules in the jet stream acquire an upward component of velocity through collisions with other molecules, while some are directed upward by turbulence and other irregularities in the vapour stream as it leaves the jet assembly. Placing a cooled cap of the type shown in Fig. 5.7 over the upper jet to intercept these wayward molecules can reduce this source of backstreaming by more than 95 percent. Varian offer an extra large cold cap, the 'Mexican Hat' cold cap shown in Fig. 5.8, for use in their VHS series pumps, and this is claimed to reduce the backstreaming rate to a level too low to be measured by the standard procedure prescribed by the AVS. Backstreaming rates for well-designed pumps equipped with cold caps are typically specified as being in the range from 10^{-6} to 10^{-5} mg/min per mm^2 of inlet area. Although low, this rate of backstreaming is still sufficiently large for a pump with an inlet diameter of 150 mm to coat a bell jar 300 mm in diameter and 300 mm high, with an internal surface area of about 0.5 m^2, with a layer of oil about 10 nm thick in an hour. This provides a basis for understanding why it is not a good

Fig. 5.8 The 'Mexican hat' cold cap for reducing the backstreaming of oil vapours in oil diffusion pumps. (Courtesy of Varian Vacuum Products.)

practice to leave specimens standing for long periods in vacuum evaporators which are evacuated by oil diffusion pumps.

5.5.2 *Baffles*

It is almost standard practice to place baffles above diffusion pumps to reduce the level of backstreaming. These are devices designed to intercept and condense upward-moving oil molecules and return them to the pump. They consist of collections of cooled apertures, plates, and fins which are said to be optically dense; that is, they are arranged so there is no straight-line path by which a beam of light, or an oil molecule, can pass through without striking a cooled surface at least once. The temperature of the elements of a baffle should be well below the ambient temperature in the vacuum system. With present pump fluids, water-cooling is normally sufficient. Whereas refrigeration obviously provides additional benefits, the temperature should not be so low that the oil freezes on the baffle and cannot return to the pump. Fig. 5.9 shows schematically the construction of a chevron baffle, which is perhaps the most popular type presently in use. Such a baffle will usually reduce the backstreaming rate by several orders of magnitude, so that levels can be attained with fluids such as Convalex-10, DC-705, and Santovac-5 that are acceptable for all but the

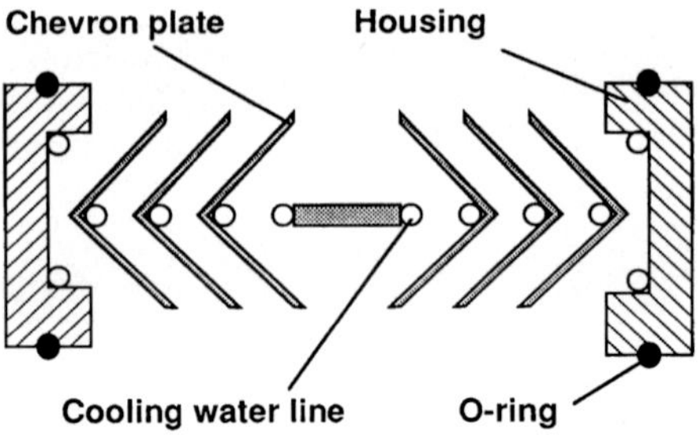

Fig. 5.9 A water-cooled chevron baffle.

most demanding applications. Again, Varian state that the rate of backstreaming above their chevron baffles is too low to be measured by the standard AVS procedures.

It should be obvious that if a baffle is effective in intercepting oil molecules moving upward from the pump into the vacuum system it will also impede the movement of gas molecules from the vacuum system downward into the pump. As a general rule, the use of a baffle decreases the effective speed of a pump by nearly a factor of two. When evaluating the specifications for a vacuum system it is important to note whether it incorporates a baffle, and whether the effect of the baffle is included in the quoted pumping speed. Thus, if an apparatus is advertised as having a 600 l/s pump but also has a water-cooled baffle, the effective pumping speed available for evacuating the apparatus will be only about 300 l/s.

Manufacturers are cognizant of the effect of baffles in reducing pumping speed, and several produce special pumps designed to minimize it. These include the 'Diffstak', 'Crystal', and 'DIFFset' series of pumps produced by Edwards, Alcatel, and Balzers, respectively. The construction of these pumps is shown in Fig. 5.10. Their unique feature is the unusual design of the upper jet and the large size of the water-cooled cold cap surrounding it. This cold cap is designed so that it functions in combination with a sharp outward bulge in the body of the pump to form an efficient water-cooled baffle system of exceptionally high conductance. Also note that this pump includes many of the other advanced design features mentioned above, such as a fractionation tube, an internal baffle, a foreline baffle, and an ejector jet. Compared with standard pumps equipped with standard baffles, these pumps provide nearly twice the pumping speed for sizes with inlet diameters up to 200 mm, and about one and one-half times

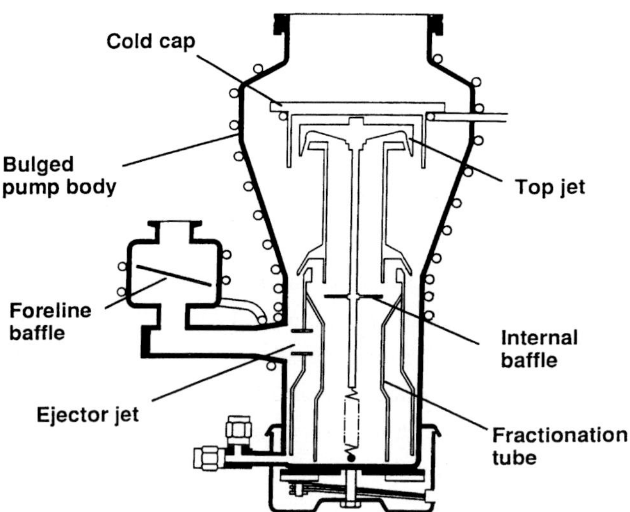

Fig. 5.10 An oil diffusion pump in which the cold cap above the top jet and a bulged pump body function as a water-cooled baffle of high conductance. (Courtesy of Balzers.)

the speed for the 250 mm size. When combined with specially-designed valves, these pumps make it possible to assemble pumping systems that are smaller, faster, cleaner, and considerably less expensive than comparable systems based on standard pumps and components. As a consequence, they have been used widely in vacuum apparatus manufactured in recent years.

Even the use of a well-designed baffle cannot totally prevent oil molecules from moving upward into the vacuum system, however. Given enough time they can still pass through the baffle by creeping along the surfaces of the baffle, by evaporating from surfaces of the baffle, and by passing through the baffle via collisions with other gas molecules. The Convalex-10, DC-705, and Santovac-5 fluids have such low vapour pressures and creep rates that the first two mechanisms make minimal contributions when these fluids are used with well-designed baffles which are properly cooled. The rate of backstreaming due to collisions of errant oil molecules with gas molecules reaches a maximum at about 0.1 Pa (10^{-3} Torr). At higher pressures gas flow enters the transitional range and acquires some viscous character, and the oil molecules tend to be swept down through the baffle. At lower pressures the density of gas molecules decreases to the point where collisions with oil molecules become very infrequent. Contributions from this mechanism are therefore very small at operating pressures in the high vacuum range, but are greatest when the pump is operating in the range of transitional flow during the pumpdown process.

5.5.3 *Liquid nitrogen traps*

In response to the residual backstreaming phenomena that cannot be controlled by other methods, it is common practice to install traps cooled with liquid nitrogen above the baffles in vacuum systems where contamination must be reduced to the lowest possible level. At the temperature of liquid nitrogen (77 K, –196°C) all pump fluids become solids with vapour pressures so low that they do not contribute in any practical way to backstreaming. For example, extrapolation of data taken at higher temperatures indicates that the vapour pressures of fluids such as Convalex-10, DC-705, and Santovac-5 are near 10^{-18} Pa (10^{-20} Torr) at 200 K (–73°C), and in the realm of 10^{-40} Pa at 77 K. Well designed liquid nitrogen traps therefore provide the best available ultimate solution to the backstreaming problem. Such traps also act as extremely effective pumps for water, which has a vapour pressure near 10^{-18} Pa (10^{-20} Torr) at 77 K. Like baffles, traps have finite conductance and therefore contribute an additional decrease of from 30 to 50 percent in the effective speed of the pumping unit.

Unfortunately, the use of a liquid nitrogen trap is not an absolute safeguard against backstreaming. There are still ways in which backstreaming can occur, even in a well-trapped system. One involves the violent ejection of flakes of frozen oil from the surfaces of traps due to thermal contraction when liquid nitrogen is added. Another involves the free evaporation of oil molecules from traps when they are warmed to room temperature. This latter mode of contamination is particularly troublesome because most traps are designed so that the cooled elements cannot be removed from them for cleaning while still at liquid nitrogen temperature. In fact, most traps cannot be disassembled in any way, and the process of removing them from a vacuum system for cleaning is a major undertaking. The usual method of cleaning a trap is to allow the liquid nitrogen to evaporate out of it, and then to pump it out for several hours while it is at room temperature. When this is done the materials collected in the trap evaporate and travel with equal probability upward into the vacuum system, contaminating it, and downward into the pump, recontaminating it. Trap heaters are usually provided to speed up this process of redistributing the contaminants. In instruments such as electron microscopes and vacuum evaporators, which cannot be cleaned by a high temperature bake-out procedure while the trap is being cleaned, this is the equivalent of 'washing your feet in your boots', cleaning a trap gets you pretty much where you would have been if the trap had not been used in the first place.

However, it may be possible to reduce the amount of contamination transferred into an instrument while a trap is warmed up to room temperature if there is a high-vacuum valve directly above the trap. To do this it is necessary to keep this valve closed whenever the trap is not at liquid nitrogen temperature. When the trap is again filled with liquid nitrogen, leave this valve closed and pump out the trap and the valve over a weekend with the aim of recapturing some of the contaminants that were deposited on the valve while the trap was at room temperature. This procedure will be even more effective if the valve, and any tubing or fixtures between it and the trap, can be heated to 150°C during this time.

Inevitably, however, someone will forget to fill the trap on schedule, the supply of liquid nitrogen will run out unexpectedly, the device for automatically filling the trap with liquid nitrogen will fail, or some other situation will arise which allows the trap to warm up while the system is in operation, whereupon oil molecules will pass from it into the vacuum system. As an extreme example, I know of one situation in which the sensor on an automatic liquid nitrogen filling device failed in such a way that the flow of liquid nitrogen into the trap was not stopped when the trap became full. Instead, the device continued to supply liquid nitrogen to the trap throughout the night until all the liquid nitrogen in a large supply tank was used up. During this time, the liquid nitrogen overflowed from the trap and ran down over a water-cooled baffle that was located below the trap and above the diffusion pump. This cooled the baffle enough so that the water in it froze, expanded, and ruptured the cooling coils inside it. Towards morning, after the supply of liquid nitrogen was exhausted, the baffle warmed up again, whereupon water leaked out of the ruptured tube and completely filled the diffusion pump and the backing line. Fortunately, the automatic valve operating system shut the vacuum system down before serious contamination of the column of the instrument occurred. However, when an attempt was innocently made to start the system up again the next morning the rotary-vane backing pump spewed forth a most uncharacteristic oil-water emulsion. It required a rather considerable amount of work, and involved some expense, to get the instrument back into operation. Admittedly, accidents such as this are rare, but hearing about them is somewhat entertaining and telling them may help someone else avoid a similar problem.

At least the use of a liquid nitrogen trap does have some rewards. It probably will slow down the rate of contamination somewhat when an instrument is new, and it will always significantly decrease pumpdown time and ultimate pressure because of the high pumping speed it provides for water, which constitutes the major component of the gas load produced by gas desorption processes (see Section 2.10.3, p. 60). Realistically, however, you need to balance these advantages against the considerable effort and expense involved in providing the liquid nitrogen needed to keep a trap in operation, without undue expectations for a long-term reduction in contamination.

5.5.4 *Operating procedures*

It is also possible to reduce the rate of backstreaming by careful system management. It is generally accepted that the rate of backstreaming is greatest during periods when the upper jet of an oil diffusion pump is overloaded and not pumping properly. This always occurs while a pump is warming up to operating temperature just after it has been turned on, and while it is cooling down just after its heater has been turned off. During these times it is very important to keep the valve between the pump and the vacuum system closed, and to be sure that traps, baffles, and the body of the pump are properly cooled. High levels of backstreaming also occur when the valve to the vacuum system has just been opened and the pump is exposed to air at pressures above its critical inlet pressure. Also, during this time the flow of gas through the pump can become transitional in nature and it is possible for some oil to be swept out of the pump into the backing line. This can lead to gradual depletion of the oil from the diffusion pump, causing the remaining oil to overheat and decompose more rapidly. It is therefore important to rough pump a system well enough before the valve above the diffusion pump is opened so that it requires no more than 15 or 20 s, and preferably less, for the diffusion pump to make the transition from the rough vacuum to the high vacuum range.

All three of the above situations arise often in the operation of vacuum evaporators, which are frequently turned on and off and cycled to atmospheric pressure, and so backstreaming in these units can be expected to be high and difficult to control. The first two occur only infrequently in the operation of electron microscopes in many laboratories, since the diffusion pumps are seldom turned off there. However, many electron microscopes are cycled to moderately high pressures rather frequently to change filaments, to load and unload film, and to change specimens, and so contributions from the third cause can become significant over a few months of operation. Unfortunately the valve systems on most instruments are automatically controlled and do not have a reliable vacuum gauge readout. Consequently, the operator has little control over, or accurate knowledge about, the evacuation process. These automatic valve operating systems should not be trusted implicitly, because the response of vacuum gauges can change over time, measurement circuits can drift out of calibration, and they may not have been adjusted to operate properly in the first place. Therefore, any indication that something is not quite right in the pumpdown process should be called to the attention of your friendly, co-operative service engineer with a request for an immediate, thorough investigation, and remedy if appropriate. It is, of course, always helpful to have a set of good vacuum gauges on your instrument, as suggested in Section 3.3 (p. 105).

5.5.5 *Nitrogen purge*

The XEI Scientific company have recently introduced an apparatus for utilizing a rather interesting approach to removing and preventing oil contamination of vacuum systems which can be left standing at a pressure in the rough vacuum range for a significant fraction of the time, such as those on most scanning electron microscopes. This apparatus, which is sold under the trade name of SEM-CLEAN, introduces a carefully controlled flow of nitrogen sufficient to maintain the pressure in the system at about 100 Pa (1 Torr). At this pressure gas flow is viscous in character, and so the hydrocarbon molecules that randomly desorb from the surfaces inside the system from normal thermal agitation have a high probability of being swept along with the flow of the nitrogen gas and removed from the system by the roughing pump. This purging process is carried out over nights and weekends when the instrument is not in use. During these times the diffusion pump is turned off and isolated from the vacuum system by closing the main and backing valves (see Fig. 5.11, Section 5.7.1, below), and only the rotary-vane roughing pump is left operating. Purging an instrument in this manner over a few weeks is said to ultimately reduce oil contamination to an insignificant level. Since the initial source of contamination appears to be the very low molecular mass hydrocarbon fragments produced by degradation of the oil in the backing pump, recontamination of the instrument when it is pumped down into the high vacuum range for normal use is substantially eliminated by a foreline trap with a copper mesh filling. The purging process, of course, also removes oil molecules from this trap and keeps it operating effectively at all times. Although not a part of the system as presently produced, it would seem advantageous to use a source of ultraviolet radiation, such as one of the Phototron devices produced by Danielson Associates, in conjunction with the SEM-CLEAN system, because photons of ultraviolet light would provide the activation energy needed to substantially increase the rate at which the oil molecules desorb from the surfaces (see Section 2.10.3, p. 63).

5.6 Ultimate pressure

One fundamental factor governing the ultimate pressure attainable with an oil diffusion pump is the overall compression ratio K it can develop between its inlet and outlet ports, which is defined in terms of the inlet and outlet pressures, P_i and P_o, as:

$$K = \frac{P_o}{P_i} \tag{5.3}$$

It is generally accepted that well-designed pumps can sustain compression ratios of the order of 10^{10} for heavy gases such as oxygen and nitrogen. On this basis, an outlet pressure of 1 Pa (10^{-2} Torr) should make it possible to achieve an inlet pressure of the order of $P_i = P_o/K \approx 1/10^{10} \approx 10^{-10}$ Pa (10^{-12} Torr). Unfortunately, compression ratios for light gases, such as helium and hydrogen, are very sensitive to pump design and prevailing operating conditions, and so may be lower by as much as a factor of 10^3 to 10^4. In fact, these gases readily diffuse back through some diffusion pumps. As a consequence, the ultimate pressure becomes a complex function of pump design and the partial pressure of these gases in the backing line. This phenomenon and methods of controlling it are discussed more fully in Section 6.1.3 (p. 234) and in O'Hanlon (p. 195).

Obviously, the vapour pressure of the pump fluid will also have a significant influence on the ultimate pressure. The pressure at the inlet of a pump cannot be less than the vapour pressure of the pump fluid at the temperature prevailing there. As indicated by the vapour pressure data given in Table 5.1 and Fig. 5.6, this should be in the range from 10^{-8} to 10^{-6} Pa (10^{-10}–10^{-8} Torr) for the majority of pump fluids, assuming an inlet temperature of 20°C. In practice, however, these values are better regarded as 'potential' levels that might be attained if other factors did not degrade the situation. The backstreaming of vapours from the mechanical pump, contamination with volatile materials from the vacuum system, and oxidative and thermal degradation, all introduce low molecular mass components into the pump fluid that raise its vapour pressure above the values for the uncontaminated fluids given in Table 5.1 and Fig. 5.6. As mentioned above, low molecular mass gases, such as hydrogen and helium, diffuse from the backing line through the diffusion pump in sufficient concentration to raise the ultimate pressure appreciably. In addition, as discussed in Section 2.10 (p. 55), gas influx processes have an over-riding influence on the ultimate pressures attainable in all vacuum systems. Considering just the pumping system itself, gas evolution from the surfaces of the valve and baffle, and from the O-rings used to seal between these units, is enough to degrade ultimate pressures significantly. Pumps with integral baffles of the type shown in Fig. 5.10 achieve a significant decrease in gas evolution by eliminating the O-rings needed for standard baffles. As was pointed out in Section 2.10.3 (p. 62), the attainment of pressures appreciably below the 10^{-5} Pa (10^{-7} Torr) range requires long-term bakeout treatments that cannot be used on standard systems which are usually assembled with O-rings. Furthermore, the levels of gas influx in systems as complex as electron microscopes can be expected to be so high that they overwhelmingly determine the ultimate pressures attainable.

If the pumping speed curves of Fig. 5.4 were extended to lower pressures they would, of course, fall rapidly to zero as the value for the ultimate pressure is approached. In characterizing their pumps, manufacturers either do not give such ultimate pressure data, or they list values that are considerably higher than the vapour pressures of the recommended pump fluids. Typically, values are given as 'better than' some value in the range from 10^{-6} to 10^{-5} Pa (10^{-8} to 10^{-7} Torr). Suppliers of pump fluids often give values for operation with a baffle only (i.e. at 20°C), which tend to be one or two orders of magnitude greater than the corresponding fluid vapour pressure. When given, values for operation with a liquid nitrogen trap are usually comparable to or only slightly less than the room temperature vapour pressures of the pump fluids. These all are probably conservative values chosen to compromise between the ultimate capabilities of the pumps and fluids and the degrading contributions of the systems on which they are likely to be used. It is generally accepted, for example, that pressures in the 10^{-9} Pa (10^{-11} Torr) range can be achieved in small, clean test systems, which have been properly baked out, when they are evacuated by diffusion pumps charged with one of the low vapour pressure fluids such as Convalex-10, DC-705, or Santovac-5, and liquid nitrogen trapping is used. Diffusion pumps are unquestionably capable of attaining ultimate pressures that are more than adequate to meet the needs of most vacuum systems used in electron microscopy. It is unfortunate indeed that oil backstreaming is so difficult to control.

5.7 Operating vacuum systems with diffusion pumps

As mentioned at the beginning of this chapter, the oil diffusion pump was the first type of high vacuum pump to be used really extensively. Consequently, most of the principles used in designing high vacuum systems, and the practices involved in operating them, were developed for systems with oil diffusion pumps. A sound understanding of these principles and practices is therefore of value, not only because of its direct applicability to systems with diffusion pumps, but also because of the insight it provides in operating systems with other types of pumps. We will describe first a traditional system with a full set of valves, and then consider some common modifications and simplifications to this system.

5.7.1 The traditional three-valve system

All the basic principles involved in the operation of vacuum systems employing oil diffusion pumps can be developed using the vacuum system shown schematically in Fig. 5.11. Here the diffusion pump is connected to the vacuum chamber through a water-cooled baffle and a large, high-conductance valve, which is usually called the 'main valve' or the 'high vacuum valve'. The diffusion pump is backed by a rotary-vane pump which is connected to it through the 'backing line' and 'backing valve' (which are sometimes called the 'foreline' and the 'forevalve'). It is helpful to have a backing gauge, which may be either a thermocouple gauge or a Pirani gauge, located between the backing valve and the outlet port of the diffusion pump, as shown, although this gauge is not critically necessary, and is therefore not included in many vacuum systems. There is also a 'roughing line', which bypasses the diffusion pump and connects the mechanical pump directly to the vacuum chamber. This contains the roughing valve and the roughing gauge, a Pirani gauge or thermocouple gauge for monitoring the pressure in the vacuum chamber when it is being rough pumped. Air inlet valves, or 'venting valves', are provided for both the vacuum chamber and the backing pump. There is also a high vacuum gauge to measure the pressure in the vacuum chamber when high vacuum conditions are attained. This is usually a cold cathode discharge gauge or a hot cathode ionization gauge.

This is actually a very important vacuum system, because it is the one found on most simple apparatus such as the vacuum evaporators used in electron microscopy laboratories, and because most of the more complex systems contain a similar arrangement of basic components, with peripheral subsystems added for special purposes. Furthermore, the sequence of operations for this system is also used on many systems with other types of vacuum pumps. If you understand the operation of this basic system you will understand, or should be able to figure out, the operation of most other vacuum systems you are likely to encounter. We will assume that the principles developed here are 'common knowledge' when we discuss more complex vacuum systems in Chapter 9.

The overall operating procedure for this basic system has several distinct stages, which can conveniently be described in terms of the four different operational configurations of the valves and gauges shown in Fig. 5.12. Each of these stages involves a group of steps which are discussed in the following subsections. The over-riding concern throughout this procedure is to avoid losing or oxidizing the oil in the diffusion pump while minimizing the opportunity for oil vapour to backstream from the pumps

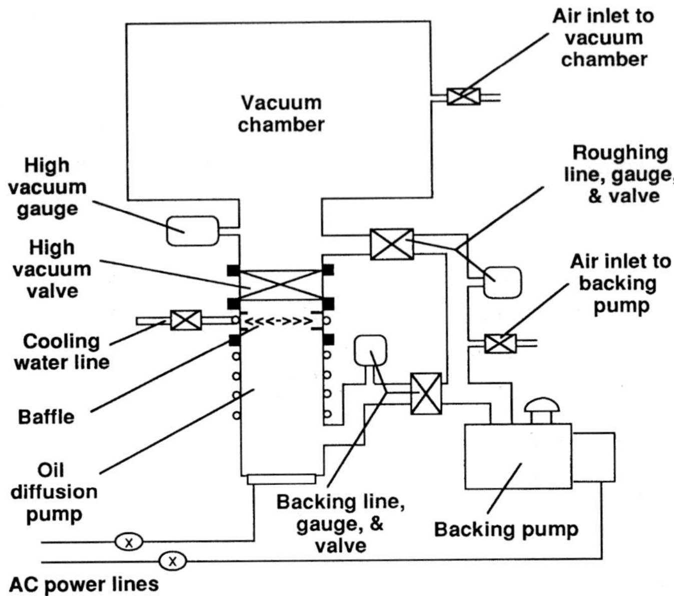

Fig. 5.11 A basic vacuum system with an oil diffusion pump and main, roughing, and backing valves. The backing gauge is often omitted.

into the vacuum system. There are four basic Rules that are inherent in this procedure:

i. The body of an oil diffusion pump must always be properly cooled when the heater is turned on.
ii. The pressure in the outlet tube of the diffusion pump should not be allowed to rise above the level of the critical backing pressure of the pump.
iii. Air at pressures near one atmosphere should never be admitted to a diffusion pump while the pump's heater is turned on and the oil in it is boiling.
iv. Whenever the diffusion pump is on it should be pumped on by the backing pump.

Watch for the methods used to comply with these rules as you read the following instructions. The consequences of violating these rules are discussed in Section 5.8 (p. 214).

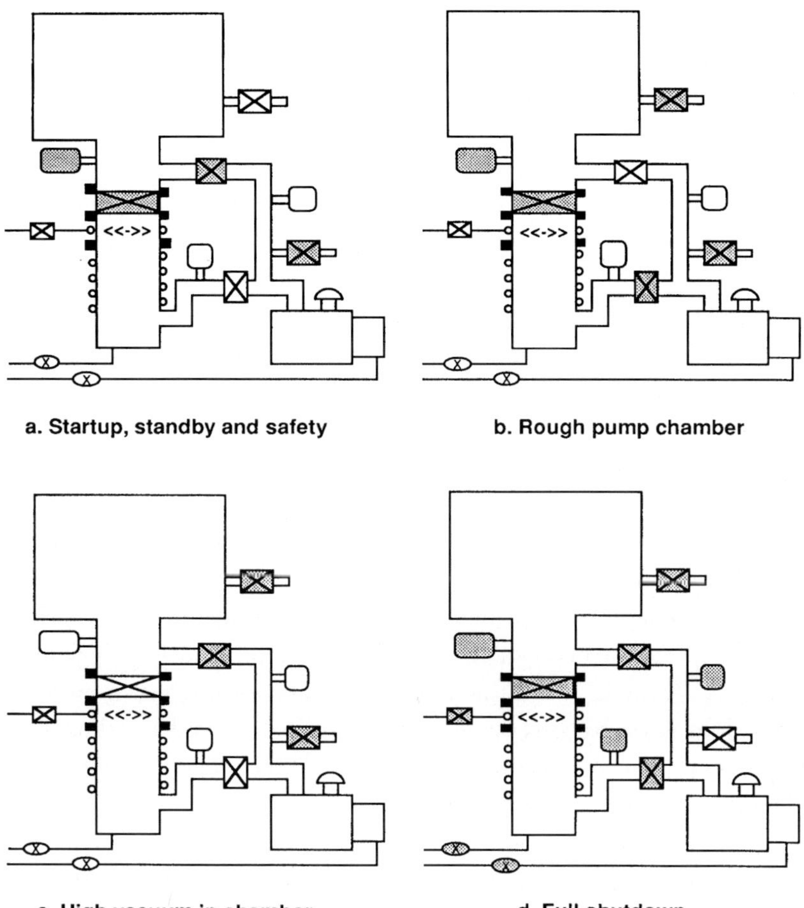

Fig. 5.12 The four operational configurations for the controls on a vacuum system with an oil diffusion pump and main, roughing, and backing valves. Shading indicates that valves are closed and that switches and gauges are turned off.

5.7.1a *Startup and standby*

When a vacuum system such as the one shown in Fig. 5.11 is shut down for a long period of time, the high vacuum, roughing, and backing valves are all usually closed, the vacuum chamber is left under vacuum, the air inlet valve to the backing pump is left open, all electrical power is turned off, and the cooling water to the diffusion pump is shut off, as indicated in Fig. 5.12d. The procedure for starting the system up from this *shutdown* configuration is as follows:

1. Close the air inlet valve to the backing pump.
2. Turn on the main power switch for the apparatus, and then turn on the switch that feeds electrical power to the backing pump.
3. Open the backing valve and turn on the roughing gauge. Turn on the backing gauge if one is present.
4. Open the valve that supplies cooling water to the diffusion pump, then turn on the electrical power to the diffusion pump heater. This should put the system in the *startup* configuration shown in Fig. 5.12a.
5. Allow from 15 to 30 min, depending on the size of the pump, for the oil in the diffusion pump to reach its boiling point and for proper pumping action to be established. During this time, monitor the pressure in the backing line to be sure that it remains well below 30 Pa (0.2 Torr), a conservative value for the critical backing pressure for most diffusion pumps.

With the system in this configuration, the diffusion pump is effectively isolated from the vacuum chamber by the main and roughing valves, while it is being backed by the backing pump through the open backing line valve. The system can be left in this configuration indefinitely without danger of damaging the diffusion pump or contaminating the vacuum system as long as the backing pump functions properly and the cooling water continues to flow. This is therefore often called the *standby* configuration. For a simple system of the type shown here, air can be admitted to the vacuum chamber through its air inlet valve and specimens can be loaded into, or removed from it, or the apparatus can be left standing idle awaiting the next use. The procedures for inserting and removing specimens from more complex systems, such as those found on electron microscopes, will be described in Chapter 9.

This is also known as the 'safety' configuration because the diffusion pump is isolated from the influence of problems that might arise in the vacuum chamber while being protected by the action of the backing pump, and oil vapour cannot backstream from either pump into the vacuum chamber because the roughing and high vacuum valves are both closed. Any time there is an indication of trouble in a vacuum system, return it to this configuration *as rapidly as possible* and then turn off the electrical power to the diffusion pump heater.

The backing and roughing gauges both read the pressure in the backing line during this stage of the operating procedure, giving some redundancy, and so the backing gauge is often omitted, as mentioned above. This is an acceptable design compromise if the main and backing valves can both be trusted to perform reliably. The advantage of having a backing gauge will become apparent in the next subsection, however.

5.7.1b *Rough pumping the vacuum chamber*

After samples have been loaded into the vacuum chamber while the system is in the standby configuration, the pressure in the chamber must be reduced to a level below the critical backing pressure of the diffusion pump before the high vacuum valve is opened and the diffusion pump is allowed to pump out the chamber. Starting with the system in the standby configuration, this rough pumping operation is carried out as follows:

1. Close the air inlet valve to the vacuum chamber.
2. Close the backing valve, keeping the high vacuum valve closed.
3. Slowly open the roughing valve to begin the rough pumping operation.
4. The roughing gauge now monitors the pressure in the vacuum chamber as it is rough pumped by the backing pump. The system should now be in the configuration shown in Fig. 5.12b.

There is some risk involved in this stage of the operating procedure. Ideally, as stated above in Rule 4, the backing pump should pump on the outlet line of the diffusion pump at all times while the diffusion pump is hot to ensure that the diffusion pump oil is protected from oxidation by a proper backing pressure. During this stage, however, the backing valve is closed while the diffusion pump is operational so that the backing pump can be used to rough pump the chamber. This is an acceptable procedure provided the system is not left in this configuration any longer than is absolutely necessary, and provided the high vacuum and backing valves both seal properly and prevent the pressure in the diffusion pump from rising above its critical backing pressure during this time. This is the point in the operating procedure where it is advantageous to have a separate backing gauge to give a direct indication of the pressure at the outlet of the diffusion pump. A noticeable increase in the reading of the backing gauge as the roughing valve is opened indicates that air is leaking through one of these valves into the diffusion pump. If this occurs, the roughing valve should be closed immediately and the backing valve opened as quickly as possible to re-establish a proper backing pressure for the diffusion pump. Then it is safe to begin investigating the source and cause of the leak. Even if the two valves appear to be functioning properly, however, some gas evolution will occur from the baffle and valve above the diffusion pump, and the backing pressure will slowly increase. It is therefore prudent not to prolong this stage of the operation any longer than necessary. It is best to proceed to the next stage of the operation as soon as the pressure in the chamber drops to a safe level for operating the diffusion pump. Don't go out for a cup of coffee while the system is in this roughing configuration.

Usually it is desirable to open the roughing valve slowly so that the pressure in the vacuum chamber decreases very gradually, a procedure often called 'soft-start roughing'. If the valve is suddenly opened fully, the air will be pumped out so rapidly that highly turbulent air currents will be generated in the vacuum system throughout the period during which the pressure is in the range of viscous flow. Under these conditions, loose particles of dust, dirt, and other debris are picked up by the air currents and blown around the vacuum system. Some particles will, of course, be carried into the roughing line where they might lodge in the roughing valve and keep it from sealing properly. There is also a very high probability that some will be deposited on specimens, in apertures, and at other sites within the vacuum system where they are basically not desired, and where they can cause a variety of problems. Since it is virtually impossible to remove all loose particles from a vacuum system, the potential for these problems is universally present. Such problems are highly likely to occur in systems with automatic valves that snap open instantly. Unfortunately valves of this kind are found on most electron microscopes.

5.7.1c *Establishing a high vacuum*

As soon as the backing pump has reduced the pressure in the vacuum chamber below 20 Pa (0.2 Torr), proceed to establish the desired high vacuum in the vacuum chamber. This is done as follows:

1. Close the roughing valve.
2. Open the backing valve.
3. Slowly open the high vacuum valve.
4. Wait a few minutes, then turn on the high vacuum gauge.

Usually there will be a small increase in the pressure in the backing line immediately after the high vacuum valve is opened. This occurs because, as discussed in Section 5.3 (p. 177), a diffusion pump has a very high throughput when the inlet pressure is between its critical backing pressure and its critical inlet pressure. When operating in this range it will usually discharge gas into the backing line faster than the backing pump can remove it, unless a very large backing pump is used. Therefore, it is normal for the pressure in the backing line to increase in the manner described for a short time, usually no more than 10 or 15 s. During this time the upper jets of the diffusion pump are overloaded and the rate of backstreaming is abnormally high. If the system is well designed, however, the diffusion pump will rapidly reduce the pressure in the vacuum system below its critical inlet pressure and into the high vacuum range. As this occurs, the backing pressure will quickly recover and steadily decrease to a final steady-state value, which should be near or below 1 Pa (10^{-2} Torr). If the pressure in the backing line rises above the critical backing pressure of the diffusion pump, all jets in the diffusion pump become overloaded and conditions for severe backstreaming prevail. Under such conditions only the backing pump contributes to the removal of gas from the vacuum chamber, and so it would have been better to have rough pumped the chamber a bit longer to begin with.

There is obviously some judgment involved in selecting the pressure at which to open the main valve and begin this final stage of evacuation. The pressure must be well below the critical backing pressure of the diffusion pump. As discussed in Section 5.2 above, the critical backing pressure of a pump is not constant, but varies with operating conditions. In particular, it decreases by about 25 percent when the pump is working near its critical inlet pressure, as it is immediately after the high vacuum valve is opened. Assuming a typical normal value of about 40 Pa (0.3 Torr), it is best to use a conservative value of about 30 Pa (0.2 Torr) as an operational guide for the maximum pressure at which the diffusion pump should be brought into use. It is beneficial to rough pump the vacuum chamber to a pressure somewhat below this to reduce the initial gas load on the diffusion pump. This will decrease the time required for the diffusion pump to bring the pressure in the chamber below its critical inlet pressure, thereby reducing the time during which severe backstreaming from the diffusion pump can occur. However, prolonging the rough pumping operation unnecessarily provides an opportunity for the pressure in the roughing line to drop into the range of molecular flow, whereupon oil vapours from the backing pump can diffuse into the vacuum chamber. Fig. 2.1 (Section 2.2, p. 31) shows that this is likely to occur if the pressure falls much below about 10 Pa (0.1 Torr).

On the basis of these considerations, and assuming there is a proper balance in the sizes of the diffusion pump, the backing pumps, and the volume of the vacuum system, a procedure that should be appropriate for most laboratory vacuum apparatus is to rough pump the vacuum chamber to a pressure in the range between 10 and 20 Pa (0.1–0.2 Torr) and then to switch over to the high vacuum configuration promptly. The quantity of gas remaining in the vacuum chamber should then be small enough so that the pressure in the backing line will not rise above the critical backing pressure of the diffusion pump, and the diffusion pump should be able to bring the pressure in the chamber down into the 10^{-2} Pa (10^{-4} Torr) range within a few seconds (20 at the maximum) after the high vacuum valve is opened. If the high vacuum valve is manually controlled it is beneficial to open it slowly. This will restrict the rate at which gas initially enters the diffusion pump and reduce the severity of the overloading phenomena just described. It will also reduce the sweeping of vapours of the diffusion pump fluid into the backing line, which otherwise can result in a significant loss of fluid from the pump over many cycles of operation.

5.7.1d *Opening the vacuum chamber*

When it is desired to admit air to the vacuum chamber that has been evacuated, proceed as follows:

1. Close the high vacuum valve.
2. Turn off the high vacuum gauge.
3. Open the air inlet valve to the vacuum chamber.
4. Open the chamber when the gas in it reaches atmospheric pressure.

The backing valve is left open, the roughing valve is left closed, and the roughing gauge is left on, returning the system to the standby configuration discussed above and shown in Fig. 5.12a.

The behaviour of the pressure in the backing line should be watched during these steps. Normally it will decrease noticeably as soon as the main vacuum valve is closed and the diffusion pump is no longer evacuating the large volume of the vacuum system, and it should not increase again when air is admitted to the vacuum system. Any such increase in the backing pressure is an indication that the main vacuum valve is leaking, a matter that should be investigated immediately. If the pressure begins to approach the critical backing pressure of the diffusion pump, turn off its heater immediately, rough pump the vacuum system again to reduce the rate of leakage, and then return the system to the safety configuration until the diffusion pump is cold.

On most apparatus the air inlet valve should be heavily throttled so that air flows in very slowly. Otherwise, the rapid influx of air will generate turbulent air streams which can pick up loose particles of dust and dirt and blow them around the system, possibly contaminating and damaging specimens, apertures, and gauges. It may also be desirable to place baffles inside the chamber, in front of the opening of the air inlet line, to ensure that the incoming air does not flow directly over the specimens and into other critical regions.

5.7.1e Full shutdown

The operation of shutting the system down completely is normally carried out by starting with it in the standby configuration and proceeding as follows:

1. Establish a high vacuum, since it is usually desirable to shut the system down with the chamber under vacuum, and then close the high vacuum valve and turn off the high vacuum gauge.
2. Turn off the electrical power to the diffusion pump heater, but allow the backing pump to continue pumping on the diffusion pump for a period of 10–30 min, depending on the size of the pump, while the oil in the diffusion pump cools well below its boiling point. Note that the high vacuum valve should always be closed while the diffusion pump is cooling down, because this is a time when the backstreaming rate is high.
3. Close the backing valve and turn off the roughing gauge, and the backing gauge if one is present.

4. Open the air inlet valve to the backing pump and immediately turn off the electrical power to this pump.
5. Shut the valve that feeds cooling water to the diffusion pump.
6. Turn off the main electrical power switch for the apparatus.

This should put the system in the shutdown configuration shown in Fig. 5.12d.

5.7.2 *Systems with buffer tanks and separate roughing pumps*

As mentioned in Section 5.7.1b above, it is fundamentally not a sound operating practice to isolate the diffusion pump and use its backing pump to rough pump the vacuum chamber. This is particularly true for large systems which require long periods for rough pumping. One way to overcome this problem is to use a separate roughing pump, leaving the backing pump to serve only the function of backing the diffusion pump. Another involves the installation of a *buffer tank* in the backing line. These modifications are shown schematically in Fig. 5.13. One function of a buffer tank, as its name implies, is to reduce the rate at which the pressure in the backing line increases during short periods of high throughput, thereby considerably decreasing the danger that it will rise above the critical level. However, buffer tanks are perhaps most often used in systems that do not have separate roughing pumps, where they serve as a reservoir for gas discharged by the diffusion pump while it is isolated from the backing pump,

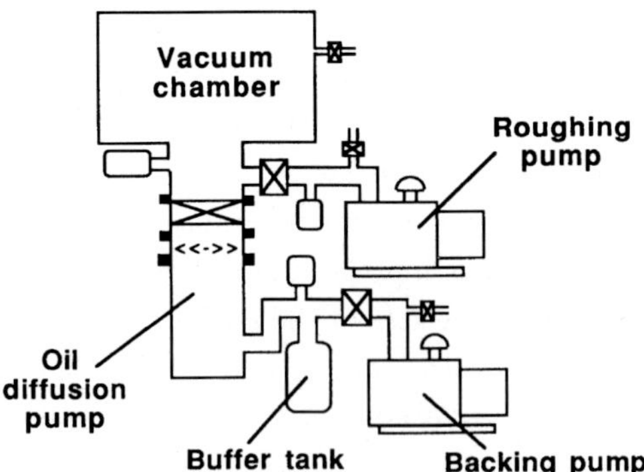

Fig. 5.13 A schematic diagram showing the addition of a buffer tank and a separate roughing pump to a vacuum system with an oil diffusion pump.

thereby prolonging the time it is safe to leave this pump isolated while the vacuum chamber is being rough pumped. For example, the quantity of gas required to raise the pressure in a 10-litre buffer tank from 1 Pa to 40 Pa is 10(40 – 1) = 390 Pa-l. A diffusion pump such as the one used in the example calculations of Section 5.3 (p.178) would certainly not have a throughput greater than about 1 Pa-l/s if it were isolated from the vacuum chamber by closing both the backing and high vacuum valves. Under these circumstances it would take 390 s, or more than 6 min, for the pressure in the buffer tank to rise to near the critical level.

5.7.3 *Valveless systems*

Some industrial processes make use of valveless vacuum systems of the type shown in Fig. 5.14. The vacuum chamber on these systems is rough pumped through the diffusion pump. When rough pumping is undertaken, the diffusion pump must be cooled down enough so that the oil in it is not vaporizing significantly to prevent the vapours from being drawn into the backing line, thereby depleting the oil in the pump. The oil should also be cool enough so that it is not in danger of undergoing oxidative degradation. The diffusion pump must be similarly cooled before air is admitted to the vacuum chamber. Diffusion pumps used on such systems usually have cooling coils around their boilers so that water can be used to speed up the cooling process, and silicone or perfluorinated polyether fluids are generally used in them, because of the superior resistance of these fluids to thermal and oxidative degradation.

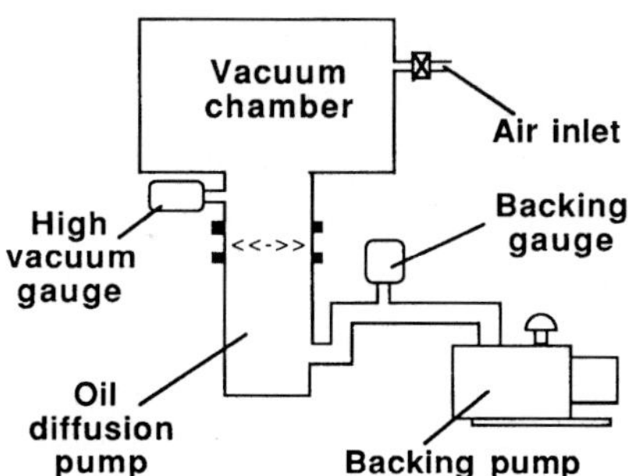

Fig. 5.14 A vacuum system for an oil diffusion pump without vacuum valves.

5.8 Potential problems

There are four very damaging events that can occur in the course of operating systems evacuated with oil diffusion pumps: (i) interruption of the supply of cooling water to the pump, (ii) failure of the mechanical backing pump, (iii) admission of air at high pressure to a hot diffusion pump, and (iv) loss of electrical power. It is extremely important to be cognizant of these potential problems and to make every possible effort to prevent them from occurring, because they can lead to levels of system contamination that are many orders of magnitude greater than those produced by the backstreaming processes that occur during normal operation. In addition, they usually cause severe degradation of the pump fluid which is accompanied by the build-up of carbonaceous decomposition products on the jet assembly and in the boiler of the pump. Fig. 5.15 shows a jet assembly badly coated in this manner. Cleaning a pump after an accident severe enough to produce this effect is a very difficult and unpleasant task indeed.

If you have the great misfortune to be confronted with a problem of this kind, it is probably easiest and safest to return the pump to the manufacturer for refurbishing. However, with sufficient courage and good abilities at tasks such as this, it is possible to do the job locally. First, all liquid oil must be removed from the pump using the solvent recommended by the manufacturer of the pump fluid, and following recommended safety procedures for handling it. About the only way to get the carbon out of the interior of the pump in an ordinary laboratory is to scrape it out with a tool such as a chisel or a screw driver, or to grind it out with a coarse grade of silicon carbide metallographic paper. Alternatively, it can be removed by sand blasting, if you can find an operator who is accustomed to working on delicate small-scale jobs of this nature. The carbon deposits on the jet assembly can also be removed by grinding with silicon carbide paper or by sand blasting, although it is easy to damage a jet assembly with these methods. Another approach is to place the pieces of the jet assembly individually in a furnace at a temperature of 500–600°C for a few minutes. *Be very careful* if you try this, because overheating can damage these parts. Use the lowest temperature that will burn the carbon off quickly. Also, this is a potentially hazardous undertaking which should be carried out only by someone experienced in working with furnaces. Most furnaces

Fig. 5.15 The jet assembly of an oil diffusion pump that was heavily coated with carbonaceous material by degradation of the pump fluid following failure of the rotary-vane backing pump.

have openly exposed electrical heating wires, and great care must be exercised not to contact these when handling parts inside the furnace with metal tongs. Afterwards the pieces can be cleaned and polished with a metal polishing compound of the type commonly marketed for use on household cooking ware (see Section 2.10.4c, p. 70). Whatever method is used, *great care must be exercised* in working on the jet assembly, because the pump will not function properly if the parts of this device are bent out of shape, misaligned, or otherwise damaged.

5.8.1 *Loss of cooling water*

If the walls of the diffusion pump are not properly cooled the oil vapours from the jets do not condense and return to the boiler, but migrate into the vacuum chamber and the backing line. As the amount of oil in the boiler decreases, the remaining oil becomes overheated and decomposes, increasing the rate of backstreaming, system contamination, and oil loss. The supply of cooling water can be interrupted by a variety of causes, most of which are usually not under the immediate control of the instrument operator, such as: plumbing activities somewhere else in the building, plant, or city; the disruption of power to a recirculating water chiller, or the failure of a fuse or other component in it; the sudden obstructing of the cooling water line by mobile particles of scale or rust. Audrey Glauert tells of an unusual situation she encountered while working at a laboratory in Africa; water was occasionally interrupted when hippos bathing in a dam in a nearby river blocked the inlet to the town's water supply. There are obvious solutions to all of such causes; the trouble is, they invariably occur unexpectedly and at a time when the instrument is unattended.

Undoubtedly the best source of cooling water for an oil diffusion pump is a refrigerated water recirculating unit. Such units, which are often referred to as 'water chillers', can be purchased with the help of the manufacturers of the electron microscopes or through companies that deal in general supplies and accessories for electron microscopy. These units should be filled with distilled water to eliminate the formation of scale deposits in the cooling water lines, and in the cooling coils on the pump. Algae, which tend to grow in these units, can cause problems by blocking water lines. Algal growth can be greatly retarded if light is excluded by keeping the reservoir covered and by using opaque tubing for the cooling water lines. The addition of substantial concentrations of ethylene glycol to the water in the reservoir is sometimes recommended as another means of controlling algal growth. I do not favour this practice because I know of several instances when water from the cooling system has found its way into the interior of an instrument. Plain distilled water will usually evaporate away before it causes serious damage, whereas the ethylene glycol is a serious contaminant in such circumstances. If the exclusion of light is not adequate, the compound Chloramine-T (the sodium salt of *N*-chloro-*p*-toluenesulphonamide), which is generally available from speciality chemical companies (e.g. Polysciences), can be used at the level of about 0.25 g/l (1 g per gallon) of water. This material dissolves in much lower concentrations than the ethylene glycol, and therefore poses a less serious contamination problem. A filter should always be installed on the intake to the water line to prevent such algae as do grow in the reservoir from entering the line. This filter must be cleaned periodically, and the water in the chiller must be replaced if algal growth becomes excessive.

If cooling water is drawn directly from a water main the minerals dissolved in it are prone to precipitate out as scale deposits when the water enters the hot coils around the barrel of the diffusion pump. To prevent blocking of these coils, it is advisable to remove these scale deposits at least yearly by dripping a scale-removal solution slowly through the coils for 10 or 15 min. Don't wait until the coils get blocked to do this or you may have a very difficult problem on your hands. Scale-removal solutions are available from hardware stores and plumbing shops. These solutions are usually strongly acidic, and so should be handled with great care (e.g. wear protective clothing, rubber gloves, and a face shield, and cover nearby objects with polyethylene sheeting to protect them if the solution is spilled or spattered around).

5.8.2 *Failure of the backing pump*

Failure of the mechanical backing pump to provide an adequate backing vacuum can occur from an equal number of unexpected events. These include: leaks in, or collapse of, backing line tubing; neglecting to maintain a proper level of oil in the backing pump; development of a leak around a shaft seal on the backing pump; a broken belt or coupling in the backing pump drive system; failure of the backing pump motor; loss of electrical power to the backing pump; etc. When a backing pump fails, the pressure in the backing line rises above the critical backing pressure for the diffusion pump causing the jets of the diffusion pump to become overloaded, develop shock fronts, and cease their pumping action. The diffusion pump will then simply sit and boil oil molecules into the backing line and any accessible part of the vacuum system. I remember one instance of backing pump failure when enough oil entered the viewing chamber of an electron microscope to fog the viewing window so badly it was impossible to see through it. As the situation deteriorates, the amount of oil in the boiler of the pump decreases, causing the oil left there to overheat, while the pressure of air in the pump increases, leading to oxidative degradation of the pump oil. The carbon-coated jet assembly shown in Fig. 5.15, one of the worst I have ever encountered, resulted because I unknowingly stepped on the power cord to the backing pump and pulled it from its socket while finishing up work on another service problem. It took several days of unpleasant work to clean up the diffusion pump, the vacuum manifold, and the main vacuum valve after this incident. Fortunately, the automatic valve operating system closed the main valve directly above the diffusion pump soon enough to prevent contamination of the column of the instrument. Needless to say, the plug on this power cord is now firmly clamped into its socket and the cord has been relocated so that it is more protected. The most serious service problems I have encountered in working for more than 30 years with instruments evacuated by oil diffusion pumps have resulted from failure of backing pumps.

5.8.3 *Inrush of air*

It should be obvious that the admission of air at high pressure into a hot oil diffusion pump will trigger a series of events substantially the same as those resulting from failure of the backing pump described above. This could arise from the development of a serious leak in the vacuum system, such as might occur if an improperly fastened vacuum coupling opens up, or if a glass vacuum gauge is accidently broken, etc. More often, however, it occurs because an inexperienced, or over-tired, or harassed operator makes a mistake in manipulating the valve system while loading or unloading a specimen or photographic film in an electron microscope, or when manipulating the valves on a vacuum evaporator. The only way to effectively prevent this is to train all operators thoroughly in the proper operation of the valve system, and to emphasize the catastrophic consequences of failure to follow proper procedures.

A particularly insidious cause of this problem is a leak in either the main valve or the backing valve, because then air is admitted into the hot diffusion pump each time the system is brought up to atmospheric pressure, if the main valve leaks, or each time it is rough pumped, if the backing valve leaks. Such a leak usually will not be detected directly, because most vacuum systems do not have a vacuum gauge located in the backing line between the diffusion pump and the backing valve, nor will it usually be indicated indirectly, because most instruments do not have a high vacuum gauge of sufficient sensitivity to register the overall degradation in performance that is occurring until it becomes a very serious problem. Instead, filament life gradually decreases, pumpdown time slowly increases, and specimen contamination rate increases until it finally becomes obvious that something is drastically wrong with the vacuum system. Such a leak can arise from various causes. A piece of loose dirt or lint can get blown around the system and become lodged on the sealing surface of a valve when air is admitted to the system or when it is being pumped down from atmospheric pressure. O-rings, gaskets, and metal bellows used in constructing vacuum valves sometimes fail. Unfortunately, some instruments have valve systems which were 'cleverly' designed to save space and manufacturing costs by people unfamiliar with good vacuum practice, but which are simply not capable of providing reliable service over long periods of use. These systems are a constant source of trouble, because it is usually impossible to adjust them so they will work properly, and there usually is not enough room to replace them with a reliable set of valves. It is particularly unfortunate when a valve system of this kind is part of an instrument that is otherwise well designed. The most effective action that can be taken to protect against problems due to valve leakage is to install a thermocouple or Pirani gauge in the backing line, between the backing valve and the outlet port of the diffusion pump, to monitor the vacuum there at all times. It is particularly valuable to use a controller for this gauge which has a set-point switch that can be adjusted to sound an alarm, and possibly activate a protective circuit, if the pressure rises above a safe level. Any time there is any indication of a leak in one of the valves the entire system should be shut down immediately and the problem should be referred to a service engineer. Above all, do not keep using the system hoping that the problem will cure itself, because it won't, and this will only lead to more severe degradation of the oil in the diffusion pump and increased contamination of the instrument. If the valve system is basically defective, it is obviously advantageous, and very satisfying, to have the instrument under service contract, because this will shift the cost of repair back on the manufacturer where it rightfully belongs. A service contract usually pays for itself promptly in these circumstances. It is also helpful to replace the oil diffusion pump with a turbomolecular pump, which is more tolerant of exposure to air at high pressures, if the instrument otherwise merits this added expense.

5.8.4 *Loss of electric power*

The loss of electric power to a vacuum system involving an oil diffusion pump can also have serious consequences, depending on how the system is configured and how its protective circuits function, if it has any. A serious situation can arise if the power loss causes the backing pump to stop without being properly vented, because this pump is then shut down under vacuum and it is possible for the oil in it to be sucked up into the backing line. This can leave the pump without sufficient oil to operate properly when it is turned on again, and can greatly increase the likelihood of contaminating the diffusion pump and the vacuum system with oil from the backing pump. This problem and its solution are discussed in some detail in Section 6.1.9b (p. 255), in connection with the management of systems with turbomolecular pumps.

5.8.5 *A protective circuit*

It should be obvious that it is extremely difficult to design an electronic circuit that will provide absolute protection against all possible causes of failure and damage to an oil diffusion pump. Actually, the automatic valve operating systems on most electron microscopes do provide a very considerable amount of protection. For example, the one on the scanning electron microscope involved in the accident that caused the problem illustrated in Fig. 5.15 did shut off the main valve above the diffusion pump in time to prevent serious backstreaming of oil into the column and specimen chamber. However, it did not shut off the electrical power to the diffusion pump heater in time to prevent severe damage to the diffusion pump (there was not a drop of liquid oil left in the pump when the problem was finally discovered). The automatic valve operating systems on most electron microscopes use a Pirani gauge located on the high vacuum side of the diffusion pump as their primary sensor, and this is often not sufficient to provide all the information needed for proper safety measures to be taken. These systems will usually turn off the filament and high voltage, and close one or more of the vacuum valves, if the pressure in the column rises significantly above the normal operating level, but they often do not operate properly to prevent damage to the diffusion pump. Vacuum evaporators, which are most subject to accidents involving operator misuse, usually do not have any safety circuits.

The consequences of diffusion pump failure in instruments used in electron microscopy are so potentially damaging that it is worthwhile to take reasonable precautions to prevent any accident that does occur from progressing to the point where a situation such as shown in Fig. 5.15 develops. The basic thing to do to accomplish this is to shut off the electrical power to the diffusion pump heater whenever there is any indication that a problem is developing, because not much happens in an oil diffusion pump when its heater is turned off. Once shut off, the power should stay off until a proper investigation has been conducted and it has been determined that it is safe to put the pump back into operation.

Fig. 5.16 is a diagram for a simple, inexpensive protective circuit for accomplishing this. This circuit can be constructed with a double-pole, double-throw (DPDT) relay R with contacts rated for at least 10 A at 120/240 V AC (e.g. Potter & Brumfield/Siemens type K10P-11A15 or KUP-11A15). General purpose relays of this kind can usually be found in stock in amateur electronics supply stores The two normally-open contacts of this relay are used in parallel to achieve a total current-carrying capacity of 20 A, which should provide a good margin of safety for the diffusion pumps on most laboratory apparatus. A power relay with a higher current-carrying capacity, which may have to be ordered specially, should be used for pumps requiring larger currents. When the solenoid of this relay S is energized the normally-open contacts NO close and current is fed to the diffusion pump heater DP. If the solenoid is not energized these contacts open, as depicted in Fig. 5.16, and the flow of current to the diffusion pump heater is interrupted. The pilot light L is then lit by current from one of the normally-closed contacts NC of the relay, signalling that the system requires attention. To prevent the diffusion pump from being shut off permanently by momentary losses of power, such as can occur during thunderstorms and similar transient phenomena, power to the solenoid of the DPDT relay is controlled by a normally-open, 20 s, time-delay relay T (e.g. Amperite 115NO20B). The contacts of this relay

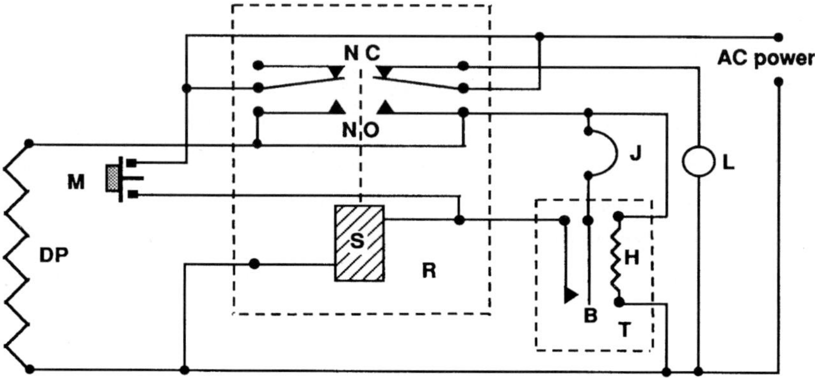

Fig. 5.16 A protective circuit for oil diffusion pumps. DP, diffusion pump heater; J, removable jumper for adding sensors; L, pilot light; M, momentary contact switch; R, double-pole, double-throw relay with normally-open (NO) and normally-closed (NC) contacts and solenoid (S); T, thermally activated time-delay relay with heater (H) and bimetallic contact (B).

are actuated by a bimetallic strip B which is warmed by a small internal heater H. Current must flow through this heater for about 20 s before the bimetallic strip warms up enough to close the contacts. It also requires about 20 s for the bimetallic strip to cool off enough for the contacts to open again when current to the heater is stopped. This heater is connected in parallel with the diffusion pump heater, so it ceases to heat the bimetallic strip any time power to the diffusion pump is interrupted. Thus, any time the current to the diffusion pump is interrupted for more than 20 s the bimetallic element cools enough to open the contacts in the time-delay relay, the current to the solenoid is cut off causing the DPDT relay to open, and the diffusion pump heater is permanently turned off. If such a power interruption does not last for the 20 s needed for the time-delay relay to open, current will be fed to the solenoid when the power comes on again, the DPDT relay will close again, and current will continue to be fed to the diffusion pump. However, once the time-delay relay opens the diffusion pump is turned off permanently, and it is necessary to deliberately start it again. This is done by holding the momentary contact switch M closed for the 20 s needed to warm up the bimetallic element in the time-delay relay so that the contacts in it close and maintain the flow of current to the solenoid of the DPDT relay. The necessity of performing this time-dependent task ensures that the manager of the instrument is aware that a significant problem has occurred and has an opportunity to investigate its cause and to take corrective measures before serious system contamination or damage to the diffusion pump occurs.

As just described, this protective circuit is designed to respond to interruptions in power to the diffusion pump initiated by safety circuits in an electron microscope or from an external power failure. However, it is easy to adapt this circuit to give positive protection against failure of the backing pump or leakage of the main and backing valves. This can be done by installing a thermocouple or Pirani gauge tube in the backing line, between the backing valve and the outlet port of the diffusion pump, using a controller for it that has a set-point switch, and adjusting this switch to open if the pressure in the backing line rises above about 40 Pa (0.3 Torr). Since these set-point switches usually cannot carry the full heating current to a diffusion pump, they must be used indirectly to provide the desired control function. This can be done by substituting the set-point switch for jumper J in the protective circuit of Fig. 5.16. Then, if the backing pressure rises enough to cause the set-point switch to open, power to the solenoid of the power-control relay will be interrupted, the power-control relay will open, and power to the diffusion pump heater will be shut off. Again, the time-delay relay will prevent the system from being shut down by momentary increases in backing pressure, such as may occur during the transition from rough pumping to final evacuation; however, the power to the diffusion pump will be cut off permanently if the increase in backing pressure lasts for more than the 20 s needed for the time-delay relay to open. I have installed such a circuit on the instrument with the pump illustrated in Fig. 5.15 to prevent the diffusion pump from cycling on and off repeatedly again while the automatic valve operating system tries vainly to re-establish an operating vacuum without a properly functioning mechanical pump.

This circuit can also be used to provide protection against loss of cooling water to the diffusion pump. This can be done by attaching a normally-open, thermally-activated switch (e.g. Alcatel No. 056497 or Edwards No. B279-03-006) to the body of the diffusion pump between the lowest two cooling water coils, and connecting it into the protective circuit at the position of jumper J. If this switch is designed to open at a temperature in the range from 50 to 75°C, power to the diffusion pump will be shut off in time for the oil to cease boiling before the upper part of the pump becomes so hot that oil condensation no longer occurs there.

Many diffusion pumps have such a thermal switch already wired into one of the power lines in series with the pump heater. Since these switches are designed to carry the full heating current to the diffusion pump they can be used in another way. This involves cutting the wire leading from the switch to the diffusion pump heater, connecting the segment that remains attached to the thermal switch to the upper AC contact in Fig. 5.16, and connecting the segment leading to the diffusion pump heater to the normally-open contacts of the DPDT relay. With this arrangement an opening of the thermal switch will have the same effect as an interruption of power from any other cause.

5.9 Routine service

In the absence of problems such as those described in the preceding subsections, oil diffusion pumps require a minimum of routine service and maintenance. Such routine service as is required consists principally of examining the condition of the oil, the jet assembly, and the interior of the pump. Most of these service procedures involve using solvents to remove residual oil from parts of the pump and associated vacuum lines, and this should be done *with great care*, following locally prescribed safety regulations faithfully. Finally, it is convenient to use a hair dryer or a heat gun (available from Fisher) to warm the parts that have been cleaned so that they are free of solvents and thoroughly dry. If this is done, however, great care must be exercised to be sure that solvent vapours are not ignited by the heating device, causing a fire or an explosion. Perform these cleaning procedures in a well ventilated area, wear rubber gloves, safety goggles, and appropriate protective clothing, and wait until all detectable traces of solvent have evaporated off the parts before the heating device is brought into use.

To service an oil diffusion pump properly it is necessary to remove it from the apparatus it serves. Once this is done the jet assembly can be taken out and inspected. If the jet assembly is clean it can be covered with aluminium foil and set aside until it is reinstalled in the pump. If there are gummy or carbonaceous deposits on it they should be removed with solvents or abrasives, whereupon the entire unit must be cleaned and dried before it is reinstalled. Solvents for the different types of pump fluids are given in the sections above which describe the fluids, although I have found a mixture of equal parts of acetone with toluene or xylene to be about as effective as anything. If solvents will not clean the jet assembly satisfactorily, it may be necessary to use a metal polish, fine steel wool, or 1200- or 600-grade silicon carbide metallographic polishing paper. If this is done, great care must be exercised not to bend or damage the unit in any way, because doing so will adversely affect its subsequent performance. If the deposits are severe it may be best to burn them off in a furnace, as described in Section 5.8 above. Finally, the jet assembly should be thoroughly rinsed with reagent grade isopropyl alcohol. After it is dry it can be covered with aluminium foil to keep it clean. Before it is reinstalled it should be warmed with a hair dryer for 15 or 20 min to drive off as much moisture as possible.

The oil should then be poured out of the pump and examined. If it is significantly discoloured, or if it is cloudy or contains particulate matter, it should be replaced. Otherwise, enough fresh oil should be added to make up the proper volume (which should be indicated on a data plate fastened to the pump, or given in the instruction manual for the pump). Actually, the pumps on most laboratory apparatus require so little oil that it is a minor expense to replace it as a routine procedure. Larger pumps may require enough oil so that the expense of replacement can make it desirable to reuse the old oil if possible. The interior of the pump should also be inspected. If gummy or carbonaceous deposits are present there they should be removed as described in Section 5.8, after which the interior of the pump must be thoroughly cleaned with solvents. This unit can also be warmed with hot air to remove as much surface moisture as possible.

The baffle, trap, and valve above the pump, and the pumping line leading from the pump to the vacuum chamber, should be cleaned as thoroughly as possible with solvents, and then carefully dried with a hair dryer or a heat gun (available from Fisher). It is also a good idea to thoroughly clean the backing line and the backing valve that serve the diffusion pump, because these will usually be coated with oil from the backing pump. The valve should be disassembled and cleaned with solvents. A pumping line that is made of rubber tubing can be cleaned to some extent by pulling a tight-fitting lint-free cloth pad, which is attached to the end of a long flexible wire, through it several times. It is not a good idea to use solvents to clean lines made of rubber tubing, because the rubber will absorb the solvent and subsequently release it into the vacuum system. Pumping lines made of flexible metal bellows can be repeatedly rinsed with solvents and then thoroughly dried with hot air. All O-rings and gaskets associated with the pumping system should also be examined carefully. If any are cracked, distorted, or otherwise damaged they should be replaced. Before the system is reassembled the gaskets can be cleaned by wiping them with a lint-free cloth that is moistened with the diffusion pump fluid. If this is done, it should not be necessary to coat them with vacuum grease.

The frequency with which such a cleaning procedure needs to be carried out will depend on a number of factors. Pumps which are cycled from high vacuum to atmospheric pressure frequently, such as those on vacuum evaporators and most scanning electron microscopes, are subject to rather severe use and should probably be serviced once or twice a year. Pumps on ion beam mills and similar apparatus that must handle high gas loads for long periods of time will require even more frequent attention, while the servicing of pumps on systems that operate continuously in the high or ultra-high vacuum range and are seldom cycled to atmospheric pressure can be usually delayed until the apparatus is shut down for other reasons.

Pumps that use hydrocarbon or synthetic ester fluids, which are the most susceptible to thermal and oxidative degradation, should be serviced more frequently than those using the more stable silicone and perfluorinated fluids. Some experience will be required to determine the optimum service interval for each apparatus and operating situation. This will be greatly facilitated if a careful, accurate record is kept of all service procedures performed, and of the general performance of the pump between times. In particular, it is important to record the type of fluid used in a pump, the amount required, and the date it was changed. I also like to record this information on a cardboard key tag that is hung by a string on the outlet tube of the pump so that it will be handily available when needed. You would be surprised how easy it is to misplace and forget such simple but essential details when things are going smoothly, and how hard it is to find them again when an emergency arises and you need them in a hurry.

5.10 Concluding remarks

This discussion of oil diffusion pumps may seem overly long and detailed; however, it is not out of proportion to the importance of these pumps in electron microscopy and vacuum technology in general. Overall, they were the workhorses of the vacuum industry for more than 50 years. Even today they are the principal type of high vacuum pump used in most industrial applications, and they are found on most high vacuum apparatus presently in use in electron microscopy laboratories. They are relatively inexpensive to purchase and to operate, they are compact and simple in design and construction, they contain no moving parts to wear out or break down, and they are noiseless and non-polluting. They require minimal service and maintenance, and what is required can be performed by most competent laboratory technicians without requiring the expensive services of factory-trained engineers. They are being replaced by other types of pumps mainly on instruments such as advanced electron microscopes and surface analysis units, and on manufacturing apparatus in the electronics industry, where oil contamination is a serious problem. However, this problem with oil contamination is so pervasive that it is almost certain that their use on electron microscopes will steadily diminish in the future.

6 *Turbomolecular and molecular drag pumps*

Turbomolecular and molecular drag pumps are mechanical high vacuum pumps which function by transferring momentum to gas molecules through collisions between the gas molecules and a rapidly rotating device. The first pumps of this kind — molecular drag pumps — were introduced several years before oil diffusion pumps, and were initially called simply 'molecular pumps', possibly in recognition of the fact that their pumping action involves interactions between the pumping mechanism and individual gas molecules, and to distinguish them from the rotary-vane and other mechanical pumps that were available then in which the pumping action could be considered to involve the co-operative flow of gas molecules in a viscous gas stream. During the period from 1920 to 1960 molecular pumps were more or less abandoned because of the overwhelming success of the oil diffusion pumps, and because of numerous serious difficulties inherent in their design and manufacture. Interest in these pumps was rekindled by the introduction of turbomolecular pumps in the late 1950s by the Pfeiffer company, because it was immediately recognized that they represented a very attractive alternative to oil diffusion pumps for attaining pressures in the high and ultra-high vacuum ranges with a drastic reduction in problems due to oil contamination. Intensive programmes of research and development over the intervening years by most major manufacturers of vacuum equipment have brought these pumps to a high state of refinement. Work at CIT/Alcatel during the late 1970s and early 1980s led to the reintroduction of commercial models of the molecular drag pump in the mid-1980s. Because they have several unusual characteristics, these new models are finding use in a wide range of applications, again where oil-free pumping is required.

The references cited in this chapter are listed in Appendix 1, and the suppliers and manufacturers of vacuum equipment referred to here are listed in Appendices 2 and 3.

6.1 Turbomolecular pumps

Turbomolecular pumps, which are also commonly called 'turbopumps' and designated as TMPs, are mechanical pumps in which momentum in a preferred direction is imparted to gas molecules by bladed disks rotating at high speeds. Pumps of this type were first produced commercially in the late 1950s. Although they were installed on electron microscopes by a few users prior to 1965, it is only relatively recently that manufacturers have designed instruments based specifically on their use. O'Hanlon (Sections 7.4, 10.2, 11.2.2, and 12.1.2) and Harris (Chapter 8) discuss the design, construction, and use of these pumps in considerable detail.

6.1.1 *Principle of operation*

Although the pumping mechanism of a turbomolecular pump resembles the drive mechanism of a gas turbine engine, its function is the reverse. In the gas turbine engine expanding gases act on the turbine blades causing the mechanism to rotate and drive an external device. In a turbomolecular pump the turbine blades are rotated by an external motor for the purpose of collecting and compressing gas. As shown schematically in Fig. 6.1, the turbine mechanism consists of two parts, the rotor and the stator. As its name implies, the 'rotor' is the part of the mechanism that rotates. It consists of a set of disks mounted on a common shaft driven by an electric motor. The disks are slotted radially and the material between adjacent slots is twisted or machined to form a small blade extending radially outward from the shaft, so each disk resembles a multi-bladed fan. There may be from as few as 20 to as many as several hundred blades per disk, depending on the diameter of the pump. Using parallel terminology, the 'stator' is the part of the turbine mechanism that is stationary. It consists of a matching set of bladed disks which are interleaved between the disks of the rotor. The stator blades extend radially inward from the barrel of the pump, and are pitched in the opposite sense to those of the rotor. Fig. 6.2 is a photograph of a turbomolecular pump with part of the barrel cut away to show the arrangement of the disks and their blades.

The pumping action results from the interactions between the gas molecules and these bladed disks. At the inlet port of the pump gas molecules wander into the the first rotor disk by virtue of their normal thermal motion. Impact with the moving blades of this disk gives the molecules a

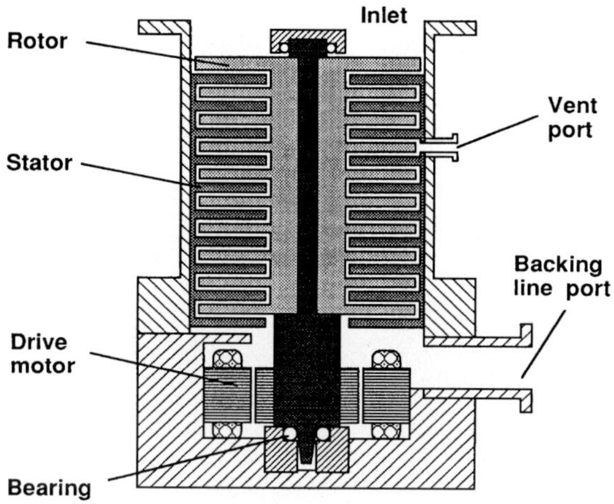

Fig. 6.1 A schematic diagram of the basic components of a turbomolecular vacuum pump.

directed velocity into the blades of the adjacent stator disk which, in turn, direct them into the blades of the next rotor disk, whereupon the process is repeated. The rotor disks transfer the energy to the gas molecules needed to move them through the pump. The stator disks reduce back-diffusion of the gas molecules, prevent them from migrating radially outwards toward the barrel of the pump, and force them to travel parallel to the axis of the pump towards its outlet. Like oil diffusion pumps, turbomolecular pumps function best at pressures in the range of molecular flow where the momentum transferred from the rotor blades can effectively propel the gas molecules into, and through, adjacent stator blades. At higher pressures the motion of the gas molecules leaving the rotor blades is randomized by collisions with other gas molecules and the effectiveness of the pumping action decreases drastically. For this reason turbomolecular pumps, like oil diffusion pumps, must be backed up by rough vacuum pumps which keep the pressure in them below the range of viscous flow.

6.1.2 *Compression ratios*

The pumping characteristics of turbomolecular pumps are usually described in terms of the compression ratios they develop for different gases. The compression ratio K for a given gas is defined by eqn. 5.3 (Section 5.6, p. 200) as the ratio of its pressure on the outlet side P_o to its pressure on the inlet side P_i (i.e. $K = P_o/P_i$). A disk on the rotor and the

Fig. 6.2 A photograph of a turbomolecular pump with part of the case cut away to show the arrangement of the internal parts. (Courtesy of Balzers.)

stator disk that follows it are usually treated as forming a single compression stage. Typically, pumps contain from 10 to 15 such compression stages, which act in series. Although the compression ratios for the individual stages may be only in the range from 5 to 10, when several stages act in series the overall compression ratio is roughly the product of the values for the individual stages. Consequently, pumps with 10 or more stages can develop overall compression ratios of the order of 10^{10}. This is sufficient to produce an inlet pressure in the ultra-high vacuum range if the pressure at the output port is maintained below about 10^2 Pa (1 Torr) (e.g. $P_i = P_o/K \approx 10^2/10^{10} \approx 10^{-8}$ Pa $\approx 10^{-10}$ Torr).

Detailed analyses of the statistical aspects of the kinetic molecular processes involved have shown that the compression ratio for a single pumping stage operating in the range of molecular flow has the following functional dependence:

$$K \propto G_b \ \exp.\left(S_b \sqrt{M}\right) \tag{6.1}$$

In this equation G_b is a geometrical factor which is dependent on the size, shape, spacing, and angle of the blades, S_b is a factor relating to the linear speed of the blades, and M is the molecular mass of the gas molecules being pumped. Thus, the compression ratios developed by turbomolecular pumps depend on blade geometry, blade speed, and the mass of the gas molecules.

Blade geometry is a critical design parameter, and manufacturers have devoted much effort to developing geometries that yield optimal combinations of properties for their pumps. Blade geometry varies from the inlet stages to the outlet stages, as shown in Fig. 6.2. Blades near the inlet usually have large pitch angles and large spacings between them. This optimizes their effectiveness in capturing the gas molecules that enter the pump and yields a high pumping speed. Blades near the outlet usually have smaller pitch angles and spacings, and produce the high compression ratios needed for reaching the ultra-high vacuum range.

The speed with which the blades travel is an equally important design parameter. Because of the exponential dependence of compression ratios on blade speed shown by eqn. 6.1, small changes in rotor speed cause large changes in pumping characteristics. Theoretical analyses, as well as practical experience, indicate that for effective pumping to occur the rotor blades must travel with linear speeds of the same order as the speeds of gas molecules being pumped, typically 400–500 m/s (eqn. 1.3 in Section 1.6, p. 20). To achieve these high linear blade speeds, rotors must turn at unusually high rates. The rotors on smaller pumps typically turn at 40 000–80 000 rpm, while those on larger pumps turn at 10 000–30 000 rpm. Because the central portions of large disks do not move with sufficient linear speeds, their blades usually extend only about one-third of the distance inwards from the periphery towards the shaft. These high rotational speeds place severe demands on the motors that turn the turbines, and on the bearings that support the turbine shafts. Again, manufacturers have done a great deal of research and development work to produce motors and bearings suitable for these pumps.

The relationships in eqn. 6.1 establish another very important characteristic of turbomolecular pumps; namely, that the compression ratio K developed by a given blade geometry and speed is exponentially dependent on the square root of the molecular mass of the gas molecules. This can be demonstrated by examining this equation in more detail. Basically, the quantity $K_M = \exp(\sqrt{M})$ which we can call the molecular component of the compression ratio equation, gives the dependence of K on molecular mass, since S_b and G_b become constants once a given pump is designed and put into operation. Table 6.1 contains values of K_M for several different gases, and shows this dependence quite clearly. For a single compression stage with a given blade configuration and speed, the compression ratios for hydrogen and helium are less than five percent of those for nitrogen and oxygen, the most abundant gases in air, which, in turn, are only of the order of one-thousandth of a percent of those for pump oil molecules. The compression ratios developed for different gases by multiple-stage pumps are proportional to these values raised to an exponential power which is of the order of the number of compression stages in the pump. Consequently, these differences grow to become several orders of magnitude. The compression ratios given by manufacturers for multi-stage turbomolecular pumps are of the order 10^9, 10^4, and 10^3 for nitrogen, helium, and hydrogen, respectively.

The compression ratios developed by turbomolecular pumps are also critically dependent on the backing pressure provided at their outlet ports. This dependence is usually documented by curves showing the variation of compression ratio with backing pressure, such as those in Fig. 6.3. Typically the compression ratios for all gases reach constant, maximum values at backing pressures below about 1 Pa (10^{-2} Torr). For the lighter gases, such as hydrogen and helium, compression ratios usually decrease significantly as the backing pressure increases above 1 Pa (10^{-2} Torr). For nitrogen, oxygen, and heavier gases, compression ratios usually do not begin to decrease until the backing pressure approaches 10 Pa (10^{-1} Torr). As backing pressures approach 100 Pa (1 Torr) the compression ratios for all gases drop precipitously.

Table 6.1 Values of K_M for different gases, showing the strong effect of the molecular mass M of the gas molecules being pumped on the single-stage compression ratio of a turbomolecular pump.

Gas	M	$\sqrt{M}$	K_M
Hydrogen	2	1.4	4
Helium	4	2.0	7
Water	18	4.1	69
Nitrogen	28	5.3	199
Oxygen	32	5.7	286
Argon	40	6.3	558
Octane	66	8.1	3375
Pump oil	300	17.3	3×10^7

$K_M = \exp(\sqrt{M})$ is the molecular component of the equation for the single-stage compression ratio of a turbomolecular pump when pumping a gas with molecular mass M.

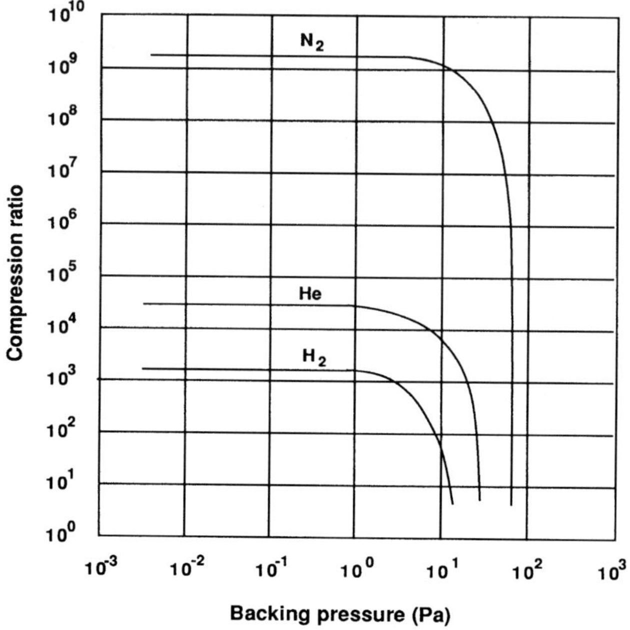

Fig. 6.3 Curves showing how compression ratios for nitrogen, helium, and hydrogen typically vary with backing pressure for multi-stage turbomolecular pumps.

6.1.3 Ultimate pressure

The basic relationship for the ultimate pressure attainable by a turbomolecular pump can be obtained by rearranging eqn. 5.3 (Section 5.6, p. 200) as follows:

$$P_i = \frac{P_o}{K} \approx P_{system} \approx \frac{P_{backing}}{K} \tag{6.2}$$

This shows that, apart from the system effects such as outgassing, virtual leaks, etc. discussed in Section 2.10 (p. 55), the ultimate pressure produced by a turbomolecular pump is determined by the compression ratio it can develop for the gases it is pumping, and by the partial pressures of these gases in the backing line. As indicated above, the compression ratios for nitrogen and oxygen, the most abundant gases in the atmosphere, are about 10^9. Since it should be easy to produce a pressure below 1 Pa (10^{-2} Torr) in the backing line with a good two-stage rotary-vane pump, this equation suggests that it should be possible to obtain a pressure at the inlet of the turbopump of at least:

$$P_i(\mathrm{N_2,O_2}) \approx \frac{P_o(\mathrm{N_2,O_2})}{K(\mathrm{N_2,O_2})} \approx \frac{1}{10^9} \approx 10^{-9}\ \mathrm{Pa} \approx 10^{-11}\ \mathrm{Torr}$$

and this is well into the ultra-high vacuum range.

In practice, however, it is usually the low compression ratio which turbomolecular pumps develop for hydrogen that determines the ultimate pressures they attain, because the decomposition of the oil in rotary-vane pumps produces a partial pressure of hydrogen in the pumping lines leading from these pumps which is of the order of 10^{-5} Pa (10^{-7} Torr). Eqn. 6.2 shows that this will lead to a partial pressure of hydrogen at the inlet of a turbomolecular pump of about:

$$P_i(\mathrm{H_2}) = \frac{P_o(\mathrm{H_2})}{K(\mathrm{H_2})} \approx \frac{10^{-5}}{10^3} \approx 10^{-8}\ \mathrm{Pa} \approx 10^{-10}\ \mathrm{Torr}$$

and this becomes the practical limiting pressure attainable by turbomolecular pumps backed by rotary-vane pumps. In fact, hydrogen is usually the most abundant gas in clean vacuum systems evacuated by turbomolecular pumps, apart from the water that desorbs from surfaces inside the system. Most manufacturers currently specify ultimate pressures of the order of 10^{-8} Pa (10^{-10} Torr) for their turbomolecular pumps.

It must be recognized that lower ultimate pressures can be attained with turbomolecular pumps, but to do so it is necessary to take special measures to control the level of hydrogen in the backing line. One such measure involves constructing the backing line of flexible stainless steel tubing. Another consists of reducing the number of O-ring seals in the backing line to a minimum. These measures are effective because significant quantities of hydrogen diffuse through the walls of rubber tubing and through rubber O-rings and gaskets. Still another approach consists of inserting a small turbopump, a small oil diffusion pump, or a molecular drag pump into the backing line between the main turbopump and the rotary-vane backing pump, thereby reducing the partial pressure of hydrogen, and all other gases, at the outlet of the main turbopump by several orders of magnitude. With such measures, ultimate pressures well into the 10^{-10} Pa (10^{-12} Torr) range can be attained. A recent development that achieves this result in a very compact design involves the addition of a molecular drag pump segment to a turbomolecular pump (see Section 6.2.3a, below). It is also advantageous to use a titanium sublimation pump (see Section 7.2, p. 296) in parallel with a turbomolecular pump to evacuate systems in which the lowest attainable pressures are required, because sublimation pumps have high pumping speeds for hydrogen and compensate for the deficiency of the turbopump in this regard.

6.1.4 Freedom from oil backstreaming

The characteristic of turbomolecular pumps which is perhaps of greatest interest in electron microscopy is their nearly complete freedom from oil backstreaming. This can also be explained on the basis of the relationships in eqn. 6.1. As shown by the data in Table 6.1, which are derived from this equation, the single-stage compression ratios turbomolecular pumps develop for gases of high molecular mass, such as mechanical pump oil vapours, for example, are several orders of magnitude greater than those for oxygen and nitrogen. This difference becomes overwhelmingly large when translated into values for multi-stage pumps. Typically, compression ratios greater than 10^{40} are cited for oil vapours, compared with values of the order of 10^{10} for nitrogen. In practice, this means *there will be no backstreaming of oil vapour molecules* from the backing line through a turbomolecular pump and into the vacuum system as long as the rotor of the turbopump is turning at nearly full speed. Spectra taken with residual gas analysers show that properly managed systems evacuated with turbomolecular pumps contain no detectable hydrocarbons.

The words 'properly managed' are critical in this context, however, because usually a turbomolecular pump can prevent backstreaming of oil vapour molecules only if its rotor is turning with at least about 60 percent of its full operating speed. Thus, if a turbomolecular pump is allowed to stop while the pressure in it and the backing line is in the region of molecular flow, oil molecules from the backing line can freely diffuse through the pump and into the vacuum system. Likewise, failure of the backing pump can lead to an increase in pressure in the region of the outlet stages of a turbomolecular pump sufficient to slow it down to the point where backstreaming of oil molecules can occur.

Whenever a turbomolecular pump is shut down it must be vented to a pressure of at least 10^2 Pa (1 Torr) so that the pressure of the gas in it and in its backing line is well above the range of molecular flow. The best venting gas is dry, oil-free nitrogen, although clean dry air is also acceptable. The venting gas must not be admitted into the backing line, because then it will sweep the oil molecules contained in the backing line into and through the turbopump, leading to rapid contamination of the pump and the vacuum system. The proper procedure is to admit the venting gas at a location such that it will flow from the turbopump into the backing line and sweep oil molecules down the backing line away from the turbomolecular pump. The usual arrangement is to install a venting valve in the vacuum chamber or in the pumping line leading from the turbopump to the chamber. However, some models of turbopumps have special venting ports located in their barrels near the middle of the turbine assembly, as shown in Fig. 6.1. This makes it possible to bake out the inlet section of the pump and the line leading to the vacuum system without requiring an expensive, bakeable, venting valve. For a turbopump to provide oil-free operation on an instrument with a vacuum system that is operated unattended for long periods of time, such as an electron microscope, it must be equipped with effective safety devices to ensure proper venting procedures are followed if the pump unexpectedly shuts down or if the backing pump fails to provide an adequate backing vacuum.

Although hydrocarbon pump oil molecules are unlikely to backstream through an operating turbomolecular pump in the vapour state, it is possible for them to creep along the surfaces of the stator in the liquid state and thereby ultimately reach the vacuum system. While this is a slow process, it is nonetheless a possibility to be considered in systems where a high degree of freedom from oil contamination is desired, and it could become significant in systems that are evacuated continuously for long periods of time, such as electron microscopes. Avoidance of this problem requires special attention to the cleanliness of the backing line to prevent liquid oil molecules from reaching the outlet section of the turbomolecular pump. Perhaps the most effective way to accomplish this is to install a good oil trap with a replaceable element (see Section 4.1.5a, p. 146) in the backing line before the system is ever turned on, and then to faithfully clean the trap and replace the filler material at intervals short enough to ensure that oil never has an opportunity to pass through it. If there is ever an indication that oil has had an opportunity to pass through the trap, then the segment of the backing line between the trap and the turbopump should be cleaned thoroughly or replaced.

6.1.5 *Pumping speed*

The pumping speeds of turbomolecular pumps vary with the molecular mass of the gas molecules in roughly the same way as compression ratios. That is, pumping speeds for light gases (e.g. hydrogen and helium) are significantly smaller than those for heavier gases (e.g. oxygen and nitrogen), because molecules of the lighter gases diffuse back through the rotating blades of the pump more readily than those of the heavier gases. This effect is shown in Fig. 6.4. For a given gas, pumping speed varies with inlet pressure in a manner very similar to oil diffusion pumps (compare Fig. 6.4 with Fig. 5.4 in Section 5.2, p. 177). For each gas, the pumping speed has a constant maximum value at pressures below about 10^{-1} Pa (10^{-3} Torr) where gas flow is molecular in character. As was true for oil diffusion pumps, the theoretical maximum possible pumping speed S_{tm} in this range is basically the conductance of the inlet port, as given by eqn. 2.13b (Section 2.5.1, p. 46) (i.e. $S_{tm} = C_{ma} = 0.09D^2$ for a pump with an inlet diameter of D mm). In practice, however, the rotor significantly reduces the conductance of the pump inlet, giving a complex dependence on the size, shape, spacing, and pitch of the blades on the disk of the first stage of the rotor, and on the linear speed with which these blades travel, so that this theoretical maximum speed is never attained. Fig. 6.5 shows approximately the manner in which maximum pumping speed for air varies with inlet diameter. Comparison of these curves with those of Fig. 5.3 (Section 5.2, p. 175) reveals that the pumping speeds of turbomolecular pumps are generally significantly smaller than those of diffusion pumps of the same inlet diameter. For pumps smaller than 200 mm in diameter, such as are commonly found on laboratory apparatus, the difference is about a factor of two. Note, however, that if the diffusion pumps are equipped with baffles or traps, as they usually are to help reduce the backstreaming of pump oil into the system, their effective speeds are cut approximately in half and then become comparable to those of the turbopumps. For the larger pumps, which are used mostly in industrial applications, the difference is somewhat greater. For example, turbomolecular pumps with inlet diameters of 750 mm typically have speeds near 10 000 l/s compared to typical speeds of 55 000 l/s for diffusion pumps of comparable size. This difference arises because, for mechanical reasons, the rotors in these large turbopumps cannot turn as fast as those in smaller pumps, and so the linear speeds near the centres of the disks are too slow for effective pumping. Therefore, the blades in these pumps usually do not extend all the way from the periphery to the centre of the disks, and so the actual acceptance area for incoming gas molecules is significantly less than the nominal area of the inlet flange.

Manufacturers currently offer turbomolecular pumps with inlet flanges ranging from 50 to 750 mm in diameter. These pumps are generally rated on the basis of the maximum pumping speeds they achieve for nitrogen. Pumps on apparatus in electron microscopy laboratories usually have inlet flanges 50–200 mm in diameter and have rated speeds in the range from 100 to 1000 l/s. Larger pumps provide speeds ranging up to 10 000 l/s. As noted above, the pumping speeds for helium and hydrogen are about 10 and 25 percent less than for nitrogen, respectively. Pump specifications usually also include speeds for these gases because of their importance in determining ultimate pressures.

At pressures slightly above 10^{-1} Pa (10^{-3} Torr) gas flow enters the transitional range, the pumping action becomes less effective, and the pumping speed begins to decrease. As pressure increases above this level, pumping speed decreases rapidly, becoming very small at pressures above

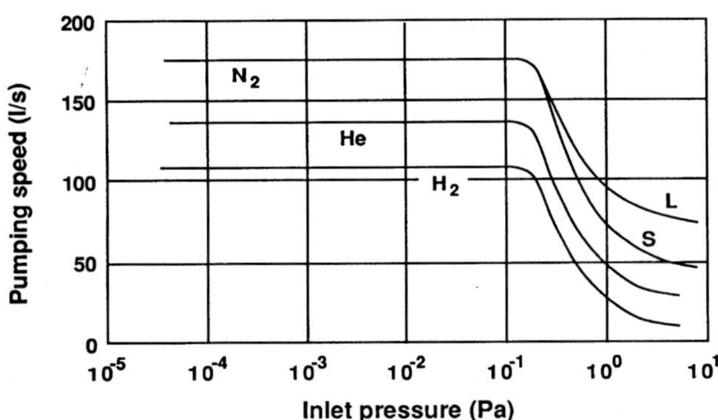

Fig. 6.4 The variation of pumping speed for nitrogen, helium, and hydrogen with inlet pressure typical of a turbomolecular pump with an inlet diameter of 100 mm. The branches L and S show the effect of large and small backing pumps on the pumping speed at pressures above 0.1 Pa (10^{-3} Torr).

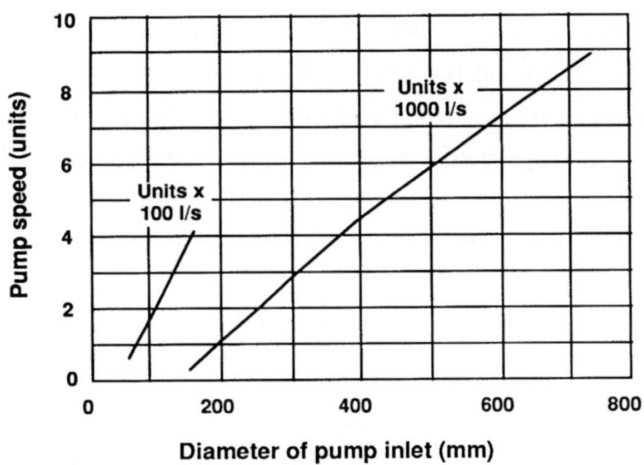

Fig. 6.5 The variation with inlet diameter of the pumping speed of turbomolecular pumps for air.

10 Pa (10^{-1} Torr) where flow becomes fully viscous in character, as indicated in Fig. 6.4. Over this pressure range, turbomolecular pumps experience overloading similar to that described for oil diffusion pumps in Section 5.2 (p. 176), and their throughput becomes essentially constant. The overall pumping speed then becomes dependent on the speed of the backing pump, as indicated schematically in Fig. 6.4.

6.1.6 Backing pumps

Since the backing pressure influences both the ultimate pressure attainable and the pumping speed at pressures above 10^{-1} Pa (10^{-3} Torr), the selection of a proper backing pump can be a somewhat critical matter. For routine applications, where the lowest possible ultimate pressures are not required, manufacturers uniformly recommend the use of two-stage rotary-vane pumps. Such pumps are capable of providing backing pressures well below the critical value of 1 Pa (10^{-2} Torr) needed to develop maximum compression ratios and pumping speeds, and thereby ensure optimal performance of the turbomolecular pumps. As described in Section 6.1.3, the rotary-vane pump may be supplemented by a small turbo or oil diffusion pump in systems where the lowest ultimate pressures are required. A recent development is the use of a molecular drag pump (Section 6.2, below) in series with a diaphragm pump (Section 4.2.3, p. 161) for backing turbomolecular pumps when a totally oil-free system is desired.

Eqn. 5.1 (Section 5.3, p. 177) can be used as a basis for determining the size of the backing pump needed for a given turbomolecular pump. This equation $Q_p = S_p P_p$ gives the relationship between the speed of a pump S_p, the pressure at its inlet port P_p, and its throughput Q_p. Since all gas molecules that enter the inlet of a pump must be discharged through its outlet, the throughput of a turbopump Q_p must be the same at the inlet and outlet ports, and so this equation can be modified as follows:

$$Q_p = S_i P_i = S_o P_o \tag{6.3}$$

or

$$S_o = \frac{Q_p}{P_o} = \left(\frac{P_i}{P_o}\right) S_i \tag{6.4}$$

where S_o and S_i are the pumping speeds and P_o and P_i are the pressures at the outlet and inlet ports, respectively. The rated speed of the backing pump must be at least as great as the value of S_o given by this equation when appropriate values of P_o and P_i are used in it. These values are generally based on the principle that the backing pump should be large enough to keep the pressure at the outlet port of the turbomolecular pump below the range of viscous flow during periods when its throughput is greatest. On this basis, Fig. 2.1 (Section 2.2, p. 31) suggests that the pressure at the outlet port of the turbopump P_o should not be allowed to rise above about 10 Pa (0.1 Torr). Likewise, a value for P_i can be obtained from Fig. 6.4, which shows that turbopumps usually develop maximum throughput at inlet pressures near 10^{-1} Pa (10^{-3} Torr). Inserting these values into eqn. 6.4 gives:

$$S_o \approx \left(\frac{0.1}{10}\right) S_i \approx \frac{S_i}{100} \tag{6.5}$$

suggesting the use of a backing pump with a speed of at least one-hundredth the rated speed of the turbomolecular pump. For example, a backing pump with a rated speed of at least (350)/100 = 3.5 l/s or 200 l/min would be selected for a turbomolecular pump with a speed of 350 l/s. In practice a somewhat larger backing pump might be used (e.g. 250 or 300 l/min), because the speed of two-stage rotary-vane pumps at 10 Pa (0.1 Torr) is usually only about 0.8 of their rated speed at atmospheric pressure (Section 4.1.2, p. 139).

Since typical turbomolecular pumping systems can be considered as two-stage systems, with the turbopump as the first stage and the rotary-vane backing pump as the second, the ratio of the speed of the turbopump S_{turbo} to that of the backing pump $S_{backing}$ is often referred to as the 'staging ratio' of the system, i.e.

$$R_s = \frac{S_{turbo}}{S_{backing}} \tag{6.6}$$

and the discussion in the previous paragraph suggests a maximum value of 125:1 for this ratio. In practice, most manufacturers recommend using two-stage rotary-vane pumps with speeds large enough to provide staging ratios in the range from 25:1 to 100:1 as the backing pumps for their turbomolecular pumps. Pumps of this size are particularly appropriate for systems that are cycled between atmospheric pressure and the high vacuum range frequently, such as vacuum evaporators, for which a major concern is the handling of the high throughput encountered during the pumpdown process. Selection of backing pumps on this basis will reduce the total time for the rough pumping operation, reduce possibilities for oil backstreaming, and minimize the time required to make the transition from viscous to molecular flow, thereby preventing the turbopump from being overloaded long enough during the pumpdown process to damage the power supply for the drive motor. Even larger backing pumps may be desirable on systems used for various physical and chemical processes which use or generate large amounts of gas and produce operating pressures near 0.1 Pa (10^{-3} Torr), where turbopumps have high throughput.

It is interesting to note, however, that the throughput characteristics of turbomolecular pumps are very similar to those of oil diffusion pumps discussed in Section 5.3 (p. 177) That is, a significantly smaller backing pump can handle the throughput of a turbopump when it is operating in the high vacuum range than is needed to handle its maximum throughput during the pumpdown process. This has interesting implications for systems that are rarely cycled between atmospheric pressure and the high vacuum range, such as many electron microscopes, because here the primary task of the backing pump is to handle the throughput of the turbomolecular pump while it is operating in the high vacuum range. The principal requirement of the backing pump then becomes maintaining a backing pressure low enough to enable the turbopump to develop maximum compression ratios for all gases during the long periods it spends operating at inlet pressures near the ultimate pressure of the system. Fig. 6.3 indicates that a backing pressure P_o of 0.5 Pa (5×10^{-3} Torr) will usually meet this requirement. Taking 10^{-3} Pa (10^{-5} Torr) as the highest inlet pressure P_i that might be expected in such long-term operation, eqn. 6.4 gives a required outlet pumping speed of about:

$$S_o \approx \left(\frac{10^{-3}}{0.5}\right)S_i \approx \frac{S_i}{500}$$

corresponding to a staging ratio of about 500:1. This is only one-fifth of the backing speed suggested by eqn. 6.5. For example, a backing pump capable of providing a speed of 350/500 = 0.7 l/s or 42 l/min could adequately serve a 350 l/s turbopump operating at 10^{-3} Pa (10^{-5} Torr). Making allowance for the decrease in pumping speed with pressure exhibited by rotary-vane pumps, one with a rated speed of 100 l/min would readily meet this requirement. Since throughput decreases linearly with pressure, a proportionally smaller backing pump would suffice at lower inlet pressures. The advantage in using a smaller backing pump is usually a significant reduction in the initial cost of the system. The disadvantage is that the pumpdown process will be prolonged somewhat and the turbopump will be overloaded longer each time the system is pumped down from atmospheric pressure, increasing the danger of overheating the turbopump's power supply. While these are matters of concern primarily to designers of vacuum systems, awareness of them should enable users to understand and manage their vacuum systems more intelligently.

One very important advantage of turbomolecular pumps is that they are considerably more tolerant of high backing pressures than oil diffusion pumps. This is primarily due to the fact that, unlike diffusion pumps, turbopumps do not contain openly-exposed, boiling oil that can be oxidized and swept into the system. As just noted, a backing pressure near 0.5 Pa (5×10^{-3} Torr) is required for optimum functioning of most turbomolecular pumps. However, most pumps are not damaged if the backing pressure rises well above 10 Pa (0.1 Torr) for short periods of time. If this happens, some of the disks encounter gas with transitional or viscous flow characteristics and the pumping mechanism becomes less effective. This causes both pumping speed and compression ratios to decrease, as shown by Figs. 6.3 and 6.4, and also usually leads to a gradual decrease in rotor speed. There is little danger of oil backstreaming from the backing line into the vacuum system as long as the rotor speed does not drop below about 60 percent of the normal value for operation in the high vacuum range. However, operation under these overload conditions places extra stress on the bearings, which can lead to reduced bearing life, and increases the load on the drive motor and its power supply, which may cause the power supply to automatically shut off. The maximum pressure at which a turbopump can safely operate for a long period of time will therefore depend on the design of its bearings and power supply. Most manufacturers give values in the range from 1 to 10 Pa (10^{-2} to 10^{-1} Torr) for the maximum backing pressures their pumps will tolerate for long-term operation.

Excessive inlet pressures cause much the same effects as excessive backing pressures, and values given by manufacturers for the maximum acceptable inlet pressure for long-term operation usually are in the range from 0.1 to 1 Pa (10^{-3} to 10^{-2} Torr). However, turbopumps can make a significant contribution to system evacuation beginning at pressures as high as 1000 Pa (10 Torr). Although their pumping speeds may be only of the order of 1 or 2 l/s at these pressures, they translate into values near

100 l/min, and are comparable to the speeds of moderately sized rotary-vane pumps. Furthermore, their pumping speed increases rapidly as pressure decreases. Thus, in the pressure range where the speed of a mechanical roughing pump starts to decrease, a turbopump can take over and markedly increase the rate of evacuation through the region of transitional gas flow. Prolonged operation at such high pressures may lead to overheating of the pump's motor or power supply; however, short periods are usually well tolerated. Consequently, a small turbopump is often used ahead of a mechanical pump to provide rapid rough pumping of a large system prior to crossing over to a diffusion, cryogenic, or ion pump, or to a larger turbopump.

6.1.7 *Design features*

The first turbomolecular pumps produced commercially were of the 'double-flow' design shown in Fig. 6.6. Such pumps have two sets of turbine blades mounted on opposite ends of a single horizontal drive shaft with a common inlet chamber between them. Gas that enters the inlet chamber is pumped laterally by the two turbines and discharged into a common duct leading to the outlet port. However, most pumps produced today are of the 'single-flow' design, shown in Figs. 6.1 and 6.2, and have only a single set of turbine blades and an inlet port concentric with the drive shaft. Double-flow pumps offer some advantages in mechanical stability, and usually provide somewhat greater pumping speeds for a given inlet size because of the co-operative action of the two turbines, although some of the potential advantage in pumping speed is lost because of the additional conductance contributed by the inlet chamber. The dual-flow design also allows the inlet port to be rotated into any desired orientation around the pump axis to accommodate input lines running from odd directions. One manufacturer (Welch) also makes dual-flow pumps with inlet chambers which have several inlet ports and so can serve more than one apparatus. Double-flow pumps are still available from at least two manufacturers (Balzers and Welch). However, single-flow pumps are lighter, more compact, and less expensive than the double-flow pumps, because they have only one set of turbine blades, and are more widely available and preferred for most applications.

In order to achieve the high linear blade speeds needed for effective pumping to occur, the rotors of turbomolecular pumps must turn at very

Fig. 6.6 A photograph of a double-flow turbomolecular pump with part of the case cut away to show the arrangement of the internal parts. (Courtesy of Balzers.)

high rotational speeds, typically near 60 000 rpm for smaller pumps and 10 000 rpm for the largest ones. Specially designed, variable-speed, high-frequency electric motors are used to drive the rotors of most pumps. The component of these motors that rotates is mounted directly on the turbine shaft, while the stationary parts of the motor are attached to the surrounding pump housing, and the entire motor is housed in a chamber on the output side of the pump which is evacuated by the backing pump, as shown schematically in Fig. 6.1. This arrangement does not require a shaft seal between the atmosphere and the interior of the pump, and so the many problems associated with shaft seal failure are eliminated. The motors are usually driven by solid-state electronic control units which provide electric current of variable frequency and variable power. When a pump is first started up the control unit provides low-frequency, high-power input to the motor. As the rotor speed increases the input frequency is increased to match, and the power level is gradually decreased. When a high vacuum is achieved the frequency and power are automatically adjusted by an electrical feedback system to maintain the rotor speed constant at the optimum level. The time needed for the turbine to reach full rotational speed varies from 2 min for the smallest pumps to more than 30 min for the largest ones, and is typically in the range from 5 to 10 min for pumps of the size found on most laboratory apparatus. Many control units provide a standby mode of operation in which the rotor speed is reduced to about 70 percent of its normal operating value. At this speed, the turbine turns fast enough to prevent oil backstreaming and to provide significant pumping action, yet bearing wear is greatly reduced. Operation in this mode may be acceptable and beneficial over weekends and holidays, and during other periods when a vacuum system is not actively in use. One of the most important functions of a control unit is to monitor critical operating parameters and to automatically shut the pump down and properly vent the system if the bearings overheat or the motor becomes dangerously overloaded. Control units may also provide a variety of other features, such as: a pump speed indicator; an overload warning; an interface for remote control of the startup, shutdown, and standby functions; an input for a control signal from a vacuum gauge; outputs for controlling vent valves and the backing pump; and a computer interface.

The high rotational speeds at which turbomolecular pumps run also place severe demands on the bearings that support the turbine shaft. Manufacturers have used a variety of different types of bearings to meet these demands, and pumps are frequently categorized on this basis. High precision ball bearings are used on most pump models. One of the latest innovations is the use of precision ceramic ball bearings, which are very resistant to wear. In some designs the ball bearings are lubricated with oil, while grease is used in others. When oil is used it is usually necessary to mount the pump in a particular orientation so that the oil circulation system will function properly. Pumps lubricated with grease can usually be mounted in any orientation from vertical to horizontal, and some can even be mounted upside down. Special grades of oil and grease are required for these bearings, and manufacturers' instructions in this regard, and in regard to lubricating practice, should be followed meticulously to minimize bearing wear. Ball bearings also generate a considerable amount of heat, because there is physical contact between the bearings and the rotating shaft, and this heat must be removed to prevent the bearings from overheating and becoming damaged. Both water- and air-cooling are used for this purpose. Water-cooling systems are the most efficient and reliable, and are therefore to be preferred in most applications. However, air-cooled pumps can be highly advantageous for use on portable systems. Pumps that require cooling for their bearings are usually equipped with built-in, thermally-activated, cut-out switches that will shut them down if the cooling system fails to function properly and the bearings begin to overheat.

Early models of turbomolecular pumps were not widely used in electron microscopes because of their higher cost, compared to oil diffusion pumps, and because of problems associated with the ball bearings used in them. As the need for an oil-free environment in electron microscopes became more critical, the higher cost became more tolerable. However, early pumps produced such high levels of mechanical vibration that high-resolution images could not be attained. This vibration problem was largely related to the quality of the bearings available and the precision with which the rotors could be balanced. Bearing wear was also a serious problem in early turbopumps. Bearings often lasted only a few months, and replacement, which usually could only be performed in a clean-room environment by factory-trained engineers, was expensive and time consuming. In addition, the bearings on many early turbopumps produced an irritating, high-pitched whine which most electron microscopists found very distracting.

In recent years these problems have gradually been overcome as manufacturers have refined their designs and manufacturing methods. Turbomolecular pumps with ball bearings are now relatively reliable and trouble free. Manufacturers now use specially-selected, high-precision bearings with refined lubricating systems, so that bearings now typically have lifetimes of from 3 to 5 years. The bearings on most present pumps are located inside the body of the pump on the rough vacuum side of the turbine. This eliminates the need for shaft seals between the pump and the atmosphere and minimizes the possibility of lubricating oil reaching the high vacuum side of the pump. The rotors of present pumps are balanced with an extreme degree of precision using high sensitivity electronic sensors, so that noise and vibration have been reduced to levels that are insignificant for most applications. As these improvements have occurred, the use of turbomolecular pumps on electron microscopes has gradually increased. The most common application to date has been on scanning electron microscopes, where some residual vibration can be tolerated. These instruments usually have rather large specimen chambers that must be opened frequently to change specimens, and turbopumps are particularly well suited for such systems.

Manufacturers have long recognized that both bearing wear and mechanical vibration could be reduced drastically by using bearings that do not involve physical contact between the rotating shaft and the body of the bearing. Air bearings were tried at one time, but proved too noisy and unreliable for general use. At least one manufacturer (Welch) uses bearings in which the shaft is supported by a hydrodynamically generated oil film that separates it from the actual metal bearing surface. The latest advance along this line is the development of magnetic bearings which float the turbine shaft in magnetic fields. Both permanent magnets and electromagnets are used for this purpose. In some pump designs both ends of the turbine shaft are supported by magnetic bearings, while in others a magnetic bearing supports the top end of the shaft while a ball bearing is used at the bottom. The use of magnetic bearings represents a very significant advance in the design of turbomolecular pumps because of the potential for virtually eliminating frictional wear, greatly extending bearing life, and drastically reducing mechanical vibration and noise. Magnetic bearings do not need to be cooled and they usually allow a pump to be mounted in nearly any desired orientation. Although they increase the cost of turbopumps somewhat, magnetic bearings are very attractive for use when it is critical to avoid oil contamination and to minimize vibration. Turbomolecular pumps with magnetic bearings are now being used very successfully to evacuate the electron gun and specimen chamber on several models of high-resolution electron microscopes.

Another relatively recent development is the use of molecular drag pumping units as the final output stage of turbomolecular pumps. Molecular drag pumps are described in Section 6.2 below, and this new type of pump is discussed there. Basically, however, these hybrid pumps have much higher output pressures than an ordinary turbopump, and so oil-free diaphragm or piston pumps can be used as backing pumps, yielding truly oil-free systems.

Turbomolecular pumps have several other characteristics that make them attractive for systems where an oil-free environment is desired. They are not damaged if the system develops a leak, and can generally operate for long periods at pressures as high as 10 Pa (0.1 Torr). Unlike cryogenic pumps (Chapter 8) and sorption pumps (Section 4.2.1, p. 155), they do not accumulate gases, and thus do not become saturated and require periodic regeneration. They generally start up more quickly than diffusion pumps and cryogenic pumps, making them well suited for use in systems such as vacuum evaporators and scanning electron microscopes that must be opened to the atmosphere frequently. They are also finding extensive use in semiconductor manufacturing processes because of their oil-free characteristics and because special models are available that can pump the reactive and corrosive gases so often encountered in these processes.

6.1.8 *Vacuum systems with turbomolecular pumps*

Vacuum systems designed with turbomolecular pumps are very similar to those used for oil diffusion pumps. In fact, a turbopump can often be substituted directly for a diffusion pump without making significant changes in the components or configuration of a vacuum system. Two basic system configurations are commonly encountered. Systems with a set of three valves are often found on vacuum evaporators and scanning electron microscopes. As shown in Fig. 6.7, the arrangement of components in these systems is essentially the same as in the diffusion pump system shown in Fig. 5.11 (Section 5.7.1, p. 204). Valveless turbopump systems generally are constructed as shown in Fig 6.8, and resemble the valveless diffusion pump system of Fig. 5.14 (Section 5.7.3, p. 213), again with the replacement of the diffusion pump by a turbopump.

6.1.8a *The operation of systems with a set of three valves*

The operating procedures for systems with the traditional full set of three valves are very similar to those for the comparable systems based on oil diffusion pumps. Since these procedures were discussed in such great detail in Section 5.7.1 (p. 202), only a brief outline will be given here. To start up the system the flow of coolant to the turbopump is started, the roughing valve and the high vacuum valve are closed, and the backing line valve is opened, putting the valves in a standby configuration comparable to that of Fig. 5.12a (Section 5.7.1, p. 205). The turbomolecular pump and its backing pump are then turned on simultaneously so that the turbopump can reach at least 60 percent of full speed before the backing pump brings the pressure in it down near to the range of molecular flow. After the turbopump reaches full operating speed and a high vacuum is established in it, the backing line valve is closed and the roughing valve is opened to rough pump the vacuum chamber. When the pressure in the vacuum chamber falls below about 150 Pa (1.5 Torr) the roughing valve is

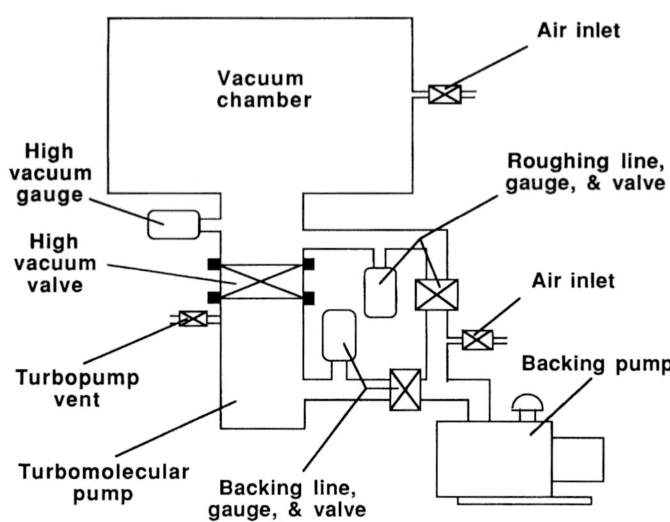

Fig. 6.7 A vacuum system with a turbomolecular pump and a full complement of vacuum valves.

closed and the backing valve is opened. The high vacuum valve is then opened and the turbopump is allowed to evacuate the chamber. Opening this valve at such a high pressure may cause the turbopump to slow down somewhat, but recovery should be prompt and rapid. There are two advantages to doing this. First, it utilizes the ability of turbopumps to operate safely for short periods at pressures well into the range for viscous flow, thereby decreasing the time needed to make the transition from the rough vacuum to the high vacuum range. Second, it also prevents backstreaming of oil because the roughing valve is closed long before the pressure in the backing line falls into the range of molecular flow.

To bring the vacuum chamber up to atmospheric pressure without shutting the pumps down, the high vacuum valve is closed, the roughing valve is left closed, and both pumps are left running with the backing valve open. This will place the system in a standby configuration similar to that shown in Fig. 5.12a, and will allow air to be admitted to the vacuum chamber without interrupting the operation of the turbopump.

To shut the turbopump off, the high vacuum and backing valves are both closed, power to the turbopump is shut off, and the pump is allowed to slow down to about 70 percent of its operating speed, whereupon air, or preferably dry nitrogen, is admitted to it very slowly through the turbopump vent. When the pump stops, the flow of coolant should be stopped to prevent condensation of moisture on and inside the pump. The air inlet valve for the backing pump is then opened and this pump is immediately shut off.

Liquid nitrogen traps are not needed to provide protection against the backstreaming of oil from the backing line into the vacuum chamber on systems of this type. As pointed out in Section 6.1.4 above, the compression ratios developed by turbomolecular pumps for gases having such high molecular masses are so large that hydrocarbon backstreaming cannot occur, provided the turbopump is always allowed to reach at least 60 percent of its operating speed before the inlet pressure drops into the range of molecular flow, and provided the rough pumping, backing, and venting processes are carried out properly. However, liquid nitrogen traps can be useful on systems that must handle large amounts of water vapour because they have very high pumping speeds for water and can provide a stage of cryogenic pumping that can be very beneficial in supplementing the action of the turbopump. When a trap is selected for this purpose it should be one that provides the highest available conductance, rather than the best features for preventing backstreaming, and it should be located between the turbopump and the high vacuum valve.

Likewise, there are no great benefits to be gained from installing traps in the backing and roughing lines of turbopump systems. Because turbopumps can tolerate such high crossover pressures, the valves that isolate these lines from the vacuum system can always be closed before conditions of molecular flow occur, and so, again, proper management of the roughing, backing, and venting processes can effectively prevent oil backstreaming. For the same reason, there is little to be gained by adding a buffer tank to the backing line of a turbomolecular pump, as was described for diffusion pumps in Section 5.7.2 (p. 212). Although the backing valve is closed and the turbopump is isolated from its backing pump during the time the vacuum chamber is being rough pumped, the fact that the chamber does not need to be pumped below 100 Pa (1 Torr), combined with the tolerance of turbopumps for high pressures, makes it unlikely that the turbopump will be damaged or that backstreaming of oil through the turbopump will occur during this time, provided the roughing operation is properly carried out, and provided the turbopump is properly used to evacuate the chamber as soon as the rough vacuum drops to about 150 Pa (1.5 Torr).

6.1.8b Operating valveless systems

Turbomolecular pumps are particularly well suited for use in valveless systems of the type shown in Fig. 6.8 because they come up to operating speed in a relatively short time and are not damaged by exposure to air for short periods at moderately high pressures. A system of this kind is evacuated by starting the flow of coolant to the turbopump, and then turning the turbopump and the backing pump on at the same time. If the backing pump is of the proper size it will reduce the pressure in the vacuum chamber to about 200 Pa (2 Torr) in about the same time as is required for the turbopump to reach about 70 percent of its full operating speed, whereupon the turbopump will gradually take over the evacuation of the chamber. In this way a short pump-down time can be attained with no danger of oil backstreaming, because the turbopump will reach a rotational speed sufficient to prevent backstreaming well before the pressure in the backing line drops into the range of molecular flow.

To shut the system down, the turbopump is turned off and allowed to slow down to about 70 percent of its operating speed, whereupon the vent valve is opened and the backing pump is immediately shut off. Dry air or nitrogen is admitted to the system at a rate adjusted to slow the turbopump to a stop rapidly but safely. When the system reaches atmospheric pressure the vent valve is closed and the flow of coolant to the turbopump is stopped.

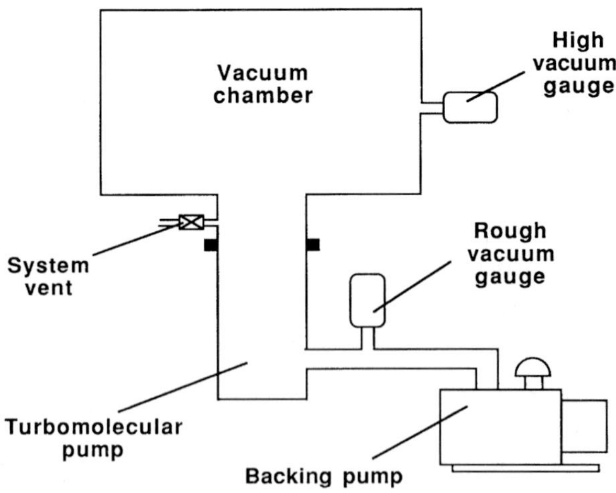

Fig. 6.8 A valveless vacuum system with a turbomolecular pump.

These operating procedures are so simple and straightforward that they can easily be controlled by a single on-off switch operating in conjunction with a programmable turbopump control unit or a simple control circuit, making the operation of these systems virtually foolproof. Valveless systems are also very economical, because they do not require an expensive high vacuum valve, nor roughing and backing valves. However, some finesse must be used in designing these systems. The size of the backing pump must be chosen properly, so that it does not evacuate the system to a pressure below about 10 Pa (0.1 Torr) before the turbopump comes up to about 60 percent of its full operating speed. The opening of the vent valve must be properly timed to minimize stress on the rotor and bearings of the turbopump. In addition, it is not convenient to add a liquid nitrogen trap to provide extra pumping speed for water vapour, because the trap would have to be warmed up to room temperature each time before the system could be let up to atmospheric pressure, and this would prolong the operation unacceptably. Because of their low cost, simplicity of construction, and ease of operation, valveless turbomolecular pumping systems are now widely used on vacuum evaporators and many other types of apparatus.

6.1.9 *Potential problems*

The major concerns associated with the operation of systems with turbomolecular pumps are preventing physical damage to the pumps, and avoiding system contamination by the backstreaming of oil from the backing line through the turbopump. While proper system design plays a major role in both of these matters, it is very important for operators to be aware of the potential sources of trouble and the ways in which they can be avoided. This will also help in evaluating systems for potential sources of trouble before they are purchased.

6.1.9a *Preventing physical damage*

A turbomolecular pump can be destroyed if a solid object or solid particulate matter is allowed to enter it while the rotor is in motion. This is unlikely to occur in a transmission electron microscope, but could readily happen in a scanning electron microscope, a vacuum evaporator, or an ion beam milling unit. A turbopump can even be damaged if a small tool is dropped into its stationary blades by someone working in an open vacuum chamber. All persons who use or work on an apparatus equipped with a turbomolecular pump should be fully aware of these dangers. One way to reduce the likelihood of incurring this kind of damage is to install a protective screen in the mouth of the turbopump. All manufacturers offer protective screens for their pumps, and such screens should be installed on all pumps used in situations where there is even a remote likelihood of solid objects or particulate matter entering the pump. The only disadvantage in using these screens is a modest decrease in pumping speed. Screens with mesh openings of the order of 5 mm in diameter are adequate for most applications and cause only a 10–15 percent decrease in pumping speed. Screens with openings as small as 0.5 mm are also available, but may cause as much as a 25 percent loss in pumping speed. However, these fine screens provide the best protection, because particles small enough to pass through them will usually also pass through most turbopumps without damaging them. Another approach that is highly effective is to mount turbomolecular pumps in a horizontal orientation so that objects cannot fall directly into them.

Manufacturers usually warn that damage to the bearings or rotor can occur if air at atmospheric pressure is suddenly admitted to a turbomolecular pump while it is running at full operating speed. Different models of pumps vary considerably in their ability to withstand such mistreatment. Some models are readily damaged, while others can stand a number of sudden exposures to the atmosphere without serious damage. Nonetheless, each such exposure places abnormal stresses on both the rotor and the bearings that support it, so every reasonable precaution should be taken to avoid such incidents. This usually involves providing proper protection for fragile windows and vacuum gauges so that they are not accidently broken while the system is in operation, and paying careful attention to operating procedures so that valves that can admit air to the turbopump are properly throttled, and are not opened at inappropriate times.

Pumps with ball bearings require periodic lubrication. Manufacturers' instructions should be followed rigorously in this matter. Even with careful maintenance the bearings of a turbomolecular pump ultimately wear out and must be replaced, typically after from 25 000 to 50 000 h of operation. Policies with regard to bearing replacement vary from laboratory to laboratory. Some users replace them routinely after 20 000 or 25 000 h of operation, while others prefer to wait until signs of bearing failure develop. If a pump begins to vibrate or to make an unusual noise, it is almost certain that a bearing is becoming worn or that the rotor has been damaged and is slightly out of balance. The pump should then be turned off immediately and the manufacturer should be contacted for advice on how to diagnose and handle the problem. Bearing replacement usually requires special equipment and expertise, and should be carried out in a clean-room environment, and so is best performed by the manufacturer.

Most turbomolecular pumps also use cooling water to keep the motor and bearings from overheating. Every possible precaution must be employed to ensure that the cooling water supply does not fail while a pump is in operation. Most of the comments made in Section 5.8.1 (p. 215) regarding problems with cooling water supplies for diffusion pumps also apply here. Fortunately, most turbomolecular pumps have built-in thermal sensors that operate through the pump's control unit to initiate appropriate shutdown and venting procedures if overheating occurs. Because the alignment of the rotor in the pump housing is highly critical, turbopumps must be mounted in such a way that they do not experience a significant load or a distorting force from the apparatus to which they are attached.

6.1.9b *Preventing oil backstreaming*

As has been noted several times in the preceding discussion, turbomolecular pumps develop such high compression ratios for molecules of the oils used in rotary-vane pumps that these molecules are totally unable to move from the backing line through a turbopump and into the vacuum system in the vapour phase, provided the turbopump's rotor is turning with at least 60 percent of its full operating speed. Therefore, operating procedures should be designed and carried out in such a way as to fully utilize this advantageous characteristic of turbomolecular pumps. General methods for doing this have been outlined in the operating procedures described above. Once the general principle is understood and appreciated, it should be possible to design, modify, or adapt procedures as needed to achieve the desired result, which basically involves turning the turbopump on soon enough so that it has an opportunity to reach the necessary rotational speed before the pressure in the backing line drops into the range of transitional flow.

It is also important to follow proper shutdown and venting procedures if the backstreaming of oil vapours is to be avoided. Turbopumps should always be vented to atmospheric pressure whenever they are shut down. The partial pressure of oil molecules from bearing lubricants and from the backing pump can be expected to be of the order of 10^{-3} Pa (10^{-5} Torr) in the outlet region of a turbomolecular pump. When the turbine of a pump is not turning at 60 percent of its full operating speed, it will not develop a sufficient compression ratio to prevent these oil molecules from diffusing from the outlet region of the pump, through the pump, and into the vacuum system. The venting gas should never be admitted on the outlet side of the pump, because doing so will physically sweep the oil molecules present there through the pump and into the vacuum chamber. Instead, the venting gas should be admitted on the inlet side of the pump so that oil molecules present in both the outlet port and the backing line are swept down the backing line away from the turbopump. In some pumps the venting port is built into the pump so that the gas is admitted in the region of the middle rotor disks, as shown in Fig. 6.1; otherwise, the venting port should be located immediately above the pump inlet as shown in Figs. 6.7 and 6.8. An additional function of the venting gas is to coat the internal surfaces of the pump and vacuum system to retard the adsorption of water onto these surfaces while the system is open. The preferred venting gas is dry, oil-free nitrogen (see Section 2.10.3, p. 64), although ordinary air can be used if it is clean and dry. If air is used some care must be exercised to ensure that it is not drawn from a region where there is a high concentration of oil vapours (e.g. near the outlet of a rotary-vane pump).

Special valves are usually used to vent turbopumped systems. These are solenoid valves designed to be operated by the pump control unit, and adjusted by the manufacturer to admit gas at a controlled rate to slow the pump down as rapidly as is safely permissible. Opening of the vent valve should be delayed until the rotor has slowed down to about 60 percent of its full operating speed to reduce the stress on the rotor and bearings of the pump. Most control units can be adjusted to provide a suitable delay. If the delay function is controlled by a thermally-activated relay (see Section 5.8.5, p. 220) or is backed by a battery, it can also prevent the system from being shut down unnecessarily due to a momentary loss of power in the supply mains.

There is one potentially serious problem that can be associated with the use of delayed venting, however, particularly in valveless vacuum systems. If a power loss occurs the rotary-vane backing pump will stop immediately, and will stand with a vacuum in its inlet port until the vent valve finally opens. Under these circumstances it is possible for air to be sucked backward through this pump, bringing considerable oil and oil vapour into the backing line, and possibly into the turbopump. Although most current rotary-vane pumps have anti-suckback devices, it is prudent not to trust these devices implicitly. The rule here is, 'If you don't want it to happen, it will, even though it can't'. We had a small turbopump badly contaminated by this suckback phenomenon when an inexperienced student shut the system down without proper venting, even though the backing pump was designed so suckback was not supposed to be possible. This turbopump had to be returned to the manufacturer to be disassembled and thoroughly cleaned before it could be used again.

It is possible to prevent liquid oil from being sucked into a turbopump by installing a buffer tank, which is large enough to contain the oil from the rotary-vane pump, in the backing line immediately above this pump. Any oil that is sucked up into this tank will eventually drain back into the backing pump when pressure equalization finally occurs. However, this will not prevent oil vapours from being drawn into the turbopump during the suckback process. The best solution is to install a normally-closed solenoid safety valve and a normally-open solenoid air vent in the backing line immediately above the backing pump in the manner shown in Fig. 6.9. It might seem satisfactory to simply wire these two valves in parallel with the backing pump motor, so that a power failure will cause the safety valve to close and the vent valve to open at the same time as the pump stops, admitting air at atmospheric pressure into the backing pump and the part of the backing line between it and the safety valve. This is not a fully satisfactory arrangement, however, because it is necessary to ensure that the safety valve closes fully before the air vent valve opens, otherwise a short burst of air can be admitted to the backing line and carry oil vapours into the turbopump. In addition, if delayed venting is also being used and the power then comes on again before the full programmed venting procedure is started, the safety valve will open before the backing pump has an opportunity to exhaust the air from the backing line. This air will then surge into the turbopump, carrying with it oil vapours from the backing line and contaminating the pump. It is therefore necessary to delay the opening of the safety valve long enough to allow the backing

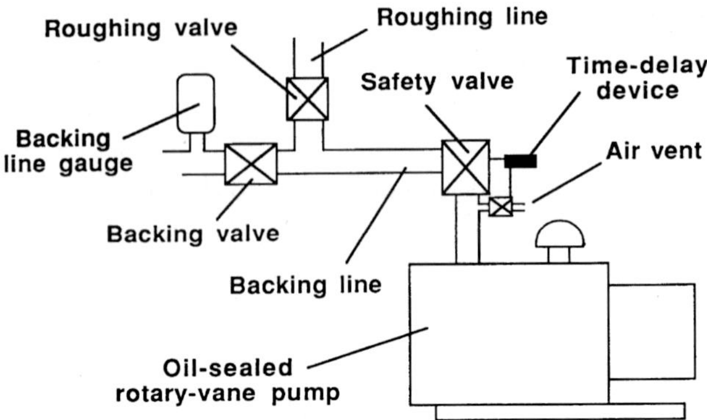

Fig. 6.9 A safety valve, air vent, and time-delay system to prevent oil from being sucked into the backing line from an oil-sealed pump.

pump to evacuate the backing line again. It is, of course, possible to design control circuits that will do this, but it may be simpler to use combination valves designed especially for this purpose. For example, the 'SECUVAC' valves sold by Leybold combine a solenoid safety valve and an air inlet valve with integral controls that automatically close and open them in the correct sequence, and with the appropriate delays, to perform this safety function properly.

6.2 *Molecular drag pumps*

'Molecular drag pumps', which are also often referred to as 'drag pumps' or MDPs, might be considered to be both the oldest and the newest type of high vacuum pump. Although they were first described prior to 1915, several years before oil diffusion pumps were introduced, it is only since 1985 that they have been refined to the point where they have begun to find significant practical use. Interestingly, molecular drag pumps are often cited as being the forerunners of turbomolecular pumps, yet it was the extensive developmental work on turbomolecular pumps that made it possible to bring molecular drag pumps to a state of practical usefulness. These pumps are of such recent development that most books on vacuum methods only mention them briefly as forerunners of turbomolecular pumps. However, Van Atta (p. 205) develops the basic theory for their function in both the viscous and molecular flow ranges in some detail and describes several early designs.

6.2.1 Function and design

The basic similarity between molecular drag pumps and turbomolecular pumps is that both use high-speed rotors to transfer momentum in a preferred direction to gas molecules; otherwise, their design, construction, and performance characteristics are appreciably different. Fig. 6.10 is a highly simplified diagram which can be used to explain the characteristics of molecular drag pumps. Here, the stator is a hollow cylinder which also forms the body of the pump, while the rotor, which turns at a very high speed, is a solid cylinder that is mounted inside the stator and is concentric with it. The stator contains a groove, or pumping channel, which runs from the inlet to the outlet of the pump. Gas molecules that wander into the inlet port by thermal motion are carried along this groove by momentum imparted to them by the rotor, and are forced to exit from the pump via the outlet port by an extension of the stator which blocks the groove in the region between the two ports. Except along the groove, the mechanical clearance between the rotor and stator is made as small as possible to minimize leakage of gas laterally out of the groove.

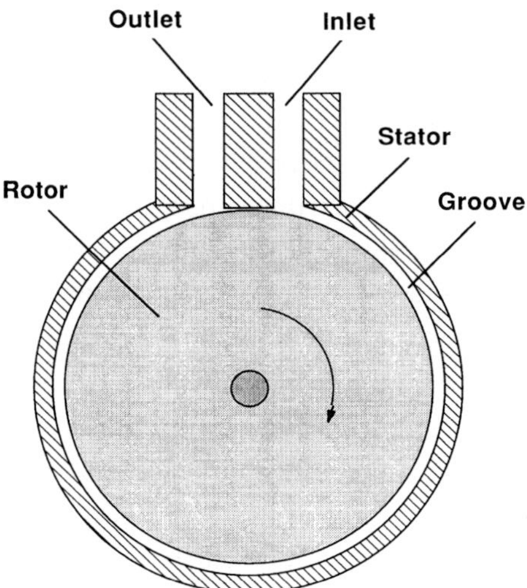

Fig. 6.10 A schematic diagram of a simple molecular drag pump.

Like turbomolecular pumps, molecular drag pumps function best at pressures where gas flow is molecular in character and the mean free path of the gas molecules is significantly greater than the depth of the groove in the stator. At these pressures, gas molecules that strike the moving rotor bounce off it with an added component of momentum in the direction in which the surface of the rotor is moving and travel to the walls of the groove without encountering other gas molecules. They then ricochet off the walls of the groove and ultimately strike the rotor again, whereupon the process is repeated until they reach the outlet port of the pump. Under these conditions, the compression ratio K of a simple pump of this kind has the following functional dependence:

$$K = \frac{P_o}{P_i} \propto \exp.\left(C \frac{sL}{d} \sqrt{M} \right) \tag{6.7}$$

Here P_i and P_o are the pressures at the inlet and outlet ports, respectively, and M is the molecular mass of the gas. L is the length of the groove and d is its depth, while s is the linear speed with which the surface of the rotor moves. C represents a collection of terms which can be considered to be constant for our purposes. This equation indicates that high rotor speeds and shallow grooves are needed to obtain high compression ratios. The basic relationship for pumping speed S is as follows:

$$S \propto sA \tag{6.8}$$

where A is the cross-sectional area of the groove, and so high rotor speeds and grooves with large cross-sectional areas are needed for high pumping speeds. Since it is physically difficult to design a groove that has both a small depth and a large cross-sectional area, it is difficult to design pumps that have both high pumping speeds and high compression ratios. A major factor limiting compression ratios is the leakage of gas laterally out of the pumping groove, a phenomenon that can be reduced by minimizing the clearance between the rotor and the stator. Both pumping speeds and compression ratios are independent of pressure during operation in the high vacuum range, and both vary with molecular mass in essentially the same manner as described for turbomolecular pumps in Sections 6.1.2 and 6.1.5 above.

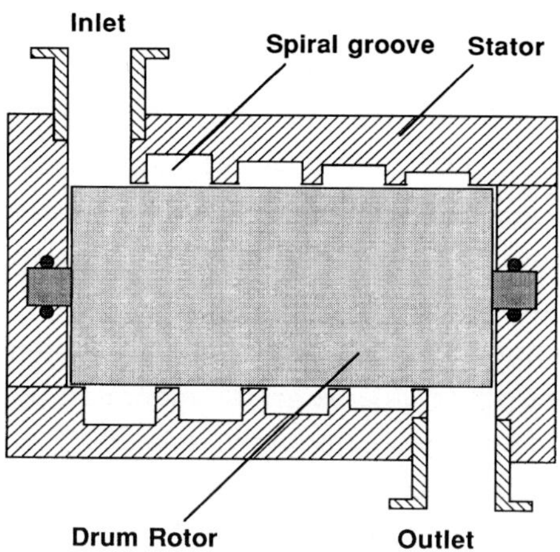

Fig. 6.11 A schematic diagram of a drum-type molecular drag pump with a spiral groove of variable depth.

Theoretical calculations, and experimental results, show that a simple pump of this kind with a groove of the order of 5 mm deep and 500 mm long, operating at about 10^4 rpm, can develop a compression ratio greater than 10^5 for nitrogen. While this is sufficient to produce pressures well into the high vacuum range, pumping speed is very limited, typically to less than 10 l/s. Several more sophisticated designs have been used to increase compression ratios and pumping speed. One that has proved very successful involves a long cylindrical rotor surrounded by a matching stator with a spiral groove, as shown schematically in Fig. 6.11. This design makes it possible to use a rotor of small diameter, which can easily be run at the high speeds required for optimal performance, while achieving the long groove length necessary for high compression ratios. Pumps of this kind are commonly called drum-type molecular drag pumps to distinguish them from early disk types. One further important refinement in design involves the use of several parallel grooves, because the pressure difference between adjacent grooves is always small and leakage of gas out of the grooves laterally is greatly reduced. This makes it possible to achieve compression ratios several orders of magnitude greater than can be obtained with single-groove pumps. It is also advantageous to vary the depth of the grooves, using a large depth near the inlet of the pump to enhance pumping speed and a smaller depth near the outlet to increase compression ratios.

As noted earlier, CIT/Alcatel pioneered in refining the design and construction of molecular drag pumps in the late 1970s to the point where a useful model could be introduced commercially. Several other manufacturers now produce similar pumps. Fig. 6.12 is a schematic cross-sectional diagram of a pump of this kind manufactured by Balzers. This is a five-stage pump, in which the rotor consists of two concentric hollow drums that intermesh with two drum-shaped extensions on the stator. Both sides of these stator drums and the chamber wall have grooves in them. These five pumping stages acting in series greatly reduce gas leakage through the pump and allow compression ratios of the order of 10^7 to be obtained with a relatively large separation (0.3 mm) between the rotor and stator surfaces. Pumping speed is enhanced by a set of impeller blades attached to the top of the rotor. These features can be seen in the detailed, three-dimensional, drawing shown in Fig. 6.13.

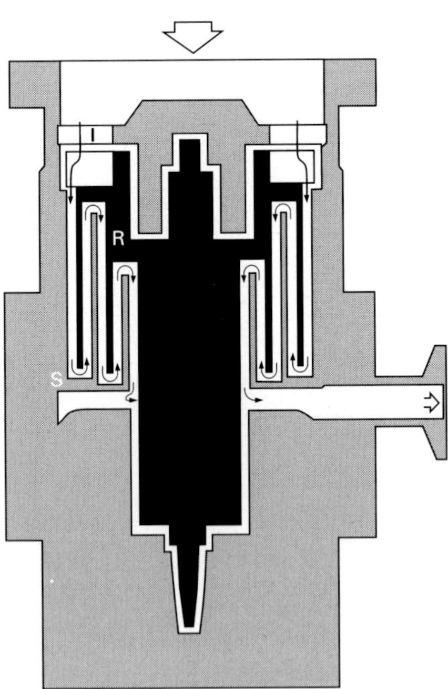

Fig. 6.12 A cross-sectional diagram of a multi-stage molecular drag pump. I, impelller; R, rotor; S, stator. Arrows show the path of gas flow. (Courtesy of Balzers.)

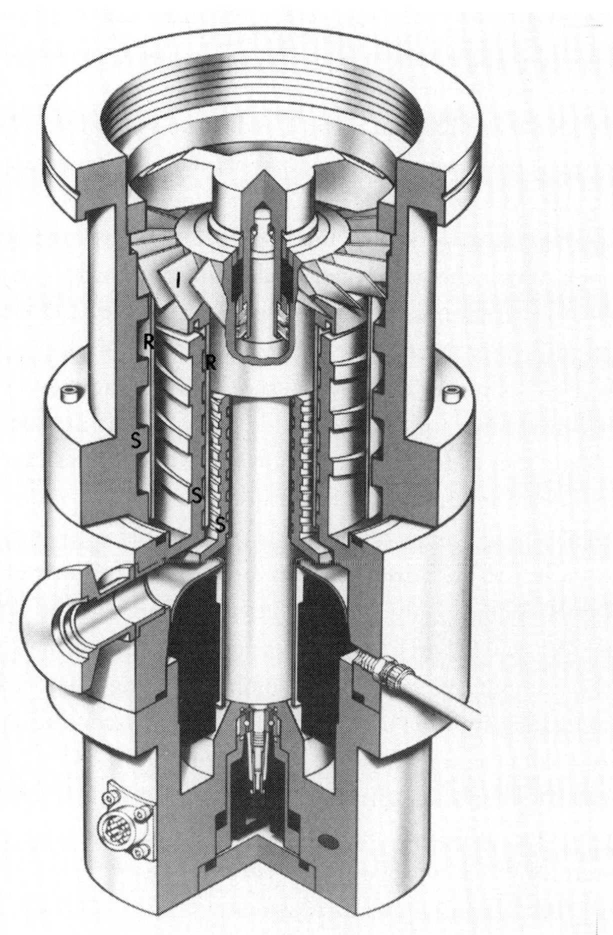

ig. 6.13 A drawing showing the details of the construction of a multi-stage molecular drag pump. I, impeller; R, rotor; S, stator. (Courtesy of Balzers.)

.2.2 Performance characteristics

One interesting characteristic of molecular drag pumps is their ability to maintain maximum pumping speed at significantly higher pressures than oil diffusion pumps and turbomolecular pumps. As shown in Figs 5.4 (Section 5.2, p. 177) and 6.4 (Section 6.1.5, above), both oil diffusion pumps and turbomolecular pumps have maximum pumping speeds that remain constant for all values of inlet pressure below about 10^{-1} Pa (10^{-3} Torr). As the inlet pressure increases above this value, however, there is a rapid drop in pumping speed caused by overloading of these pumps. Molecular drag pumps exhibit a similar constant maximum speed at low pressures, but do not show a significant loss of speed until the inlet pressure rises to near 10 Pa (0.1 Torr), and retain as much as 10 percent of their maximum speed even at 100 Pa (1 Torr). This occurs because conditions of molecular flow, which are critical to the pumping mechanism of all these types of pumps, persist to much higher pressures in the small grooves through which the gas moves in the molecular drag pumps, than in the more open structures of the other pumps (see Fig. 2.1, Section 2.2, p. 31). Practically, this means that molecular drag pumps can provide useful, high-speed pumping at inlet pressures two to three orders of magnitude higher than either turbomolecular or oil diffusion pumps can tolerate.

The other characteristic of molecular drag pumps that makes them of great current interest is the high outlet pressures they develop. Typically, molecular drag pumps produce outlet pressures in the range from 1000 to 5000 Pa (10 to 50 Torr), whereas both oil diffusion pumps and turbomolecular pumps only sustain outlet pressures in the range from 10 to 50 Pa (0.1 to 0.5 Torr). Again, this arises because of the small dimensions of the pumping grooves in the molecular drag pumps. Even when the pressure rises into the range of viscous flow, viscous interactions occur in these small grooves which are sufficient to produce a significant pumping action. The important practical significance of this is that it is not necessary to use oil-sealed pumps to back molecular drag pumps. Instead, various combinations of oil-free piston and diaphragm pumps can be used, giving fully oil-free pumping systems.

These advantages of molecular drag pumps are somewhat offset by two characteristic limitations: they have appreciably lower pumping speeds and significantly higher ultimate pressures than other high vacuum pumps. The low pumping speeds arise because of the limited conductance of the small grooves that are a critical feature of these pumps. Even when stators with several variable-depth grooves are used, it is difficult to attain maximum pumping speeds greater than about 50 l/s. These speeds are indeed small compared to those of turbomolecular and oil diffusion pumps, which range in the hundreds and thousands of litres per second. The best ultimate pressures reported for molecular drag pumps are in the 10^{-5} Pa (10^{-7} Torr) range, several orders of magnitude greater than the ultimate pressures attainable with most other types of high vacuum pumps. This rather modest performance apparently results from high levels of outgassing from components within the molecular drag pumps and a limited ability to handle the evolved gases because of their low pumping speeds.

There is not a great deal of information available on commercial models of molecular drag pumps. At present, only Alcatel, Balzers, and Osaka Vacuum list this type of pump in their catalogues. The basic Alcatel model (MDP 5010) has an inlet diameter of 63 mm, and develops pumping speeds of 7.5, 4.0, and 3.0 l/s, and compression ratios of 10^9, 10^4, and 10^3, for nitrogen, helium, and hydrogen, respectively. It can produce an ultimate pressure below 10^{-4} Pa (10^{-6} Torr) with a backing pressure of 4000 Pa (40 Torr), and can safely be operated continuously at inlet pressures up to 10 Pa (10^{-1} Torr). The construction of this pump is based heavily on technology developed for turbomolecular pumps, and so, like many models of turbopumps, it has a rotational speed in the range from 25 000 to 30 000 rpm and employs grease-lubricated bearings which permit it to be mounted in any desired orientation (even upside down) in use. Alcatel also produce a larger model (MDP 5030 with an inlet diameter of 100 mm) which has generally similar characteristics, but a rated speed of 27 l/s for nitrogen. This pump also contains features that make it suitable for handling many of the corrosive gases used in chemical and semiconductor processes. The Model TPD-020 molecular drag pump produced by Balzers, which is shown in Fig. 6.13, has an inlet diameter of 63 mm, and provides compression ratios of 5×10^7, 5×10^3, and 6.5×10^2, and pumping speeds of 18, 12, and 9 l/s, for nitrogen, helium, and hydrogen, respectively. It has rated operating and standby speeds of 90 000 and 60 000 rpm, and can produce ultimate pressures in the 10^{-5} Pa (10^{-7} Torr) range when backed by a single diaphragm pump with a capacity of about 8 l/min. Its critical backing pressure is 1300 Pa (13 Torr). The Model TS440 helical groove pump produced by Osaka Vacuum has an inlet diameter of 160 mm and a rotational speed of 19 800 rpm, and provides pumping speeds of 440 and 320 l/s, and compression ratios of 2×10^6 and 1×10^3 for nitrogen and hydrogen, respectively. Its critical backing pressure is 2000 Pa (15 Torr), and its ultimate pressure is 7×10^{-4} Pa (5×10^{-6} Torr). The limited information currently available indicates that service life and maintenance requirements for these pumps are very similar to those for turbomolecular pumps.

6.2.3 *Applications of molecular drag pumps*

It is, of course, possible to use molecular drag pumps, with rotary-vane backing pumps, as the principal high vacuum pumps on vacuum systems in the same manner as oil diffusion and turbomolecular pumps. There is little advantage in doing this, however, because turbomolecular pumps offer much higher pumping speeds and much lower ultimate pressures at about the same initial cost, and otherwise have characteristics similar to those of molecular drag pumps. Instead, molecular drag pumps are being used principally in two other ways that take advantage of their tolerance for high inlet pressures and their unusually high outlet pressures.

6.2.3a *Hybrid pumps*

One innovative use of molecular drag pumps involves combining a turbomolecular segment with a molecular drag segment on a common rotor to form a hybrid turbomolecular-molecular drag pump of the type shown in Fig. 6.14. This is a very effective design, because the resulting hybrid pump has the best characteristics of both the turbomolecular and the molecular drag segments. The input characteristics are determined by the turbomolecular segment, and so the pumping speed of the hybrid pump is comparable to that of a turbopump of the same inlet diameter, and many times greater than that of a standard molecular drag pump. The hybrid pump also has the potential for attaining an ultimate pressure much lower than can be attained by either standard type of pump. This arises because the input pressure of the molecular drag segment, which could easily be as low as 10^{-2} Pa (10^{-4} Torr), becomes the effective backing pressure for the turbomolecular segment. Since the turbomolecular segment can easily produce a compression ratio K of the order of 10^8, then according to eqn. 6.2 its inlet pressure P_i, which is the ultimate pressure P_u of the hybrid pump, can reach the 10^{-10} Pa (10^{-12} Torr) range (i.e. $P_u = P_i = P_{backing}/K \approx 10^{-2}/10^8 \approx 10^{-10}$). This is a significantly lower ultimate pressure than is normally attainable with standard turbomolecular and molecular drag pumps. Perhaps the most important characteristic of the hybrid pump, however, is the high outlet pressure it can develop. This pressure is produced by the molecular drag segment, and typically has a maximum value above 1000 Pa (10 Torr). This is much greater than the outlet pressure produced by a standard turbopump, and is high enough so that the hybrid pump can be backed by a diaphragm pump, rather than a rotary-vane pump, for totally

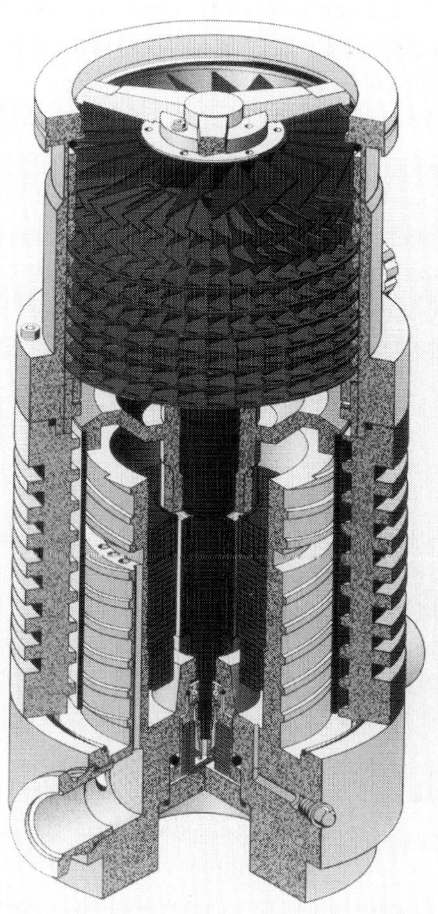

ig. 6.14 A drawing showing the details of the construction of a hybrid turbomolecular-molecular drag pump. (Courtesy of Balzers.)

oil-free pumping. The molecular drag segment also brings to the hybrid pump its characteristic high throughput in the pressure range from 1 to 10^2 Pa (10^{-2} to 1 Torr), and so the time required to make the transition from viscous to molecular flow conditions during pumpdown operations is significantly less than for a standard turbomolecular pump of comparable rated speed.

Hybrid pumps of this kind are relatively recent developments. Alcatel have produced one such pump (Model ATS) with a rated speed of 165 l/s since about 1985. Balzers only recently announced the pump shown in Fig. 6.14 (Model 180 H) which has a rated speed of 180 l/s. Osaka Vacuum also recently introduced hybrid pumps with speeds of 200, 550, 1000, 1300, and 1800 l/s for nitrogen. At the present, the cost of these pumps is about one and a half times that of a standard turbomolecular pump of comparable rated speed. This extra cost can easily be justified by the several advantages these pumps offer in many applications. In particular, when used in combination with a diaphragm backing pump they make it possible to assemble a totally oil-free pumping system with a significantly higher throughput and a potential for reaching much lower ultimate pressures than can be achieved with standard turbomolecular pumps of comparable size.

6.2.3b *Oil-free combination pumping units*

To date the most important and the most widespread use of molecular drag pumps is as components of oil-free combination pumping units. Typically, these units consist of one of the standard molecular drag pumps connected in series with two or three diaphragm pumps (Section 4.2.3, p. 161), all compactly assembled in a single cabinet which also contains all necessary electrical power supplies and controls. The molecular drag pump serves as the input module, while the diaphragm pumps, which are connected in a series or parallel configuration, serve as the backing pump for the molecular drag pump and as the output for the combination unit. Currently Alcatel produce one combination unit (Drytel 30) based on their MDP 5010 molecular drag pump and another (Drytel 100) based on the MDP 5030 pump. Danielson Associates manufacture 10 different models under the Tribodyn trade name. Four of these are based on the Alcatal MDP 5010 pump, three on the Alcatel MDP 5030 pump, and three on a molecular drag pump with an ultimate pumping speed of 12 l/s obtained from an unrevealed source. Fig. 6.15 is a photograph of one of the Tribodyn units.

The pumping characteristics of these combination diaphragm-molecular drag pump (DP/MDP) units are an interesting mixture of the characteristics of the individual pumps that comprise them. Performance at pressures below about 10 Pa (10^{-1} Torr) is largely determined by the characteristics of the molecular drag pump used in the unit. Since there is only a limited number of molecular drag pumps currently available, performance

Fig. 6.15 A combination molecular drag-diaphragm pump unit. (Courtesy of Danielson Associates, with permission.)

at low pressures is relatively limited. Thus, the ultimate pressure for all units is in the 10^{-5} Pa (10^{-7} Torr) range, which is typical for all current molecular drag pumps. DP/MDP units based on the Alcatel MDP 5010 pump all have a maximum pumping speed of 7.5 l/s, while units based on the MDP 5030 pump all have a maximum speed of 27 l/s. However, pumping speeds at pressures above about 1 Pa (10^{-2} Torr) are determined largely by the diaphragm pumps used in the units, and it is here that a considerable range of performance characteristics become possible. Fig. 6.16 is a plot of data published by Danielson Associates for the four Tribodyn units they manufacture using the MDP 5010 molecular drag pump. As just noted, all these units have the same maximum speed of 7.5 l/s at low pressures. However, pumping speeds at pressures above 1 Pa (10^{-2} Torr) differ widely, because of differences in size of the diaphragm pumps and in the way they are interconnected (i.e. in parallel versus in series, etc.).

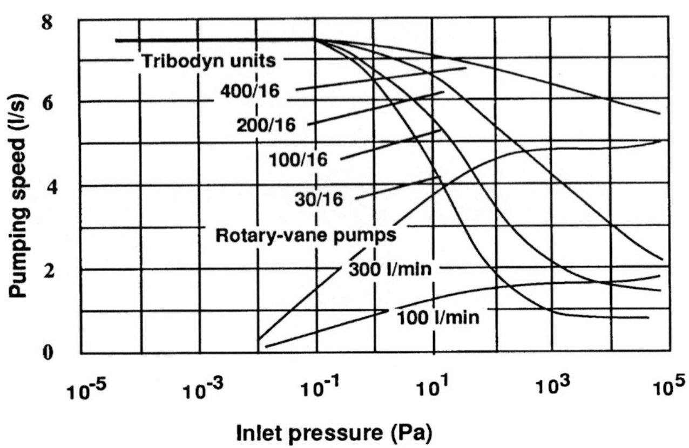

Fig. 6.16 The variation of pumping speed with inlet pressure for the series of Tribodyn compound pump units based on the Alcatel MDP 5010 molecular drag pump, and for two rotary-vane pumps. (From data of Danielson Associates, with permission.)

Since their diaphragm pumps work effectively at atmospheric pressure, these DP/MDP units can be used to evacuate systems from atmospheric pressure to pressures in the high vacuum range without a separate backing pump. However, their performance characteristics when pumping down from atmospheric pressure are markedly different from those of rotary-vane pumps, the pumps most commonly used for this purpose. As shown in Fig. 6.16, and discussed in Section 4.1.2 (p. 139), rotary-vane pumps exhibit maximum pumping speed at atmospheric pressure and rapidly lose speed as the pressure drops below about 10^2 Pa (1 Torr). Conversely, these combination pump units have rather low speeds near atmospheric pressure where the diaphragm pumps carry most of the pumping load, but show a rapid increase in speed as the pressure drops below 10^2 Pa (1 Torr) and the molecular drag pump begins to make a significant contribution. The consequence of this is that it may require somewhat longer to rough pump systems from atmospheric pressure to 10^2 Pa (1 Torr) with these DP/MDP units than with rotary-vane pumps. This is not totally disadvantageous, however, because it provides 'natural' soft-start rough pumping (see Section 5.7.1b, p. 207), which is usually desirable to prevent particles from being blown around inside the vacuum system. As the pressure drops further and approaches 1 Pa (10^{-2} Torr) the DP/MDP units reach nearly maximum speed, while the speeds of rotary-vane pumps typically fall to about one-half their maximum value. The performance of the combination units in this range is quite remarkable. For example, the four Tribodyn units characterized in Fig. 6.16 have speeds near 7 l/s, which is about 420 l/min, at 1 Pa (10^{-2} Torr). It would require a rotary-vane pump with a rated speed (at 1 atm) of nearly 1000 l/min to provide a comparable speed. This is a very large rotary-vane pump — about the largest made, in fact, and much too large for most laboratory apparatus. On the other hand, the two smallest of these Tribodyn units are only about 300 mm square and 200 mm high, and weigh less than 25 kg. Because the molecular drag pumps in these units can operate at pressures above 10 Pa (10^{-1} Torr) for several hours without damage, these DP/MDP units can be used effectively to rough pump rather large systems.

The size of the diaphragm pumps used in these DP/MDP units is a very important feature in their design. The power supplies of the molecular drag pumps become overloaded and automatically shut off if these pumps are exposed to inlet pressures above about 500 Pa (5 Torr) for more than about 10 min. Therefore, to permit convenient, valveless, single-switch operation, the diaphragm pumps must be large enough to evacuate the system to this pressure before an overload condition develops. Furthermore, a large fraction of the time involved in pumping a system down from atmospheric pressure to the high vacuum range is taken up by the initial stage of the process, when the diaphragm pumps are carrying the load. Once the pressure drops to about 100 Pa (1 Torr) and the molecular drag pump takes over, the evacuation proceeds more rapidly. Since it is annoying and inconvenient to wait for a seemingly endless period for the rough pumping operation to be completed, different sizes and configurations of diaphragm pumps are used in the various DP/MDP units to make them suitable for use on differently sized vacuum systems. Since this is an important and unique feature of the DP/MDP units, it is usually reflected in their ratings and model designators. For example, the Tribodyn 30/16 unit will evacuate systems with volumes up to 30 l (e.g. a bell jar 300 mm in diameter and 600 mm high) from atmospheric pressure to below 10^{-1} Pa (10^{-3} Torr) in a 'reasonable' time of about 10 min, while the Tribodyn 400/16 unit will evacuate systems with volumes up to 400 l (e.g. a tank 600 mm in diameter and 1400 mm high) in a 'reasonable' time of about 30 min. (An acceptable or 'reasonable' pumpdown time for large systems is naturally somewhat longer than for small ones.) The number 16 in these model designators indicates that the maximum pumping speed of the units is 16 cfm; that is, they are based on the MDP 5010 molecular drag pump which has a maximum speed of 16 cfm (7.5 l/s or 450 l/min). Units of cubic feet per minute (see Table 2.1 in Section 2.1, p. 28), and the abbreviation 'cfm', are used here because these are the conventions followed in manufacturing industries in the United States where these DP/MDP units presently find greatest use. Monetary considerations usually prevail over scientific rigour. The largest Tribodyn pump, Model 500/57, is capable of evacuating a 500 l system from atmospheric pressure to below 10^{-1} Pa (10^{-3} Torr) in less than 30 min, and has a maximum speed of 57 cfm (27 l/s or 1600 l/min, provided by the MDP 5030 molecular drag pump). Likewise, the Drytel 30 and Drytel 100 units are designed for systems with volumes up to 30 and 100 l, respectively. Timers are available for most units that can be used to delay the startup of the molecular drag pump until after the diaphragm pumps have been in operation for a selected period of time, making it possible to use the units on larger systems by prolonging the initial stage of the pumpdown process.

As noted above, these compound pump units are surprisingly light and compact. Most of them can be mounted in any orientation, even upside down, and most are light and rugged enough so that they can be suspended directly from a sturdy apparatus. Since no valves, traps, or baffles are required, conductance losses can be minimized and the full pumping speed of the unit can be utilized directly to evacuate the system. Most of the units are designed for 'single-switch' operation in which the molecular drag and diaphragm pumps are turned on and off simultaneously by a single main power switch. Because of the wide range of pumping characteristics they provide, these DP/MDP units have a number of interesting uses. Obviously, they can serve as complete, oil-free pumping systems, capable of evacuating apparatus from atmospheric pressure well into the high vacuum range. Use in this way is most advantageous when oil-free pumping is critical, when it is not necessary to reach pressures below the 10^{-4} Pa (10^{-6} Torr) range, and when the relatively low pumping speeds of these units in the high vacuum range can be tolerated. Applications of this kind fall into two principal categories: evacuation of apparatus with a small volume where high pumping speeds are not required, and evacuation of systems of limited conductance which inherently do not justify high pumping speeds. Examples of the first category include the evacuation of desk-top evaporators, portable leak detectors, and small vacuum ovens. Examples of the second category include industrial systems where several objects, such as television tubes, are evacuated simultaneously through a manifold of low conductance. They could, for example, be used advantageously to ensure oil-free evacuation of the specimen airlock chamber on electron microscopes, which are usually pumped through long tubes of small diameter, or for pre-pumping photographic film.

Interestingly, the most advantageous use of these units is in applications where they can serve as oil-free roughing pumps. They are widely used to provide oil-free backing for turbomolecular pumps. In fact, Danielson Associates add turbomolecular pumps to their various Tribodyn units and market the resulting combination under the Tandem trade name. The DP/MDP units are particularly useful for providing oil-free roughing of systems that are pumped into the ultra-high vacuum range by ion (Section 7.1, p. 276) or cryogenic pumps (Chapter 8). One of the weakest features in the design of current high-resolution transmission electron microscopes is the use of oil diffusion pumps to evacuate the column before the ion pump is turned on, because this procedure introduces the strong possibility of contaminating the column with pump oil. Replacing the diffusion pump with one of these combination units would completely eliminate this danger. The DP/MDP units are recommended for pumping on cryogenic and sorption pumps during regeneration treatments. Since they have high pumping speeds at intermediate pressures and do not cause condensation of water vapour they are better than rotary-vane pumps for rough pumping operations that involve handling large amounts of water vapour, such as, for example, desiccating photographic film for use in electron microscopes.

6.3 Concluding remarks

Turbomolecular pumps represent a very attractive alternative to cryogenic, ion, and oil diffusion pumps for many applications in electron microscopy. Although early turbomolecular pumps had a number of serious shortcomings, those presently produced by most major manufacturers are highly reliable, have excellent performance characteristics, and afford a number of advantages over other types of high vacuum pumps. They are available in a convenient range of sizes with control units that provide safe, easy, nearly-automatic operation. They are capable of operating at inlet pressures approaching 100 Pa (1 Torr) without being damaged, at least for short periods of time, yet can produce ultimate pressures in the ultra-high vacuum range in systems that are properly designed and managed. Because smaller turbopumps reach operating speed in a few minutes and tolerate exposure to high pressures well, they are ideally suited for use in valveless vacuum systems. Such systems are fast, convenient, and nearly foolproof to use, and are becoming common features of compact vacuum evaporators and similar apparatus. If venting and roughing operations are carried out properly, turbopumps can provide an oil-free, ultra-clean vacuum without the use of baffles or liquid nitrogen traps. These features are being steadily enhanced by the rapidly developing technology of using turbopumps in combination with molecular drag units. The principal disadvantages of turbomolecular pumps are their higher initial cost, the need for costly, periodic replacement of bearings, and the potential for producing residual levels of mechanical vibrations. These last two disadvantages are rapidly being reduced in importance as magnetic bearings undergo refinement, although these bearings currently increase the initial cost of the pumps considerably. Turbomolecular pumps are now being used on several models of electron microscope, and on a number of other kinds of apparatus commonly found in electron microscopy laboratories, such as vacuum evaporators, ion mills, and sputter coaters. They are performing so well in these applications that they can be expected to become increasingly popular and to find even wider use in the future. In fact, one major manufacturer of scanning and transmission electron microscope has recently announced the intent to use turbomolecular pumps on most future models of these instruments.

Although molecular drag pumps were introduced considerably before turbomolecular pumps, their development and refinement to a state where they are practical and useful is so recent that they are not yet being used to their full potential. They have so many unique advantages and characteristics, however, that they will undoubtedly see wide use in the future, particularly in providing a solution to the nagging problem of developing oil-free roughing systems. We will undoubtedly see them appearing in this role on many of the next generation of electron microscopes. A custom-designed vacuum system for an ultra-high resolution transmission electron microscope that uses both turbomolecular and molecular drag pumps is described in Section 9.5 (p. 410).

7 Ion, sublimation, and bulk getter pumps

This chapter might well have been entitled 'Getter pumps', because all the pumps discussed here rely on the use of getters for their pumping action. Getters are highly active substances that remove gas molecules from the volume to be evacuated by reacting with them chemically to yield stable products with negligible vapour pressures. Because some form of chemical interaction occurs between the sorbed gas molecules and the getter material this gettering process is also called 'chemisorption' to distinguish it from the physical sorption processes involved in cryogenic pumps. Since the gases removed from the vacuum system are retained inside these pumps in or on the active gettering material, getter pumps are often classified as 'entrainment pumps'. In contrast, diffusion pumps, rotary-vane pumps, and turbomolecular pumps are called 'displacement pumps', because they move the pumped gas from the vacuum system more or less directly to the surrounding atmosphere.

Gettering was one of the first practical methods used to achieve a high vacuum. Many years before the introduction of oil diffusion pumps, which were the first effective vacuum pumps for producing a high vacuum, red phosphorus was used as a getter to reduce the pressure inside incandescent light bulbs into the 10^{-5} Pa (10^{-7} Torr) range. In this application the phosphorus was coated on the filament, and the bulb was rough pumped, baked out, and sealed off. The filament was then heated electrically, causing the phosphorus to evaporate and form a thin invisible layer on the inside of the bulb. The pressure inside the bulb was then reduced and maintained in the high vacuum range by reaction of gas molecules with this chemically-active phosphorus film. Phosphorus is still used as a getter in a few applications. Barium, a very reactive metal, is now widely used in a similar manner to maintain the vacuum inside sealed cathode ray tubes, X-ray tubes, television tubes, and similar devices. Getters have been used in the vacuum systems of mass spectrometers, gas sampling devices, and in similar instruments on several space vehicles, because they require little power to operate, have no moving parts, and have relatively low mass and volume compared to most other high vacuum pumps. Guthrie (p. 116), Roth (p. 264) and Weston

(p. 99) give interesting descriptions of commonly used getters and their applications.

The references cited in this chapter are listed in Appendix 1. The suppliers and manufacturers referred to in this chapter are listed in Appendices 2 and 3.

7.1 Sputter-ion pumps

Sputter-ion pumps are used on most recent models of electron microscopes designed specifically for high-resolution and microanalytical applications, and on many other types of vacuum apparatus, to avoid the problems of oil contamination associated with oil diffusion pumps. These pumps, which are often simply called 'ion pumps', are special types of getter pumps in which high-energy gas ions sputter the getter onto surfaces inside the pump so that a fresh film of the active material, usually titanium, is continuously available for the pumping process. Interestingly, ion pumping effects were first noted in the 1850s in early studies of gas discharge tubes. The effect was reported again by Penning in 1938 in his work on the cold cathode discharge gauge (Section 3.2.2, p. 99); however, true ion pumps were not introduced until the late 1950s when they were developed by Varian for use on microwave tubes. Although several different types of ion pumps have been developed (see Weston, Section 3.7), the only type that will be discussed here is the 'sputter-ion pump', which is also often referred to as the 'ion-getter pump', because this is the type most widely in use today. Van Atta (Section 9.6) gives an interesting discussion of the work that led to the development of these pumps, and O'Hanlon (Section 9.2.2) describes the physical principles underlying their operation in some detail.

7.1.1 Principle of operation

The simplest type of sputter-ion pump is the diode pump. The operation of this pump can be described by reference to the single Penning cell shown schematically in Fig. 7.1. This cell consists of a metal case which encloses a cylindrical anode and a pair of titanium cathode plates which are located opposite the open ends of the cylinder. The anode cylinder is usually made of stainless steel, and is maintained at a positive direct current

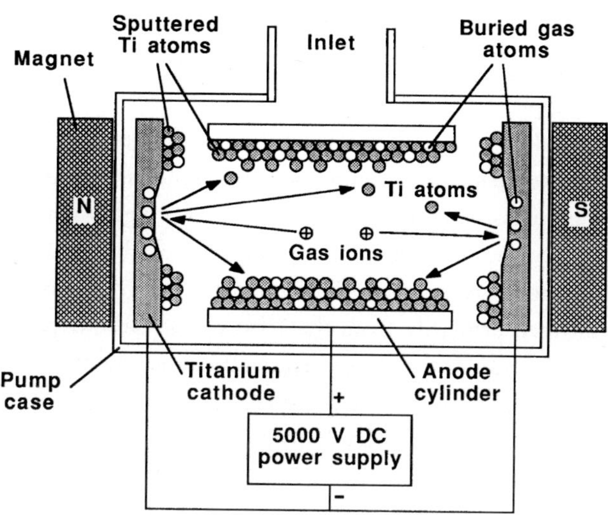

Fig. 7.1 A schematic diagram of the Penning cell of a diode sputter-ion pump depicting the sputtering process and the burial of gas atoms.

(DC) potential of from 3000 to 7000 V with respect to the cathodes and the case. A permanent magnet located outside the case provides a strong magnetic field parallel to the axis of the anode cylinder.

Although the construction of the Penning cell is simple, the processes that produce the pumping action in it are rather complex. These processes critically involve free electrons, which are initially generated by cosmic rays. These electrons are accelerated towards the anode by the large positive DC potential applied to it. Rather than travelling to the anode directly, however, the strong magnetic field that permeates the cell forces them to follow long, complex, spiral paths. As a consequence, they are 'trapped' inside the anode cylinder for a significant time and therefore have a high probability of colliding with any gas molecules present in the cell. Such collisions produce positively charged gas ions and additional electrons. These additional electrons also undergo acceleration by the DC field, and are also trapped inside the anode cylinder by the magnetic field. This trapping action can be so effective that there can be more than 10^5 high-energy electrons per cubic millimetre moving around inside the anode cylinder of a properly designed cell, even during operation at pressures in the high vacuum range. As a consequence, most gas molecules which enter the cell are struck by electrons and ionized, and many are dissociated.

The positive gas ions formed in this ionization process are drawn directly to the negative titanium cathodes, because they are so heavy that they are not appreciably deflected by the magnetic field. Under the influence of the applied potential of several thousand volts they gain sufficient kinetic energy to dislodge titanium atoms from the cathode plates. It is this ion sputtering process, which constantly deposits highly reactive titanium atoms on the interior surfaces of the pump, that is chiefly responsible for the pumping action. At pressures in the high vacuum range and lower, titanium is sputtered primarily from the areas of the cathode plates opposite the centre of the anode ring. At these pressures the rate of sputtering from areas of the cathodes around the periphery of the anode is negligible, and deposits of titanium atoms sputtered from the opposite cathode accumulate there. At pressures above the high vacuum range the sputtering action spreads outward towards the periphery of the anode cylinder and may dislodge some of this deposited titanium. At all pressures, however, most of the sputtered titanium is deposited on the interior surface of the anode cylinder. This titanium is not subsequently sputtered away, because the anode is not subjected to gas ion bombardment. The electrodynamics of the ionization, electron-trapping, and sputtering processes are such that optimal pumping occurs when the Penning cell is 10–15 mm in diameter and 15–20 mm in length.

Several phenomena are involved in this pumping action, and these contribute differently to the pumping of different gases. Any gas atom or molecule that comes to rest on a surface where titanium is being deposited can be buried there by the arriving titanium atoms. This process is called 'surface burial'. Ultimately, of course, all the gas pumped by an ion pump is buried under the sputtered titanium. Highly reactive gases, such as oxygen, nitrogen, carbon monoxide, and hydrogen, also react directly with the fresh titanium surfaces to form stable oxides, carbides, nitrides, and hydrides. This process, which is called 'chemisorption', is enhanced if the gas molecules are first dissociated by electron bombardment. Molecular dissociation is generally considered to be critically involved in the chemisorption of less reactive gases such as hydrocarbons, water, and carbon dioxide. Pumping also occurs by the mechanism of 'ion implantation', which takes place because the imposed electric potential gives some positive gas ions sufficient kinetic energy to penetrate into the surfaces of the cathode plates. While this process occurs for all gases, it is virtually the only mechanism that contributes to the pumping of the inert gases helium, neon, and argon in a simple diode pump. Finally, when some gas ions strike the cathode surfaces they acquire an electron and become neutral atoms. If these atoms rebound with enough kinetic energy they can become embedded in the surface of the anode or the opposing cathode. This process of 'neutral atom implantation' is not a prominent one in simple diode pumps, because the ions strike the cathode surfaces with such high angles of incidence that most of the neutral atoms formed are reflected with energies too low for implantation.

7.1.2 *Argon instability*

Gas atoms that have been embedded in the cathode plates of a diode pump at one time by the processes of ion and atom implantation and atom and molecule burial can later be released again as the titanium layer in which they are buried is sputtered away. This does not present a problem for the active gases since they usually react with the titanium cathodes to form compounds that are stable when subsequently sputtered, or they are quickly pumped again by the highly effective chemisorption process. The inert gases do not react to form stable compounds, however, and so the release of inert gas atoms from the cathodes, and the re-pumping of these atoms by ion implantation, occurs constantly in an operating diode pump, limiting both the pumping speed and the ultimate pressure attainable. Argon is the gas most prominently involved here, because it constitutes nearly one percent of the atmospheric gas mixture. This process also gives rise to a phenomenon called 'argon instability' in which the pressure in a pump rises due to the release of bursts of argon from the cathodes, and then recovers as the argon is pumped again. These pressure fluctuations may rise to levels as high as 10^{-2} Pa (10^{-4} Torr), and once started they generally occur at fairly regular intervals for as long as an appreciable quantity of argon is being pumped.

Argon instability usually does not occur when diode pumps are used on ultra-high vacuum systems, because these system are essentially leak-free and so there is no significant continuing load of argon to be handled. Problems can arise with ordinary systems, however, because they usually have leaks that constantly introduce argon, together with other atmospheric gases, over long periods of time. Since this phenomenon initially limited the usefulness of ion pumps, manufacturers devoted considerable effort to developing pumps which are less subject to argon instability than the simple diode pumps. This work is reviewed by O'Hanlon (p. 219) and Weston (p. 114). Two relatively simple modifications to the basic diode design have proved quite effective. Perhaps the simplest of these, and the least expensive, involves making one of the cathode plates of tantalum. Because tantalum has a much higher atomic mass than titanium (180 versus 48 g/mol), the neutral gas atoms formed when gas ions collide with the tantalum cathode rebound with sufficiently high energies that they readily become permanently embedded in the anode and in areas of the opposing cathode where titanium sublimation does not occur. In this way the mechanism of neutral atom implantation is caused to make a major contribution to the pumping of argon, and the problem of argon instability is virtually eliminated. Physical Electronics/Perkin Elmer, the pioneers of this approach, advertise that their Differential Ion Pumps, which are based on this design, can pump pure argon at pressures as high as 10^{-5} Pa (10^{-7} Torr) for several hundred hours without exhibiting instability.

The other simple approach that has been used to reduce argon instability in diode pumps involves the use of titanium cathode plates with square grooves in their surfaces. With this grooved diode design a significant fraction of the accelerated gas ions strike the walls of the grooves at low angles of incidence. If these ions are neutralized, the neutral atoms formed reflect with high energies and therefore have a high probability of becoming embedded in the bottoms of the grooves. This greatly increases the rate of neutral atom implantation, and so this mechanism makes a significant contribution to pumping the inert gases. Ions that strike the walls of the grooves also sputter titanium atoms into the bottoms of the grooves, making the gettering and burial processes very effective there. Apart from this, sputtering of titanium occurs mostly from the tops of the ridges between the grooves. Therefore, atoms that become buried at the bottoms of the grooves are effectively protected from subsequent release by the sputtering process, and so diode pumps with grooved cathodes seldom exhibit argon instability.

A third widely-used approach to improving the pumping of the inert gases involves the use of the triode pump design shown in Fig. 7.2. The cathodes in this design have an open grid-like structure, and are maintained at a negative potential with respect to both the anode and the case of the pump. With this arrangement the accelerated gas ions have a high

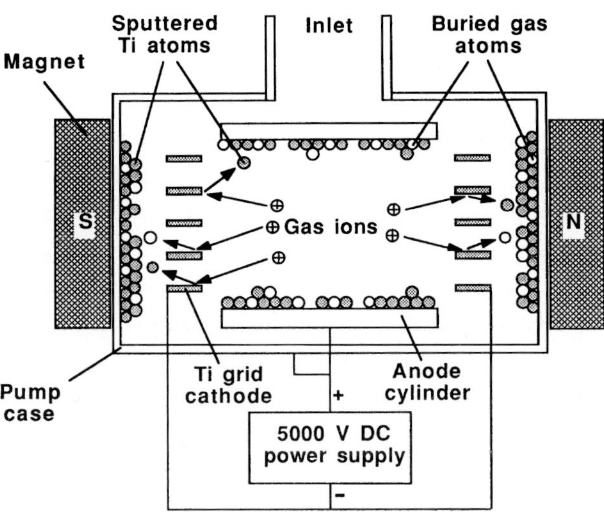

Fig. 7.2 A schematic diagram of the Penning cell of a triode sputter-ion pump depicting gas ions striking the cathode grid bars at low angles of incidence forming high-energy neutral atoms and sputtering titanium onto the pump case.

probability of striking the sides of the grid bars with a small angle of incidence. This enhances the sputtering efficiency, virtually eliminates ion and atom implantation in the cathodes, and causes most of the sputtered titanium atoms to be deposited on the interior walls of the pump case where the gettering and surface burial processes occur as described above. In addition, most ions that are neutralized during collision with the cathode grid leave the grid with high energies, because of the low angles of incidence and reflection involved, and become embedded in the titanium layer that covers the inside of the pump case with a high degree of efficiency. This greatly enhances the pumping of the inert gases by causing the mechanism of neutral atom implantation to make a significant contribution. It also virtually eliminates argon instability, because the case of the pump, which is at the same potential as the anode, is not subjected to ion bombardment and sputtering.

7.1.3 *Pumping speed*

Sputter-ion pumps are usually rated on the basis of the maximum pumping speed they develop for air. The pumping speed of an individual Penning cell is determined by its geometry and dimensions, the value of the DC potential, and the strength of the magnetic field. As noted above, optimization of these variables severely limits the acceptable cell size, and consequently the pumping speed for a single Penning cell is only of the order of 1 or 2 l/s. Higher pumping speeds are obtained by assembling a number of cells into an array, as indicated schematically in Fig. 7.3. However, this approach is limited to arrays with 30–60 cells, which have pumping speeds of 30–60 l/s, because the gas conductance to the interior cells of larger arrays is very limited. Pumps with speeds greater than about 50 l/s are usually produced by assembling several such arrays around a common inlet chamber in an arrangement that optimizes the access of gas molecules to all the cells in all the arrays and minimizes the size and number of magnets needed. Fig. 7.4 is a photograph of an ion pump that has 10 such arrays assembled around a large cylindrical inlet chamber. Each array contains about 30 individual Penning cells and provides a pumping speed of about 30 l/s, giving a total pumping speed of 300 l/s. The magnets are located between adjacent cell arrays, which share their fields, so that only 10 magnets are needed for the 10 array units. Pumps of this type are widely used on surface analysis instruments and similar ultra-high vacuum apparatus. While the pumping speeds of most other types of high vacuum pumps are directly related to their inlet diameters, as shown by Fig. 5.3

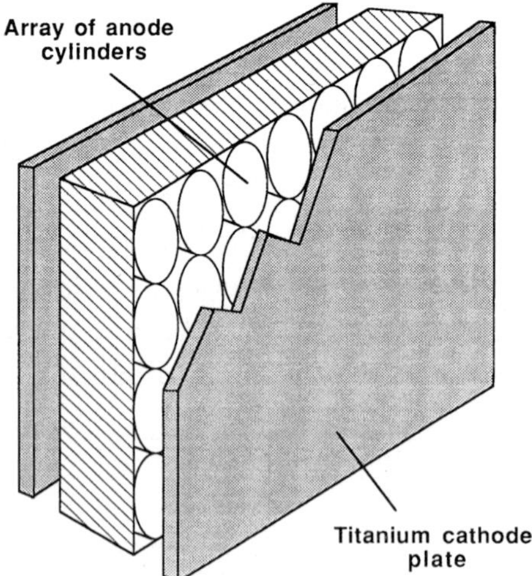

Fig. 7.3 The arrangement of Penning cells into an array to achieve high pumping speeds.

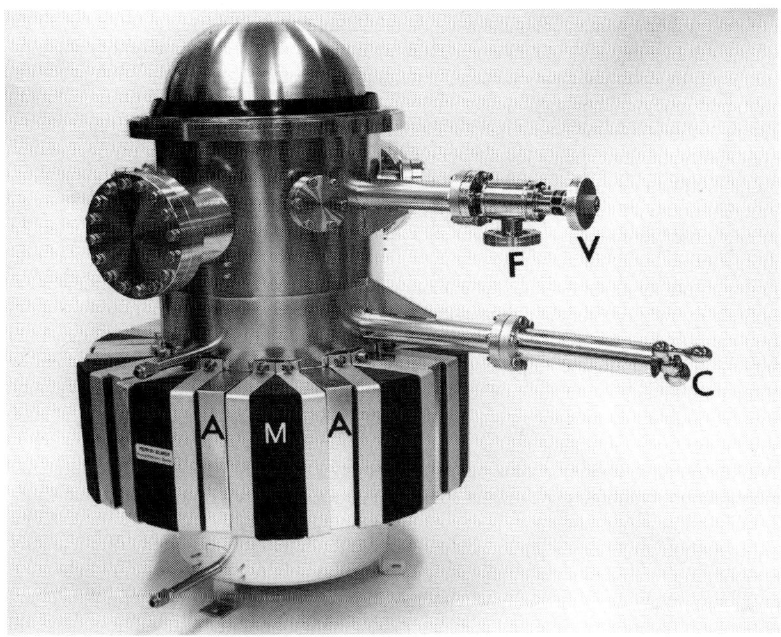

Fig. 7.4 A versatile, high-speed, compound pumping unit consisting of 10 ion pump arrays (A) and associated magnets (M) assembled around a large cylindrical inlet chamber which also contains a titanium sublimation pump and a liquid nitrogen baffle. The upper appendage provides a flange (F) and a valve (V) for attaching a roughing pump. The crank (C) operates a large isolation valve. (Courtesy of the Physical Electronics Division of the Perkin Elmer Corporation.)

(Section 5.2, p. 175), Fig. 6.5 (Section 6.1.5, p. 239), and Fig. 8.3 (Section 8.6, p. 349), the speeds of ion pumps depend more critically on the number and arrangement of the Penning cell arrays that are attached to their inlet manifolds. For example, ion pumps with 150 mm diameter inlet ports are available with pumping speeds of 60, 80, 120, 150, 220, 270, 400, and 500 l/s. Most manufacturers do not produce standard models of ion pumps with rated speeds greater than 1000 l/s.

Ion pumps usually exhibit maximum pumping speed when operating at an inlet pressure near 10^{-4} Pa (10^{-6} Torr) where conditions for the ionization and sputtering processes are optimal. At lower pressures pumping speed decreases gradually, as shown in Fig. 7.5, because the number of gas ions produced per second decreases, causing the sputtering rate to decrease. At pressures above 10^{-4} Pa (10^{-6} Torr) pumping speed decreases rapidly, becoming essentially zero near 10^{-1} Pa (10^{-3} Torr), because the

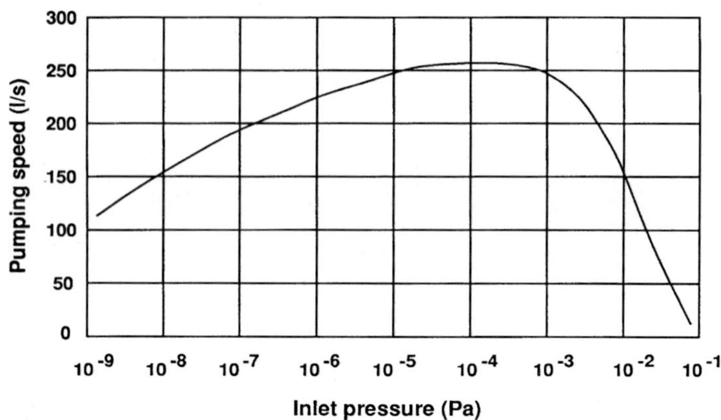

Fig. 7.5 A curve showing how pumping speed for air typically varies with inlet pressure for diode sputter-ion pumps with rated speeds of about 250 litres per second.

concentration of gas molecules becomes so large that interactions between the gas ions and the gas molecules seriously interfere with the sputtering process.

Pumping speed varies considerably for different pure gases. Active gases, such as nitrogen, carbon monoxide, carbon dioxide, water, and low molecular mass hydrocarbons, are pumped with about the same speed as air. The speed for pure oxygen is usually about one-half that for air, apparently because the cathodes become covered with a resistant oxide film that reduces the rate of sputtering of titanium. The problems associated with pumping argon and the other inert gases have been discussed above. Typically, the pumping speed for argon is less than 5 percent of that for air in simple diode pumps, and only from 10 to 20 percent of that for air in the modified diode and triode pumps. Pure hydrogen is pumped more than twice as fast as air, because the hydrogen molecules have such high speeds that they diffuse into the pump more rapidly than molecules of most other gases, and because they react so vigorously with titanium. When hydrogen molecules come in contact with clean titanium they apparently dissociate into atoms, which then diffuse rapidly into the titanium and react with it to form aggregates of atoms that resemble titanium hydride in composition and crystal structure. This process distorts the titanium crystal lattice. Although the distortion is negligible in most applications, it can lead to macroscopic warping, cracking, and flaking of the cathode plates sufficient to cause pump failure, particularly when large amounts of hydrogen or water are pumped. Because of this, most manufacturers produce special pumps with heavy, reinforced cathodes for use in such applications. When a mixture of gases, such as air, is pumped, the overall speed is a complex average of the speeds of the individual gases in the mixture.

There is an interesting secondary benefit that can be derived from the fact that ion pumps exhibit different pumping speeds for different gases; that is, ion pumps can be used as leak detectors. When the air that normally enters a system through a leak is replaced by a gas with significantly different pumping characteristics, such as hydrogen, helium, argon, or oxygen, a small change occurs in the current flowing through the pump. Thus, it is possible to locate very small leaks by watching for such current fluctuations while probing around the exterior of a vacuum system with a fine jet of one of these gases, as described in Section 10.6 (p. 437). This capability is very advantageous because it eliminates the need for a separate, expensive leak detector, and for a valve and port for attaching it to the vacuum system. Some manufacturers make special provisions for using their pumps as leak detectors. For example, Varian sell an attachment for their pump power supplies (the DetecTorr unit) and Physical Electronics/Perkin Elmer incorporate a special circuit into their microprocessor-based control units for this purpose.

7.1.4 Ultimate pressure

Sputter-ion pumps are true ultra-high vacuum pumps. Their cases and anode cylinders are usually constructed of stainless steel, the preferred material for ultra-high vacuum systems. The titanium and tantalum used in their cathodes have vapour pressures that are well below 10^{-9} Pa (10^{-11} Torr) at 1000°C, and virtually immeasurable at room temperature. Likewise, the titanium carbides, oxides, and nitrides formed by the reactive gases have negligible vapour pressures at temperatures below 1000°C, and the buried inert gases yield similarly low vapour pressures at normal operating temperatures. Thus, ion pumps are inherently capable of reaching pressures well into the ultra-high vacuum range. To facilitate this, ion pumps are constructed so that they can be baked out at temperatures in the range from 250 to 350°C while in operation, and at temperatures above 400°C with the magnets and power cable removed. Most manufacturers list ultimate pressures for their pumps that are in the 10^{-9} Pa (10^{-11} Torr) range, and several imply that lower pressures can be reached in systems that are properly designed, constructed, and managed.

7.1.5 Design features

As implied by the above discussion, four different types of sputter-ion pumps are available. Standard diode pumps, which have two flat titanium cathodes, are the most widely used because they are the least expensive and are entirely satisfactory for most ultra-high vacuum applications. They are not recommended for systems in which large amounts of argon, helium, hydrogen, or water must be pumped, nor for systems that may have significant air leaks. The remaining three types of pumps are designed for use in these situations where the plain diode pumps are not satisfactory. Diode pumps that are modified to increase their effectiveness for pumping the noble gases helium, neon, and argon, which do not react readily with titanium, are commonly called 'noble diode pumps'. As described above, these pumps usually have one titanium and one tantalum cathode. Typically, their pumping speed for air is about 5 percent lower than that of comparable plain diode pumps because tantalum is a less reactive getter than titanium, and they have a lower capacity for pumping hydrogen and water because hydrogen is less soluble in tantalum than in titanium. Pumps designed for applications in which large amounts of water or hydrogen must be pumped are called 'hydrogen diode pumps'. These usually have two titanium cathodes which are thicker than normal to increase their total capacity for dissolving hydrogen, and which are structurally reinforced to resist the warping that can occur when a large amount of hydrogen dissolves in the titanium. These pumps are also useful in applications where a long lifetime is desirable (e.g. in inaccessible locations such as sealed nuclear reactors), or where a large total gas capacity is advantageous [e.g. operation at pressures in the 10^{-3} Pa (10^{-5} Torr) range or higher], because the heavy cathodes provide extra titanium to be consumed in the pumping process. In addition, 'triode pumps' are available from most manufacturers. These usually provide speeds for inert gases that are somewhat greater than the noble diode pumps, and are generally considered to offer the greatest freedom from argon instability. In some designs the cathode lifetime is less than that of diode pumps, because the grid cathodes do not provide as much titanium for sputtering as the solid cathodes used in diode pumps. However, Varian produce a series of pumps with a proprietary cathode design (the StarCell series) that is advertised as providing greater pump life and higher throughput than diode pumps. Triode pumps exhibit significantly higher pumping speeds at pressures above 10^{-2} Pa (10^{-4} Torr) than diode pumps, and are therefore advantageous for use on systems where easy startup or rapid pumpdown is desirable.

All four types of pumps discussed above are available in standard models that range from 'appendage' pumps, which have only a few Penning cells and speeds of 2–10 l/s, to multi-array pumps with speeds of about 1000 l/s. Some manufacturers will produce larger pumps to special order. Typical speeds for pumps found on electron microscopes and other laboratory apparatus are 10, 25, 50, 100, 150, and 200 l/s. Fig. 7.6 is a photograph of a series of pumps manufactured by the Physical Electronics Division of the Perkin Elmer Corporation. In addition to standard pumps, such as those shown in this photograph, a number of special types are also available. Some manufacturers offer pumps with two inlet ports. These 'double-ended' pumps can be used advantageously in a number of

Fig. 7.6 A photograph showing the range of sizes in which sputter-ion pumps are commonly available. (Courtesy of the Physical Electronics Division of the Perkin Elmer Corporation.)

situations. For example, on an electron microscope one port could be connected directly to the electron gun chamber and the other to the specimen chamber, thereby considerably reducing the length of pumping manifold required and increasing the pumping speed to both chambers. Combination pumps that combine sputter-ion arrays with liquid nitrogen baffles, titanium sublimation units, or non-evaporable getter elements, are also available. These interesting innovations, which provide higher pumping speeds and greater capacities than standard ion pumps for reactive and condensable gases, are discussed in more detail in Section 7.3 below.

A power supply and control unit is an essential component of a sputter-ion pump. This has to be a highly sophisticated electronic device, because of the range of power requirements it must meet and the range of control functions it must perform. At pressures near 10^{-1} Pa (10^{-3} Torr), which are encountered when a pump is started up during a pumpdown process, a high-intensity glow discharge of very low impedance (i.e. high current-carrying capacity) develops, and so the power supply must be current-limited to prevent damage to itself and to the pump. Under these conditions the voltage across the pump will be only a few hundred volts, while the current through it may approach 1 ampere. At pressures below 10^{-2} Pa (10^{-4} Torr) the discharge becomes high-impedance in character. Here a potential of several thousand volts is needed to sustain the discharge, but the current decreases to a few milli- or micro-amperes. To make it possible to monitor the operation of a pump over such a wide range of conditions, most power supplies have a meter with a multi-range, multi-function circuit and selector switch for reading either the voltage or the current output. Since current varies linearly with pressure below 10^{-2} Pa (10^{-4} Torr), this meter usually also has a scale calibrated directly in pressure units for monitoring pressure. As noted in Section 3.2.3 (p. 104), this is a very useful feature on electron microscopes that are equipped with ion pumps, because these instruments usually do not have other reliable vacuum gauges. Recently, some manufacturers have taken advantage of the versatility of microprocessors to introduce microprocessor-based power supplies with a number of special features, such as digital display of current, voltage, and pressure, automatic programming of startup and shutdown procedures, programmable setpoints for control functions, diagnostic routines, and interfaces for computer monitoring and control. Power supplies capable of operating several ion pumps simultaneously are also available from some manufacturers. All ion pump power supplies should be treated with *a very high degree of caution* and respect. These units can deliver an open circuit potential of more than 5000 V, and a closed circuit output of several tenths of an ampere at several hundred volts. As a minimum, a shock from one of these units is very unpleasant; under unfavorable circumstances, it can be lethal.

The magnets that provide the magnetic field which permeates sputter-ion pumps are also important components of the pumps, because the strength of the magnetic field they provide critically influences the pumping speed attainable. Typically, field strengths in the range from 0.15 to 0.2 tesla (1500 to 2000 gauss) are needed for optimum performance. Ceramic ferrite magnets are now used on most pumps because they are lighter than metal magnets and have very favourable magnetic properties. Nonetheless, the magnets cause sputter-ion pumps to be significantly larger and heavier than most other types of high vacuum pumps. Ion pumps of the size used on most electron microscopes, which usually have speeds of only 100–150 l/s, typically weigh from 50 to 75 kg (100 to 150 lb). In contrast, turbomolecular and diffusion pumps with inlet ports of comparable diameter, which typically have speeds near 400 and 1000 l/s, respectively, usually weigh only about 25 kg (50 lb). It is not an easy task to lift an ion pump into position, or to remove it from its place, behind the column of an electron microscope. Magnetic field leakage can also be troublesome in some applications, although manufacturers are aware of this and usually design their pumps to minimize this problem. Presumably, the pumps used on electron microscopes are of special low-leakage design. For some service operations it is necessary to remove the magnets from ion pumps. If this must be done, *great care should be exercised* in handling the magnets, because they produce such strong magnetic fields that they can be unexpectedly attracted to a nearby steel object, or to another magnet, with a force great enough to break a brittle ceramic magnet, or to seriously injure fingers that happen to get caught in the way.

Most ion pumps have built-in heaters for use in performing the bakeout procedures needed to desorb gases from the case and other inactive internal parts so that pressures in the ultra-high vacuum range can be attained. Considerable care should be exercised when using these heaters, and generally when performing a bakeout procedure on the rest of a system that has an ion pump. Most ion pumps will tolerate temperatures up to about 250°C without permanent damage, and so can remain in operation during bakeout treatments up to this level. However, power cables may be damaged and magnets may lose a significant fraction of their strength, either temporarily or permanently, at the high temperatures used for baking out many ultra-high vacuum systems. Therefore, the safest approach is to follow the manufacturer's instructions rigorously when carrying out any bakeout procedure.

7.1.6 *Operating characteristics at high pressures*

Sputter-ion pumps perform effectively only at pressures below about 10^{-2} Pa (10^{-4} Torr) where the magnetic field confines the gas discharge to the central region of the Penning cells. At higher pressures the gas discharge tends to spread, becoming a generalized glow discharge at pressures above about 10^{-1} Pa (10^{-3} Torr). If a generalized glow discharge develops in a pump its current-carrying capacity becomes very large, its electrical resistance decreases correspondingly, and it enters the low impedance mode of operation. Then the current flowing through the pump increases markedly, possibly to the maximum value allowed by the power supply, the power dissipation in the pump becomes abnormally large, and the pump can heat up rapidly. This, in turn, can cause gases such as hydrogen, argon, and helium, which are physically sorbed or which do not form highly stable compounds, to be released inside the pump, thereby further increasing the pressure and exacerbating the situation. The release of gas in this manner tends to increase as the age of a pump increases, because the film of sputtered titanium that is deposited inside the pump tends to become spongy as its thickness increases, and this increases the ease with which physically buried gases are released.

Overheating in this manner is very likely to occur when an ion pump is first started up during the process of pumping a system down from atmospheric pressure. Most manufacturers indicate that it is permissible to start their pumps at pressures as high as 1 Pa (10^{-2} Torr), but usually state that starting at pressures below 5×10^{-2} Pa (5×10^{-4} Torr) is preferred and recommended. If the pressure is low enough when a pump is started, the pressure and the pump current will decrease steadily, the voltage across the pump will rise above 2000 V in a few minutes, stable pumping will become established quickly, and there will be negligible overheating or release of gases within the pump. If less favourable conditions prevail, the pump may have difficulty handling the release of gases and a slow startup may result. If a pump is started at a pressure that is much too high, the overheating and rate of gas release can become so high that a run-away condition develops, whereupon the pump can be physically damaged if it is not turned off within a few minutes.

Overheating during startup is particularly likely to occur if a pump has previously been used to pump large quantities of water, hydrogen, or inert gases, because hydrogen and the inert gases are readily released from the titanium film by heating. Leaving a pump standing open to a humid atmosphere will also promote overheating by allowing it to adsorb large quantities of atmospheric gases, and particularly water vapour, which are subsequently released when an attempt is made to start it up. Old pumps are particularly likely to overheat during startup, because, as mentioned above, they usually contain spongy deposits of titanium which readily adsorb and release large quantities of gases.

The overheating of ion pumps during startup poses a particularly serious problem in the operation of electron microscopes. Several microscopists have privately reported the development of instabilities in their microscopes after an ion pump overheated slightly during a pumpdown process. These instabilities appear to arise from high-voltage microdischarges in the electron gun, and are presumed to be caused by contamination of the surface of the gun insulator by materials released from the overheated pump. It was even necessary to replace the electron guns on some of these instruments to restore stable operation. This problem can be very serious and quite expensive to overcome, and can put an instrument out of operation for a considerable time while repairs are being made. It is therefore advisable to exercise great care to avoid it. For this reason it appears advisable to pump electron microscopes down below 10^{-2} Pa (10^{-4} Torr) before the ion pump is first turned on. Even then, it is highly advisable to further guard against overheating by operating the pump intermittently, on for two or three minutes and then off for a similar period, until the pump voltage rises to at least 2500 V. Then continuous operation can be implemented, but the pump should still be watched closely until it is clear that the danger of severe outgassing and overheating is past. The power supplies of some pumps begin to emit an abnormal hum just as the generalized discharge begins to develop, and this can serve as a warning that trouble is imminent. John Mardinly, of the Intel Corporation, who has had considerable experience with this problem, suggests that when an ion pump is installed directly on an electron gun chamber, contamination of the high-voltage insulator with material released by the pump can be prevented by inserting a short tube with an elbow between the ion pump and the gun so that there is no direct 'line of sight' path available for the material to follow from the pump onto the insulator.

Obviously, overheating of an ion pump can also occur during normal operation if the inlet pressure rises significantly above about 10^{-2} Pa (10^{-4} Torr). This could occur from such causes as the development of a leak in a vacuum system, or from improper use of the specimen airlock system of an electron microscope. Most manufacturers do not recommend operating their pumps for extended periods at pressures above about 5×10^{-2} Pa (5×10^{-4} Torr), and so the power supplies of most ion pumps contain a safety circuit that will shut off the power to the pump if the current drawn by the pump rises significantly above the value normally required for operation at this pressure. Microscopists who have experienced problems with the ion pumps on their electron microscopes emphasize the value of such a protective circuit, and warn against accepting a power supply that does not have one. Unfortunately, this circuit will not protect against the startup problems described in the preceding paragraphs, because it must be turned off to allow the power supply to deliver the abnormally large currents needed at the beginning of the startup process. Overheating, and even a run-away condition, can also develop during a bakeout treatment if the temperature is increased so rapidly that the rate of gas desorption exceeds the ability of the ion pump to handle it.

7.1.7 *Operating characteristics at low pressures*

At pressures below 10^{-2} Pa (10^{-4} Torr) the gas ionization, sputtering, and gettering processes become well controlled, and sputter-ion pumps exhibit their normal operating characteristics. The limited electrical discharge that occurs in this pressure range has sufficient electrical resistance to support several thousand volts potential and to limit the ion current flowing in the pump to a few milli- or micro-amperes. In this high impedance mode of operation the ion current flowing through a pump is proportional to the number of gas ions produced per second, which itself is proportional to the concentration of gas molecules in the pump, and so the ion current becomes a linear function of the pressure. This relationship is shown for several different sizes of pumps in Fig. 7.7. This makes it easy to use the ion current as an indicator of the pressure, and to use an ion pump as a vacuum gauge, as suggested in Section 3.2.3 (p. 104). When the characteristics of pumps of different size are compared it is found that, for operation at a given pressure, the ion current is roughly proportional to the speed of the pump. This is to be expected, because pumping speed is proportional to the number of Penning cells in a pump, and all cells draw about the same current. However, the ion current is slightly different for diode, noble ion, and triode pumps of a given speed, and also for pumps of a given speed and type produced by different manufacturers. Because ion pumps draw such little current when operating at pressures in the high and ultra-high vacuum ranges, their power consumption is small and operating costs are minimal. For example, the 200 l/s pump of Fig. 7.7 draws only 0.4 mA when operating at 10^{-5} Pa (10^{-7} Torr). For a typical operating voltage of 5 kV, this corresponds to a power consumption of only about 2 W, which is less than is required to run many portable battery-powered radios. Power consumption is even smaller at lower pressures. In contrast, oil diffusion pumps of the size used on electron microscopes typically have heaters that draw from 500 to 1000 W at all pressures. Unlike diffusion and turbomolecular pumps, ion pumps do not require a supply of cooling water. Note, however, that the power consumption of ion pumps does increase at higher pressures. For the pump just mentioned it would be about 20 W at 10^{-4} Pa (10^{-6} Torr), and it could increase to the maximum allowed by the

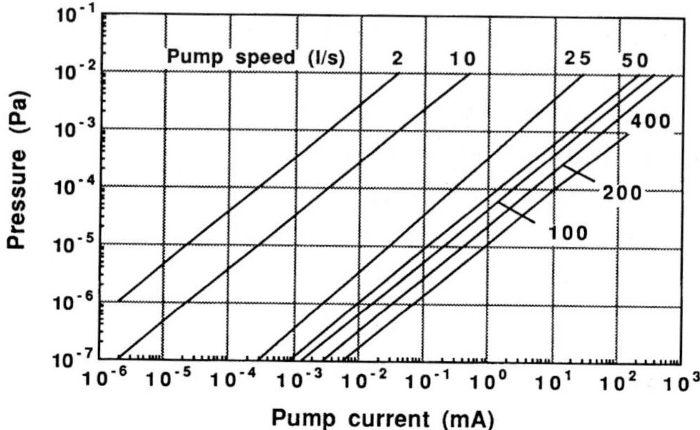

Fig. 7.7 A graph showing how the ion current in sputter-ion pumps typically varies with inlet pressure and the speed of the pumps.

power supply as the impedance of the gas discharge decreases at higher pressures.

7.1.8 Operating life

Because the gas molecules that are pumped by an ion pump are captured and retained inside the pump by the gettering action of the sputtered titanium, these pumps have a limited operating life, which is reached when the titanium available for the sputtering and gettering processes is consumed. Manufacturers usually rate pump life by giving an estimate of the time needed to exhaust the titanium when a pump is operated constantly at a pressure of 10^{-4} Pa (10^{-6} Torr). Typical values are 45 000–50 000 h for diode and noble diode pumps and 75 000 h for hydrogen diode pumps, which have heavier cathode plates than standard diode pumps. The open, grid-like cathodes found in most triode pumps provide less titanium for sputtering than the solid cathodes used in diode pumps, and so allow a pump life of only about 35 000 h. However, the unique cathode design used in the Varian StarCell triode pumps provides an exceptionally large amount of titanium for sputtering and a life of 80 000 h.

The rate at which titanium is sputtered depends directly on the ion current that flows in a pump. As shown in Fig. 7.7, this current varies linearly with pressure, and so pump life also varies linearly with pressure. Thus, a diode pump with a life of 50 000 h (5.7 years) at 10^{-4} Pa (10^{-6} Torr) would have an expected life of 500 000 h (nearly 60 years) at 10^{-5} Pa (10^{-7} Torr). When used in vacuum systems that operate constantly in the ultra-high vacuum range, ion pumps effectively last forever. On the other hand, the expected life for the same pump would be only 5000 h (about 6 months) if it were used constantly at a pressure of 10^{-3} Pa (10^{-5} Torr). Ion pumps are therefore not well suited for use on systems that require them to operate above 10^{-4} Pa (10^{-6} Torr) an appreciable fraction of the time. Interestingly, however, very little titanium is consumed by the very high ion current associated with the generalized discharge that develops when a pump is being started up at pressures above 10^{-1} Pa (10^{-3} Torr), because the low impedance of the pump under these conditions prevents it from supporting a voltage high enough to give the gas ions sufficient kinetic energy for effective sputtering of the titanium.

Three phenomena can cause the life of an ion pump to be less than the rated value. The most common of these involves the electrical shorting out of a pump by a flake of sputtered titanium that breaks off the anode and lodges between the anode and cathode. This is most likely to occur in an old pump in which the layer of sputtered titanium is quite thick, particularly if the pump has been used to pump large amounts of water or hydrogen. It is also possible for a pump to be shorted out by the growth of a whisker of titanium between the anode and cathode structures. In addition, the impedance of a pump can be greatly reduced by the deposition of sputtered titanium or a carbonaceous contaminant on the ceramic insulators that support the anode cells and isolate the electrical feedthrough. This last problem is not frequently encountered, however, because most pumps are constructed so that these insulators are well shielded. All of these conditions can sometimes be relieved temporarily by turning the power to the pump on briefly (only a few seconds) three or four times when the pressure in the pump is above 1 Pa (10^{-2} Torr). However, this may overload the power supply and damage it, and so the manufacturer's instructions for testing for these conditions and for dealing with them should be followed rigorously. Ultimately, of course, it will be necessary to recondition the pump, or to replace it. Reconditioning involves disassembling a pump, chemically dissolving all sputtered titanium off the case and anode cells, cleaning electrical insulators, and replacing the cathodes and any other defective parts. Most manufacturers offer reconditioning services, and there are independent companies (e.g. Duniway) that specialize in this.

7.1.9 General features

Sputter-ion pumps offer several advantages, the chief one being absolute freedom from oil contamination. In addition, ion pumps have no moving parts to wear out mechanically or to produce noise and vibrations, there is no need for cooling water, they can be mounted in any orientation, and their operation is relatively simple and readily adapted for automation and computer control. Ion pumps do not require valves, baffles, cold traps, or backing pumps, and elaborate safety circuits are not needed to protect against power failure, loss of cooling water, loss of backing pressure, etc. If air is accidentally admitted to the system and causes the pressure to rise to an unsafe level, the power supply will automatically switch off so that the ion pump is not damaged and does not cause contamination of the system. If an electrical power failure occurs an ion pump simply stops operating, and then automatically starts again when power is restored. Ion pumps are admirably suited for use on ultra-high vacuum systems, but are not well suited for use on vacuum systems that are frequently cycled to atmospheric pressure, nor on systems that make frequent or prolonged excursions to pressures above 10^{-4} Pa (10^{-6} Torr). Sputter-ion pumps are used on many models of electron microscopes presently being manufactured for ultra-high resolution and analytical applications.

7.2 Titanium sublimation pumps

Sublimation is the physical process in which a material changes directly from the solid to the gaseous state without melting and then condenses to form a solid again. This is the process that is used to produce the titanium film that serves as the getter in titanium sublimation pumps (sublimation pumps, or TSPs). These pumps were developed in the early 1960s, shortly after the introduction of sputter-ion pumps, and have been used extensively since to supplement the action of sputter-ion pumps in ultra-high vacuum systems. O'Hanlon (Section 9.1.1) gives a detailed discussion of the physical processes that occur in these pumps, while Van Atta (Section 9.8) discusses their development and early use.

7.2.1 Construction

A titanium sublimation pump consists of two major components — a titanium sublimation source and the pumping surface, an arrangement of surfaces surrounding the source onto which the vaporized titanium is deposited. A widely-used design for the source is shown schematically in Fig. 7.8. In this 'standard' design three or four titanium filaments are attached to a vacuum flange by a thermally- and electrically-insulated supporting structure. To produce the getter film, one of the filaments is heated directly by electric current through a multi-conductor electrical feedthrough in the vacuum flange, and the titanium that vaporizes from it is allowed to condense on the pumping surfaces that surround the source. With this arrangement, one filament is used until it burns out, and then another is put into use. When all the filaments on the assembly are used up, the assembly is removed from the vacuum system and new filaments installed. These filaments are usually made of an alloy consisting

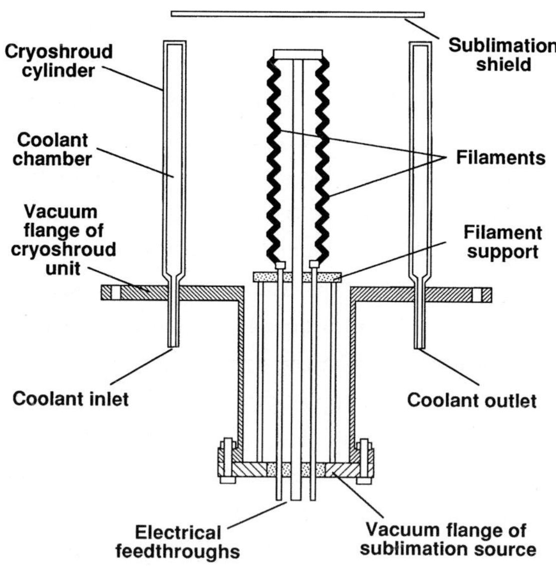

ig. 7.8 A titanium sublimation pump consisting of a sublimation source installed in a cryoshroud unit.

of 85 percent Ti and 15 percent Mo. This alloy does not readily develop hot spots, and so it sublimes more uniformly and yields more vaporizable titanium per gram than pure titanium. Typically, each filament weighs about 3.5 g and provides about 1.5 g of titanium for sublimation, and so an assembly with four filaments affords a total of about 6 g of titanium getter. The Ti-Ball source, patented by Varian, is based on a somewhat different design in which a hollow titanium sphere surrounds an independent electrical heater and is heated to the sublimation temperature indirectly by energy radiated from the heater. The sphere and heater are attached to a vacuum flange by a supporting structure in a manner similar to that shown for the filaments in Fig. 7.8. The great advantage of this design lies in the fact that one sphere can yield substantially more vaporizable titanium than the four filaments of the standard design, at about the same cost. Two models are available, one providing 15 g and another 35 g of vaporizable titanium. These afford substantially greater gettering capacities, and allow longer periods of operation before replacement becomes necessary, than most standard sources.

A number of approaches are used to provide surfaces for the the titanium vapour to condense on. Most manufacturers produce a 'cryoshroud' unit, of the general design shown in Fig. 7.8, for this purpose. The essential component of these units is a double-walled stainless steel cylinder which is mounted on a vacuum flange for attachment to a port in the vacuum system. A second smaller vacuum flange, attached to the end of an extension tube, is provided for mounting the titanium sublimation source on the axis of the cryoshroud cylinder so that the titanium vapour is deposited on the inner wall of the cylinder. Feeder tubes lead through the vacuum flange into the chamber formed by this double-walled cylinder so that water, refrigerant from a compressor, or liquid nitrogen can be fed into the chamber to cool the surface on which the titanium is deposited. A shield plate may also be installed over the top of the cylinder to prevent titanium from being deposited on other components of the vacuum system. Many other arrangements are possible, however, and so manufacturers usually sell the source as a separate item, and leave it up to the designer of the vacuum system to decide how it will be used. The simplest approach, and the one that usually yields the greatest pumping speed, is to install the source so that it extends directly into the vacuum chamber and deposits titanium on the chamber walls. Alternatively, the source can be installed in a tube or pumping manifold that is attached to the vacuum chamber. It will be obvious that a number of other arrangements are possible, depending on the configuration of the vacuum system. This versatility is one of the many advantageous features of titanium sublimation pumps. In all arrangements, shield plates may be needed to prevent titanium from being deposited on sensitive components of the vacuum system.

The power supplies for titanium sublimation pumps must provide accurate control of the power that is fed to the titanium filaments or heaters. Most of these supplies operate at a voltage in the range from 6 to 20 V AC (alternating current), and provide an output current that is adjustable from 0 to about 60 A. For effective sublimation to occur, titanium must be heated in the range from 1000 to 1550°C, where its vapour pressure varies exponentially with temperature from about 10^{-6} Pa (10^{-8} Torr) to nearly 10^{-1} Pa (10^{-3} Torr). The upper end of this operating temperature range is very close to 1675°C, the melting point of titanium (where its vapour pressure is about 0.5 Pa). The standard filaments provided by most manufacturers require about 500 W of power to reach the operating temperature range. This corresponds to a current of about 50 A for a power supply operating at 10 V AC. Because of the exponential relationships involved, however, the acceptable operating current range is usually only a few amperes, typically about five. Thus, for example, while a current of 50 A might be the minimum for obtaining a useful sublimation rate, a current of 55 A might be the maximum allowable without damaging the source. Because the current output usually must be set manually by the operator, the manufacturer's recommendations should be noted carefully and followed closely to obtain optimal performance without damaging the source. Power supplies also have a timer that can be set to allow the source to be heated continuously or intermittently. In the latter mode the heating power is repetitively turned on for a sublimation period (usually adjustable from 1 to 5 min), and then turned off for a cycle period (usually adjustable from a few minutes to a few hours). The utility of these two modes of operation will be discussed shortly.

The pumping action of titanium sublimation pumps results primarily from the chemisorption processes that occur because of the high chemical reactivity of a freshly deposited titanium film. As described in Section 7.1.1 above, the active gases nitrogen, oxygen, hydrogen, carbon monoxide, carbon dioxide, and water vapour, react chemically with the titanium atoms in the surface of the getter film, forming nitrides, oxides, hydrides, and carbides of titanium, and then are buried as additional titanium is deposited over them. However, chemically inert gases such as helium, neon, and argon are not pumped because they do not react with the getter film. For this reason, titanium sublimation pumps are almost never used as the primary pumps on vacuum systems; instead, they are employed to supplement the action of other types of pumps that handle these inactive gases.

2.2 *Pumping speed at high pressures*

As is true for oil diffusion, turbomolecular, and sputter-ion pumps, the pumping speed of titanium sublimation pumps shows a strong dependence on the operating pressure. Titanium sublimation pumps usually are not used at pressures above about 1 Pa (10^{-2} Torr), because at such high pressures gas molecules strike the hot filament so frequently that it becomes coated with stable compounds which severely retard the sublimation of titanium. Effective sublimation begins as the pressure drops below about 10^{-1} Pa (10^{-3} Torr). From here down to about 5×10^{-5} Pa (5×10^{-7} Torr) the gas molecules collide with the deposited titanium film so frequently that they react with the titanium atoms as fast as they are sublimed. Pumping speed in this range is therefore determined primarily by the titanium sublimation rate ξ, which manufacturers usually specify in units of grams per hour.

It is possible to estimate the maximum speed a pump can develop in this high pressure end of the operating range. The number of moles of titanium being sublimed each second n_s can be obtained by dividing ξ (g/h) by the number of seconds per hour (3600) and the grams of titanium per mole (48), giving $n_s = \xi/(3600\times48) = 5.8\times10^{-6}\xi$ moles of Ti sublimed per second. The gettering reaction for nitrogen is thought to be $2Ti+N_2\rightarrow2TiN$. Since nitrogen is the most abundant atmospheric gas, this suggests that each mole of titanium pumps about half a mole of air, and so the subliming titanium will pump about $n_p \approx 2.9\times10^{-6}\xi$ moles of air per second. The volume V of air pumped per second, which is actually the pumping speed for this high-pressure range S_{hi}, can now be estimated by using the ideal gas equation (eqn. 1.1, Section 1.3, p. 14; i.e. $V = nRT/P$) as follows:

$$S_{hi} \approx \frac{n_p RT}{P} \approx \frac{2.9 \times 10^{-6} \ \xi \ (8314)293}{P} \approx 7\frac{\xi}{P} \ \ 1/s \qquad (7.1)$$

where ξ is the titanium sublimation rate (g/h), P is the gas pressure (Pa), R is the ideal gas constant, and it is assumed that the gas is at room temperature (i.e. $T \approx 293$ K) . This equation is interesting because it indicates that pumping speed varies inversely with pressure in this high-pressure range. Values cited by manufactures for ξ range from 0.2 to 0.5 g/h. Using an average value of 0.35 g/h, for example, gives $S_{hi} \approx 2.5/P$ l/s, which corresponds to a pumping speed of about 2500 l/s at 10^{-3} Pa (10^{-5} Torr), 250 l/s at 10^{-2} Pa (10^{-4} Torr), and 25 l/s at 10^{-1} Pa (10^{-3} Torr). As noted at the start of this calculation, these are certainly maximum pumping speeds, because it was assumed that the gettering reactions occur in an ideal way and with 100 percent efficiency. In practice, such conditions may not prevail and speeds could be somewhat less than predicted by this equation. Nonetheless, this calculation indicates that a titanium sublimation pump can make a very substantial contribution to an evacuation process in the pressure range from 1 Pa down to about 10^{-4} Pa (10^{-2}–10^{-6} Torr), where the speeds of most roughing pumps become very low. It is also interesting to note that over this high-pressure range the throughput Q_{hi} does not vary with pressure, but depends only on the titanium sublimation rate. This can be shown simply by using the definition of pumping speed given by eqn. 2.9 in Section 2.5 (i.e. $S = Q/P$), as follows:

$$Q_{hi} = S_{hi}P \approx 7\left(\frac{\xi}{P}\right)P \approx 7\xi \quad \text{Pa}-\text{l/s} \tag{7.2}$$

This is as expected, considering the physical characteristics of the process. As was shown in Section 1.3 (p. 16), a pascal-litre of gas represents a fixed number of gas molecules (2.47×10^{17}). Under the conditions of concern here the number of gas molecules pumped per second is equal to about half the number of titanium atoms sublimed per second, which depends only on the temperature of the filament and is not influenced by the pressure of the gas.

7.2.3 *Pumping speed at low pressures*

At pressures below about 5×10^{-5} Pa (5×10^{-7} Torr) it is possible to sublime titanium appreciably faster than the gas molecules collide with the titanium film and react with it, and so pumping characteristics differ somewhat from those at higher pressures. Again, a limiting maximum value can be calculated by considering the instantaneous interaction between gas molecules and a freshly deposited titanium film. The number of moles of gas molecules that collide with each square millimetre of the film per second n_c can be calculated by dividing γ, the number of molecules that strike each square millimetre of surface per second (eqn. 1.5, Section 1.8, p. 23) by Avogadro's number N_A (i.e. $n_c = \gamma/N_A = 2.86 \times 10^{16} P/6.023 \times 10^{23} = 4.75 \times 10^{-8} P$ mol/s·mm^2, for pressure in pascals), and so the number of moles of gas that strike a film with an area of A mm^2 each second will be An_c. It is also reasonable to assume that only a fraction f of the gas molecules that strike the getter film are captured by it. Again, the ideal gas equation ($V = nRT/P$) can be used to estimate the volume of gas removed from the system per second, and this is the pumping speed S_{lo} for this low-pressure range:

$$S_{lo} \approx \frac{n_c ARTf}{P} \approx \frac{4.75 \times 10^{-8} P\,(8314)293 fA}{P} \approx 0.12 fA \quad \text{l/s} \tag{7.3a}$$

This is an interesting result, because it shows that in this pressure range the pumping speed does not vary with the pressure, but is primarily determined by the area of the getter film A (mm^2) and the fraction of incident gas molecules that 'stick' to it. The area of the getter film will depend on the arrangement of surfaces surrounding the titanium source, and can vary widely, depending on the design of a particular vacuum system, as noted above. Sticking coefficients f are difficult to evaluate accurately. They also vary with temperature, depend on the character of the surface and the fraction of it already covered with gas molecules, and are different for different gases.

It is generally accepted that at the temperature of liquid nitrogen (–196°C, 77 K) $f \approx 1.0$ for all atmospheric gases. Eqn. 7.3a suggests that at this temperature each square millimetre of getter film, which is actually a rather tiny area, has the potential for pumping one-tenth of a litre of gas each second. This is enough to give very substantial pumping speeds. For example, a cryoshroud of the type shown in Fig. 7.8, which might typically have a diameter of about 150 mm and a length of about 200 mm, would provide an area of more than 90 000 mm^2 for titanium deposition, and would give a pumping speed greater than 10 000 l/s. This is a rather impressive result, and it shows why it could be advantageous to use a cryoshroud cooled with liquid nitrogen as the substrate for the getter film in a titanium sublimation pump. Notwithstanding this, it is usually preferable to use air- or water-cooled pumping surfaces, particularly in systems that are frequently cycled to atmospheric pressure or that are required to pump appreciable amounts of water. With liquid nitrogen cooling, the condensation of water on the pumping surface causes the sublimed titanium to adhere poorly. This causes the titanium to flake off the surfaces, and also leads to pressure instability. If a surface cooled with liquid nitrogen is needed to assist in pumping water, it is usually considered desirable to keep it well separated from the surface onto which the titanium is sublimed.

For this reason the titanium is usually sublimed onto surfaces that are near room temperature, and here the sticking coefficients are somewhat less than unity. Pumping speed data published by several manufacturers indicate that near room temperature f has a value of about 0.8 for oxygen, carbon monoxide, and carbon dioxide, and about 0.3 for hydrogen, nitrogen, and water. Using a weighted average of these values for oxygen and nitrogen suggests that a reasonable estimate of the pumping speed for air at room temperature is:

$$S_{lo} \approx [0.8(0.3)+0.2(0.8)]0.12\,A \approx 0.05A \quad \text{l/s} \tag{7.3b}$$

Although this predicts a considerably smaller pumping speed than that estimated above for surfaces cooled with liquid nitrogen, it still implies that each *square millimeter* of titanium is capable of gettering more than 10^4 mm^3 of gas per second, and this is certainly enough to give very useful pumping speeds in practical systems. For example, if it were used at room temperature the pumping speed of the cryoshroud unit just described would still be greater than 4 000 l/s. By comparison, the largest standard models of sputter-ion pumps have speeds of only about 1000 l/s. This equation appears to be in reasonable agreement with such manufacturers' data as are available. For example, it predicts a pumping speed of about 25 l/s per square inch (645 mm^2) of titanium surface, which is comparable to the value of 20 l/s per square inch cited by Varian as being a conservative estimate. Thermionics give 40 to 90 l/s per square inch as the range to be expected in practice, which tends to confirm the conservative nature of the Varian value, and which suggests that eqn. 7.3b is also conservative.

In contrast to the relation described by eqn. 7.2 for operation at higher pressures, the throughput in this low pressure range:

$$Q_{lo} = S_{lo}P = 0.12fAP \quad \text{Pa-l/s} \tag{7.4}$$

is directly proportional to the operating pressure.

7.2.3a *Intermittent sublimation*

The area A in the above calculation is the area of clean, unreacted titanium surface available at any instant for gettering. However, the pumping process itself covers an existing titanium surface with the products of the chemisorption reactions, and so the area available for gettering decreases as pumping progresses. In addition, the sticking coefficient for nitrogen, the most abundant atmospheric gas, depends strongly on the fraction of the titanium surface that is covered with sorbed gas molecules. As noted above, it has a value of about 0.3 for a fresh surface, but decreases by a factor of almost 10 when about 10 percent of the surface becomes covered.

Both of these effects cause a decrease in pumping speed with time, unless the titanium film is constantly renewed. Renewal of the film can, of course, be accomplished by constantly subliming titanium. However, it is very difficult to adjust the rate of sublimation to match the rate of gas sorption, especially at pressures in the ultra-high vacuum range where the rate of titanium consumption is very low, because the sublimation rate varies exponentially with the heating current fed to the sublimation source. As a consequence, constant sublimation usually deposits titanium much more rapidly than it is actually needed. This uses up the titanium available from the source more quickly, and makes it necessary to replace the source more frequently, than a more optimal method would. Therefore, intermittent sublimation is generally used to compromise between pumping speed and source life. That is, the filament is heated for a sufficient period to coat the pumping surfaces with a fresh layer of titanium, and is then turned off long enough for a significant fraction of this titanium to be used up by the gettering process. The question now arises as to what the appropriate times might be for these two sequential steps, and it will be instructive to make a few calculations in this connection.

7.2.3b *Sublimation time*

As a basis for estimating the sublimation time t_s it is convenient to assume a geometry similar to that shown in Fig. 7.8 in which the titanium is sublimed onto a cylindrical surface that surrounds the titanium source in a symmetrical manner. If the surface of this cylinder were atomically smooth it would require about $\omega \approx 1.3\times10^{13}$ titanium atoms/mm^2 to cover it with a monolayer of titanium atoms. This corresponds to about $n = \omega/\mathrm{N}_A \approx 2.2\times10^{-11}$ moles of titanium/mm^2, or 1.0×10^{-9} g of titanium/mm^2 (since $M = 48$ g/mol for Ti). A sublimation rate of ξ g/h corresponds to a rate of $\xi/3600 = 2.8\times10^{-4}\xi$ g/s, and so the time needed to sublime a monolayer of titanium on a surface with an area of A mm^2 under such idealized conditions would be about:

$$t_s \approx \frac{1.3 \times 10^{-9} A}{2.8 \times 10^{-4} \xi} \approx 4.7 \times 10^{-6} \left(\frac{A}{\xi}\right) \ \mathrm{s} \qquad (7.5a)$$

Using this equation to calculate the sublimation time for the cryoshroud cylinder described in the examples above (with $A \approx 90\,000$ mm^2), assuming

$\xi \approx 0.35$ g/h, gives $t_s \approx 1.2$ s. This is a surprisingly short sublimation time, which may be unrealistic for practical applications for at least two reasons. First, most surfaces on which titanium is likely to be sublimed are unlikely to be atomically smooth. As mentioned earlier, the microscopic area for real surfaces can be from 10 to 100 times the apparent macroscopic area. Second, it is usually desirable to deposit several (perhaps from 3 to 5) monolayers of titanium to ensure complete burial of all gas molecules adsorbed on the existing getter surface. Taken together, these effects will increase the sublimation time by a factor U. In addition, it usually requires about 15 s for a filament to heat up to the sublimation temperature. Therefore, a more practical form for the above equation is:

$$t_s \approx 15 + 4.7 \times 10^{-6} \left(\frac{A}{\xi} \right) U \text{ s} \tag{7.5b}$$

Applying this equation to the calculation of the sublimation time for the cryoshroud cylinder, with $A \approx 90\,000$ mm^2 and $\xi \approx 0.35$ g/h, as above, gives $t_s \approx 30$ s if $U \approx 10$, and 75 s if $U \approx 50$. These results suggest that a total heating time of about 1 min should be sufficient for most practical situations.

7.2.3c Cycle time

A rough estimate of the cycle time t_c (i.e. the time the filament can be turned off between sublimation treatments) can be obtained using eqn. 1.6 in Section 1.9 (p. 24), which indicates that the time required for the sorption of a monolayer of gas molecules on a clean surface is of the order of $\tau \approx 3.3 \times 10^{-4} g/Pf$ seconds, in which P is the pressure in Pa, g is a surface roughness factor, and f is the sticking coefficient. This would be an approximate value for t_s for an operating strategy in which the pumping speed is allowed to drop to zero before more titanium is sublimed onto the pumping surface. For practical purposes it is reasonable to assume that the pumping speed would decrease to an undesirable level well before the surface becomes covered with a full monolayer; that is, when some fraction x of the surface is covered. Making allowance for this gives the following equation for the time required for the pumping speed of the getter film to decrease to an unacceptable level:

$$t_c \approx x\tau \approx \frac{3.3 \times 10^{-4} gx}{f \cdot p} \quad \text{s} \tag{7.6a}$$

It is a bit difficult to know what values to use for g, x, and f in this equation for practical applications. However, for a rough estimate it can be assumed that the surface onto which the titanium was deposited is well polished, as is usually true for surfaces in ultra-high vacuum apparatus, so that g has a value of only about 2. It also seems reasonable to assume that there would be a significant decrease in pumping speed if $x \approx 0.25$, because then one-quarter of the gas molecules striking the pumping surface would encounter sites that have already participated in the gettering reaction. If the pumping surface is near room temperature, $f \approx 0.3$ for nitrogen on a clean surface, but decreases by nearly a factor of 10 for 10 percent surface coverage, while for oxygen f remains relatively constant at about 0.8, even when surface coverage approaches a full monolayer. Using a value of $f \approx 0.03$ for nitrogen gives a weighted average of $f \approx 0.2(0.8)+0.8(0.03) \approx 0.2$ for air, and so a very rough estimate of the cycle time for operation at room temperature is:

$$t_c \approx \frac{0.25(2)\tau}{0.2} \approx \frac{8 \times 10^{-4}}{P} \quad \text{s} \tag{7.6b}$$

where P is the pressure in pascals. This equation gives $t_c \approx 16$ s when $P = 5\times10^{-5}$ Pa (5×10^{-7} Torr). This is of the order of the time required to heat up a source to the sublimation temperature, which is consistent with the fact that it is common practice to use continuous sublimation to achieve maximum pumping speed when the pressure is at this level or higher. This equation also shows that at lower operating pressures t_c increases, becoming of the order of:

a few min	at	5×10^{-6} Pa (5×10^{-8} Torr)
half an hour	at	5×10^{-7} Pa (5×10^{-9} Torr)
a few hours	at	1×10^{-7} Pa (1×10^{-9} Torr)
a day	at	1×10^{-8} Pa (1×10^{-10} Torr).

Eqn. 7.6a suggests that these times would be less by a factor of about five for a pumping surface cooled with liquid nitrogen, because at this

temperature $f \approx 1$ for all atmospheric gases. These values of t_c are significantly longer than the time required to heat a source up to the sublimation temperature, and so it becomes practical to use intermittent heating to attempt to achieve an optimal compromise between pumping speed and source life at these pressures.

7.2.3d *Practical considerations*

Although the above calculations and equations have obvious shortcomings because of the various assumptions and approximations used in them, they are nonetheless useful because they indicate the physical phenomena involved in determining t_s and t_c, and give an indication of the general range of values to be expected. For example, they show that a titanium sublimation pump will not make a maximum contribution to the evacuation of a system operating in the 10^{-6} Pa (10^{-8} Torr) range if it is only turned on for a few minutes once or twice a day. On the other hand, it clearly would be inefficient to cycle a pump on and off every half-hour in an ultra-high vacuum system operating in the 10^{-8} Pa (10^{-10} Torr) range. Likewise, they show that a sublimation time of about a minute will probably be adequate unless a system has an unusually large pumping surface area.

It usually will not be a straightforward matter to calculate values for t_s and t_c in practice, because the physical characteristics of the system, such as the titanium sublimation rate and the area and geometry of the pumping surface, generally will not be known accurately. It must also be recognized that t_s and t_c probably cannot be sharply delimited for many vacuum systems. Nonetheless, it is worthwhile to use such manufacturer's data as are available to calculate rough values for t_s and t_c. In addition, empirical methods can be employed to obtain values that are directly influenced by the characteristics and operating requirements of a particular application. For example, the cycle time can be arbitrarily taken as the time needed for the pressure to begin to increase at an unacceptable rate, or to rise to an unacceptable level, after completion of a sublimation treatment. Likewise, the optimum sublimation time can be estimated by starting with the minimum value allowed by the power supply, gradually increasing it for successive sublimation treatments, and noting the value of the cycle time after each treatment. Increasing the sublimation time in this way will increase the amount of titanium sublimed onto each unit area of the pumping surface. This should cause the cycle time to increase, reach a maximum value when optimal surface coverage is achieved, and then remain roughly constant. The optimum sublimation time then is the minimum time needed to produce this maximum cycle time.

7.2.4 *Applications and general characteristics*

As noted above, titanium sublimation pumps were first developed as supplemental pumps for use with sputter-ion pumps in ultra-high vacuum systems. This continues to be the major application for them, and they perform three basic functions in this role. First, they assist in getting the ion pump started properly during the pumpdown process. For this purpose the sublimation pump is turned on and operated in the continuous sublimation mode as soon as the roughing pump brings the pressure well below 1 Pa (10^{-2} Torr). The high pumping speed of the sublimation pump in this range helps bring the pressure into the 10^{-2} Pa (10^{-4} Torr) range quickly, whereupon the ion pump can be turned on. The sublimation pump then helps handle the gas released as the ion pump starts up, reducing chances of developing overheating or a run-away discharge in the ion pump. When the two pumps bring the pressure down into the 10^{-6} Pa (10^{-8} Torr) range, the sublimation pump is changed over to intermittent operation. Second, sublimation pumps are used in the intermittent mode of operation to provide additional pumping speed for ion-pumped systems operating in the ultra-high vacuum range. If the sublimation and cycle times are properly chosen, the use of a sublimation pump in this way can produce improvement by a factor of 10 in the vacuum over that attainable with an ion pump alone. Finally, sublimation pumps are used on a demand basis to handle large intermittent gas loads, such as might arise when a specimen airlock is operated, when a system is heated for outgassing, or when a filament is heated during an evaporation process. It will be obvious that a sublimation pump could perform similar functions in supplementing other types of high vacuum pumps.

One disadvantage associated with a titanium sublimation pump is the increase in pressure that always occurs in the system when the heating current to the source is first turned on. This is caused initially by the release of gases from the source, but may also involve desorption of gases from surfaces surrounding the source if the heating is carried on long enough for these surfaces to be warmed up appreciably. To minimize this second source of gas evolution when a pump is being used in the intermittent mode of operation, it is advantageous to use the shortest sublimation time needed to restore the getter film to full activity. If this is done a minimal amount of gas will be released into the system, and this will quickly be pumped by the fresh getter film, and the pressure will rapidly decrease again as soon as the power to the source is turned off. If a sublimation pump must be operated in the continuous sublimation mode frequently or for long periods of time, it is advantageous to cool the surfaces that surround the source with circulating water to control this gas evolution problem. Sublimation periods should also be carefully controlled if the vacuum system contains appreciable amounts of hydrogen or water in the presence of carbon monoxide or carbon dioxide, because catalytic effects which occur on the surface of the hot titanium allow these gases to react together to form methane. Since methane is not pumped readily by the sublimation pump, this can lead to a considerable increase in the gas load that must be handled by the main high vacuum pump.

Another disadvantage of titanium sublimation pumps is the need to replace the titanium source periodically. This involves bringing the system up to atmospheric pressure, removing the source assembly, and replacing the burned out elements. How often this must be done depends on the manner in which the pump is used. Obviously, the total sublimation time T_s for a given source is equal to the weight of titanium available in the source for sublimation w (g) divided by the sublimation rate ξ (g/h):

$$T_s = \frac{w}{\xi} \text{ h} \tag{7.7}$$

For example, the standard 85Ti-15Mo alloy filaments usually provide about 1.5 g of titanium for sublimation. The sublimation rate can be varied by changing the heating current supplied to the source. For a current at the lower end of the operating range ξ typically has a value of about 0.05 g/h, for which $T_s \approx 30$ h, while the maximum recommended operating current usually gives $\xi \approx 0.5$ g/h, for which $T_s \approx 3$ h. Thus, it would be reasonable to expect to replace all three filaments on a source of the type shown in Fig. 7.8 after a total of about 10 h of sublimation, if the pump is used mostly in the continuous sublimation mode at the maximum sublimation rate. For a pump that is used primarily to assist in rough pumping a system down from atmospheric pressure this might correspond to only a few days or a few weeks of system operation, depending on the size of the system and how frequently it is cycled to atmospheric pressure. On the other hand, the same three-filament source would have a life of several months if it were used in an ultra-high vacuum system that operates in the mid 10^{-8} Pa (10^{-10} Torr) range, where it might be turned on for only 2 min every 5 h, and where a lower sublimation rate of 0.25 g/h might be adequate. Source life can, of course, be increased by not making use of a sublimation pump at times when the vacuum system is not being actively used, such as over nights and weekends, and relying on the main high vacuum pump to maintain an adequate standby vacuum during these periods. It is usually advisable to avoid operating a source at too high a temperature, because this can lead to microstructural changes in the filament that can shorten its life. If a source can be seen directly it is possible to estimate its temperature by noting its colour. A bright orange colour usually corresponds to a safe temperature near 1200°C, while a bright yellow colour indicates a temperature near 1500°C, which may be too high for longest source life. The Ti-Ball sources produced by Varian, and the heavy-duty sources produced by several other manufacturers, provide substantially more titanium for sublimation, and thus last correspondingly longer than those with filaments of the standard design.

One very important factor that influences the pumping speed of a titanium sublimation pump in practical situations, which was not taken into account in the development of eqns. 7.1 and 7.3, is the effect of surrounding fixtures on the ease with which gas molecules can reach the getter film. For example, placing a shield above the cryoshroud cylinder in the manner shown in Fig. 7.8 could significantly restrict the access of gas molecules to the titanium film deposited on the interior of the cylinder. Installing the sublimation source in a pumping manifold, or in a tube attached to a vacuum chamber, would have a similar effect. In such arrangements the effective speed of the pump is given by $S_e = S_pC/(S_p+C)$ (see eqn. 2.12 in Section 2.5, p. 40), in which S_p is the speed of the pump estimated on the basis of eqns. 7.1 or 7.3, and C is the conductance of the fixtures between the getter surface and the rest of the vacuum system.

Overall, titanium sublimation pumps are very useful and have many advantageous characteristics. A major advantage is that they are totally oil free, and so will not contribute to contaminating a vacuum system with hydrocarbons. They can make a useful contribution to the evacuation of a system at pressures all the way from the rough vacuum range to deep into the ultra-high vacuum range. They are inexpensive to purchase and operate, and provide relatively high pumping speeds at relatively low cost. They have no moving parts to wear out or produce mechanical vibrations, and they are light in weight, and can be mounted in any orientation.

7.3 Combination ion-sublimation pumps

Titanium sublimation pumps work so well in conjunction with sputter-ion pumps that most manufacturers produce combination ion-sublimation pumps by adding a titanium sublimation source to one of the larger ion pumps (usually those with speeds greater than 100 l/s). This can be done in a variety of ways. One compact arrangement involves the insertion of a cryoshroud sublimation pump unit, like the one shown in Fig. 7.8, into one of the flanges of a double-ended ion pump. The sublimation pump can also be installed in a flange that is added to the body of a standard ion pump. In another common arrangement the sublimation pump is installed along the axis of a large tube and the ion pump units are welded around this tube. Many of these combination pumps can also be obtained with a cryoshroud, cryopanel, or baffle which can be cooled with liquid nitrogen to enhance the speed of the sublimation pump and to provide extra pumping speed for water vapour. These combination units are usually designed to optimize the performance of each of the components, and so they provide a compact and efficient means of incorporating the capabilities of all the component pumps into a vacuum system. For example, the multiple-array ion pump shown in Fig. 7.4 optimally accommodates both a sublimation unit and a liquid nitrogen-cooled baffle. The ion pump arrays provide a basic pumping speed of 300 l/s, the sublimation pump gives a supplemental speed of 600 l/s for reactive gases, and the baffle will give a speed of 600 l/s for water if it is cooled with liquid nitrogen. This unit also has a high-conductance valve located below the inlet flange, and the overall design is such that conductance is not a factor in limiting the pumping speed of any of the component pumping units. In addition to providing these pumping capabilities, this module is mechanically strong enough to serve as the base for a rather sizeable vacuum apparatus.

7.4 Bulk getter pumps

Bulk getter pumps differ from the types of pumps discussed above in the way in which the getter material is used. Gettering is basically a surface phenomenon. The chemical reactions that produce the gettering effect occur only when gas molecules contact the getter material as a result of their random, thermally-activated motion, and only those atoms which form the outermost atomic layer of the surface of the getter participate in these reactions. Once a gas molecule reacts with a site on the getter surface that site is 'used up' and can no longer contribute to the pumping process. Thus, the pumping capacity of a getter steadily decreases as the pumping process progresses, and so it becomes necessary to renew the getter surface in order to sustain the pumping action. Two fundamentally different methods are used for this purpose. In the sputter-ion and titanium sublimation pumps discussed above, which are often classified as 'evaporable getter pumps', this is done by vaporizing a few atom layers of fresh getter material (usually titanium) onto the surfaces where the gettering reactions occur. For obvious reasons, these pumps are also sometimes called 'thin-film getter pumps'. In bulk getter pumps, which are the topic of this section, the gettering reactions occur directly on the surface of bulk pieces of getter material (wires, rods, sheets, or powders) and the surface renewal process is effected by heating these bulk pieces to temperatures high enough to cause the sorbed gas atoms to diffuse into the body of the getter material. Since vaporization of the getter material is not involved, these pumps are also called 'non-evaporable getter pumps'.

7.4.1 Bulk getter materials

The high chemical reactivity of the getter materials in these pumps and the nature of the solid-state diffusion processes used for surface renewal make it necessary to follow special procedures in working with them. During preparation, the pieces of the getter materials are treated so that they become covered with a dense oxide layer which protects them from reaction with the atmosphere during subsequent processing, handling, and installation. After installation, however, this oxide coating must be removed so that reactions can again occur on the getter surface to produce the desired pumping action. This *activation* process is accomplished by heating the getter in a vacuum to a high temperature for a period long enough to allow the protective coating to diffuse into the interior of the getter. It is also necessary to heat a bulk getter during operation so that the sorbed gases constantly diffuse into the getter and the surface is continuously renewed. However, this operating temperature is usually significantly lower than the temperature required for activation.

The most widely used bulk getter materials are active metals and alloys. The earliest of these were pure metals such as titanium, tungsten, tantalum, zirconium, hafnium, and thorium. The dense crystal structures of these pure metals severely restrict the diffusion processes, however, and so they must be heated to such high temperatures that their practical usefulness is severely limited. Typically, full activation requires heating at 1000–2000°C, while operating temperatures are in the range from 700 to 1000°C. A major step in overcoming this limitation was achieved in the early 1970s with the development of the proprietary 84Zr-16Al alloy (trade name St 101) by SAES Getters. After proper metallurgical processing this alloy becomes a mixture of several Zr-Al intermetallic phases, and it appears that the relatively low density of atomic packing in the boundary regions between these phases allows diffusion to occur more readily than in the pure metals. This alloy can be activated by heating for 30 min at 750°C, and it pumps atmospheric gases effectively at 300–400°C. Following this breakthrough, further research and development work has yielded ternary and quaternary multi-phase alloys with even better properties. An example is the 70Zr-24.6V-5.4Fe multi-phase alloy (SAES trade name St 707), which can be fully activated by heating for 10 min at 450°C, and which can be used effectively at 250–300°C. Most of these alloys are very brittle and can easily be ground to fine powders to obtain the large surface areas needed for effective gettering. These alloys are the getters most commonly used in bulk getter pumps.

The gettering reactions for these alloys are generally similar to those described above for the titanium films in the titanium sublimation and sputter-ion pumps. When molecules of oxygen, nitrogen, carbon monoxide, and carbon dioxide, adsorb on such alloys, they dissociate and react chemically to form oxides, nitrides, and carbides. If the temperature is high enough, these reaction products then diffuse into the body of the alloys, renewing the getter surface for continued pumping. The compounds formed in these reactions are so stable that these reactive gases are not released even if the alloys are subsequently heated to temperatures near their melting points. Helium, neon, and argon do not react chemically with the getter alloys, and so they are not pumped. Hydrogen exhibits rather unusual behaviour. Hydrogen molecules dissociate on contact with the getter surface and form individual hydrogen atoms which diffuse into the interior of the getter alloys very easily. In fact, the diffusion process occurs so readily that hydrogen is pumped at room temperature. However, the chemical interaction between the hydrogen atoms and the atoms of the alloys is very weak, and so hydrogen forms only a solid solution in the getter alloys rather than a stable chemical compound. Because of this, most of the absorbed hydrogen is released if the alloys are heated significantly above the temperature at which the absorption took place. Water dissociates into hydrogen and oxygen atoms, and the oxygen atoms react to form stable oxides while the hydrogen atoms are dissolved in solid solution.

For use, the alloys are reduced to fine powders which are usually compressed into porous pellets or compression-bonded onto thin metal sheets or strips. SAES Getters produce a series of pumps under the SORB-AC trade name that use the coated metal strips as the getter element. For this application the strips are pleated like the bellows of an accordion and then bent into cylinders that are mounted around a central heating element. The getter strips and the heating element are mounted on a vacuum flange that also carries the insulators and connectors for supplying electrical power to the heater, and a thermocouple for measuring the getter temperature. Fig. 7.9 shows a pump consisting of one of these getter units mounted inside a water-cooled housing. Pumps of this design are available with either the St 101 or the St 707 alloy as the getter material, and with pumping speeds for nitrogen ranging from 20 to 400 l/s. Alternatively, these pumping units can be mounted nude inside a vacuum system, increasing pumping speed by as much as a factor of two. Danielson Associates have recently introduced a series of pumps under the Sorbodyn trade name that use pellets of a proprietary quaternary alloy as the getter material. Fig. 7.10 shows the way these pumps are constructed. The getter pellets are contained in the flanged canister which carries a heating element for activating the getter and for heating it during use, and which is enclosed during use in an insulating protective housing. Pumps containing from 100 to 2000 g of pellets are available, providing pumping speeds for nitrogen that range from about 15 to 500 l/s. Power supplies

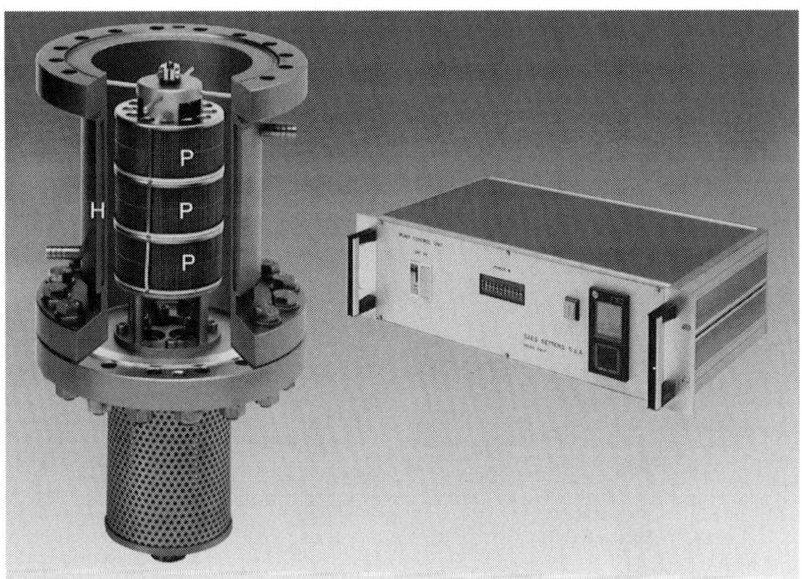

Fig. 7.9 A bulk getter pump in which the pumping elements (P) are cylinders made of pleated metal strips that are coated with particles of a getter alloy. A heater is mounted along the axis of the cylinders, and the pump unit is surrounded by a water-cooled housing (H). The control unit for the pump is also shown. (Courtesy of SAES Getters S.p.A.)

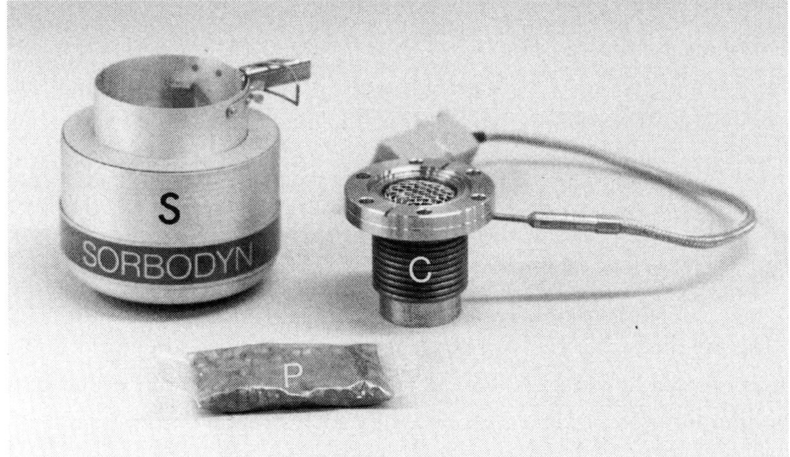

Fig. 7.10 A bulk getter pump in which the pumping element is a charge of pellets (P) of a getter alloy. When assembled, the pellets are placed inside the heated canister (C), which is then installed in the protective, insulating heat shield (S). (Courtesy of Danielson Associates.)

furnished with bulk getter pumps provide electrical power at the proper levels for the activation process and for normal operation, and meters for reading the output of the thermocouples which monitor the getter temperature.

Standard operating procedures for non-evaporable getter pumps are relatively simple and straightforward. After installation, the system is evacuated to below 10^{-1} Pa (10^{-3} Torr) by another type of pump, and the initial activation treatment is carried out. As soon as a significant fraction of the protective oxide coating is dissolved the pump will begin to contribute to the evacuation process, whereupon the pressure in the vacuum system should decrease rapidly. At the end of the activation period the temperature of the getter is decreased to the operating level and the normal pumping process occurs. In general, it is not considered advisable to expose the hot getter to reactive gases at pressures above about 1 Pa (10^{-2} Torr) so that it does not become covered with such a heavy layer of reaction products that the pumping process is impeded. In particular, before the system is let up to atmospheric pressure it is important to allow the pump to cool to room temperature, or to isolate it by means of a vacuum valve. However, if reaction products do build up to the point where the pumping speed decreases significantly, normal performance can be restored by carrying out another activation treatment.

7.4.2 *Performance characteristics*

The pumping speed of bulk getter pumps is determined in a complicated way by at least three important factors: the total area of getter surface presented for gas molecules to react with; the ease with which gas molecules can reach the getter surface; and the rate at which the getter surface is renewed by diffusion of the products of the gettering reactions into the bulk of the getter. The first two of these depend on such features in the design and construction of a pump as the total amount of getter alloy provided and the way it is distributed inside the pump, the amount of surface area per gram of getter, the conductance of the inlet port of the pump, and the amount of impedance introduced by such structural items as retaining screens, baffles, and heaters. These features are essentially fixed when a pump is manufactured, and so are beyond the control of the user. The third factor, the rate at which the gas molecules diffuse into the bulk of the getter particles to renew the active getter surface, can be controlled to a considerable extent by the operator, however. This diffusion rate R_d varies exponentially with temperature T (K) according to an equation of the form:

$$R_d = C \exp(-J/T) \tag{7.8}$$

Here C and J are constants, the values of which depend on the properties of the particular getter alloy being used. Such a relationship implies a very strong dependence on temperature. For many diffusion processes of the kind involved here a change of 50–100 kelvins can cause the diffusion rate to change by as much as a factor of 10. It is therefore important to be sure that bulk getter pumps are maintained at the optimum operating temperature recommended by their manufacturer if it is desired to achieve maximum pumping speed for long periods. During use, diffusion rates decrease due to the build-up of reaction products just beneath the surface of the getter particles, and this causes a decrease in pumping speed. Increasing the operating temperature will increase the diffusion rates, however, and maintain the desired pumping speed for a longer time. If the pumping speed drops to an unacceptable level, it can be restored by performing an activation treatment to distribute the reaction products uniformly throughout the getter. At some point, however, the capacity of the getter will become exhausted, whereupon the activation process will no longer be effective in restoring pumping speed and the getter will have to be replaced. As noted above, bulk getter pumps are presently available with speeds ranging from about 15 to 500 l/s for air.

The ultimate pressure attainable when a bulk getter pump is the only high vacuum pump in use on a vacuum system is determined largely by its inability to pump helium, neon, and argon. Since these gases comprise about one percent of the normal atmospheric mixture, the lowest pressure a bulk getter pump can produce is about one-hundredth of the pressure to which the system is rough pumped before the roughing pump is turned off and the bulk getter pump is brought into use. For example, if the total pressure in a vacuum system is first reduced to 5×10^{-1} Pa (5×10^{-3} Torr) with a rotary-vane pump, the partial pressure of non-reactive gases remaining in the system would be about one-hundredth of this, or about 5×10^{-3} Pa (5×10^{-5} Torr), and this would be the best pressure attainable with the bulk getter pump. Pumping to lower rough pressures would allow correspondingly lower ultimate pressures to be reached. For example, Danielson Associates report reaching pressures near 10^{-5} Pa (10^{-7} Torr) in systems that are rough pumped with a Tribodyn molecular drag-diaphragm combination pump (Section 6.2.3b, p. 268). Using dry nitrogen to fill a vacuum system when it is let up to atmospheric pressure or flushing with dry nitrogen prior to beginning the rough pumping operation would further reduce the partial pressure of the inert gases, making even lower pressures attainable. The gettering process requires only that gas molecules have an opportunity to come in contact with the getter surface, and this can occur at any pressure. Therefore, a bulk getter pump can make a significant contribution to reaching pressures well into the ultra-high vacuum range if it is used in conjunction with another pump that can remove the inert gases from the system. Danielson Associates produce bulk getter pumps with an ion pump attached for this purpose, as shown in Fig. 7.11.

As noted above, bulk getter pumps, like other entrainment pumps, have a finite operating life, which is determined by the capacity of the getter material for absorbing the reactive gases. Fortunately, it is usually a simple matter to restore the pumping capacity of a pump in which the getter has become saturated with absorbed gases merely by replacing the spent getter material. The frequency with which this must be done depends, of course, on the conditions under which a pump is used. Roughly, each gram of the getter alloys can absorb a total of about 1300 Pa-l of air. This is equivalent to about 3×10^{20} molecules, 5×10^{-4} moles, or 1.5×10^{-2} g of air, and is enough to provide rather impressive capacities in practical designs. For example, the Model SD-100 Sorbodyn pump, which contains 300 g of getter alloy, has a capacity to pump a total of about 4×10^{5} Pa-l of air. Operating at a pressure of 10^{-5} Pa (10^{-7} Torr), where its pumping speed is about 100 l/s, this pump would absorb only $10^{-5}\times100 = 10^{-3}$ Pa-l of air per second, and so its operating life at this pressure would be 4×10^{8} s or more than 10 years. The operating life would, of course, decrease at higher operating pressures, becoming about:

1 year	at	10^{-4} Pa (10^{-6} Torr)
1 month	at	10^{-3} Pa (10^{-5} Torr)
a few days	at	10^{-2} Pa (10^{-4} Torr)
half a day	at	10^{-1} Pa (10^{-3} Torr).

Fig. 7.11 A bulk getter pump with a sputter-ion pump attached to it to remove the inert gases which are not pumped by the getter. (Courtesy of Danielson Associates.)

Although the operating life would be quite short if used for steady-state pumping at 10^{-1} Pa (10^{-3} Torr), this pump would give rather surprising service in evacuating a vacuum system into the high vacuum range after it has first been rough pumped to this pressure. For example, a vacuum chamber with a volume of 100 l would contain 10 Pa-l of free air at 10^{-1} Pa (10^{-3} Torr). Allowing a safety factor of 10 to take care of additional gas that might desorb from surfaces inside the chamber, evacuation of this chamber into the 10^{-5} Pa (10^{-7} Torr) range would require the sorption of about 100 Pa-l of gas, and the pump would have a capacity to do this nearly 4000 times.

Although manufacturers do not generally mention doing so, it is possible to start using a bulk getter pump at pressures considerably above 10^{-1} Pa (10^{-3} Torr) by merely increasing the temperature of the getter, perhaps to the value used for reactivation, at the start of the process, and then decreasing it to the normal operating value as the pressure drops below 10^{-1} Pa (10^{-3} Torr). Such an approach would make a bulk getter pump very useful in rough pumping many vacuum systems, because it would then be possible to use an inexpensive multi-stage diaphragm pump of the type described in Section 4.2.3 (p. 161) to evacuate a system from atmospheric pressure to 400 Pa (4 Torr), whereupon a bulk getter pump could take over and bring the pressure into the high vacuum range. If the vacuum system were an electron microscope with a volume of about 25 l, for example, the bulk getter pump would need to handle about $25\times400 = 10^4$ Pa-l of gas in this process. The Sorbodyn SD-100 pump described above should have the capacity to handle this gas load about 40 times. High-resolution and analytical microscopes, which operate in the 10^{-5} Pa (10^{-7} Torr) range, are usually let up to atmospheric pressure only once or twice a year, and so this pump could serve this pumpdown function, and also contribute to the evacuation process during operation of the microscope, for a number of years before being exhausted.

The traditional procedure of heating bulk getter pumps to several hundred degrees during operation could be objectionable in some applications. For example, it might be undesirable to have a hot pump located near the column of an electron microscope at times when the microscope is being used for high-resolution applications. Conversations with Phil Danielson, of Danielson Associates, reveal that this problem is relatively easy to overcome, however. His experience indicates that the chemical reactions responsible for the pumping action of the getter alloys occur quite readily at room temperature, and that fresh alloy surfaces exhibit pumping speeds for atmospheric gases at room temperature which are nearly as great as those exhibited at the traditional operating temperatures. This means that bulk getter pumps can be used without heating above room temperature. The difference is that the products of the gettering reactions do not diffuse into the alloys at room temperature, and so pumping speed gradually decreases, and pumping life is shorter, because the surface of the getter gradually becomes covered with these products. The operational consequence of this is that activation treatments must be performed more frequently when a bulk getter pump is used at room temperature than when it is used at the traditional operating temperature. However, the period between activation treatments is very acceptable for many applications, as can be seen from a simple calculation. For example, manufacturers' data indicate that the bulk getter alloy powders have accessible surface areas in excess of 1.5×10^5 mm^2 per gram. As noted above, full surface coverage requires the sorption of about 10^{13} gas molecules/mm^2. Assuming that pumping performance might fall to an unacceptable level when the surface becomes half covered, the sorptive capacity would be about $1.5\times10^5\times0.5\times10^{13} \approx 7.5\times10^{17}$ molecules per gram. Since there are 2.47×10^{17} molecules per pascal-litre, this corresponds to a pumping capacity of about 3 Pa-l per gram of getter alloy. On this basis, the Sorbodyn SD-100 pump, which contains 300 g of getter alloy, would be able to adsorb about 900 Pa-l of gas between activation treatments. If this pump, which has a pumping speed of about 100 l/s, were used on an electron microscope operating at 5×10^{-5} Pa (5×10^{-7} Torr) it would pump $100\times5\times10^{-5} = 5\times10^{-3}$ Pa-l of gas per second, and the time between activation treatments would be about $900/5\times10^{-3} = 1.8\times10^5$ s or 50 h. In such an application an automatic timing circuit can be used to perform an activation treatment after midnight each night. The pump would then cool to room temperature by morning, and would have sufficient capacity to operate at room temperature all the following day with substantially full effectiveness.

7.4.3 *Applications*

Bulk getter pumps are inexpensive to purchase, operate, and maintain. They are completely free of oil, noise, and vibrations, and they do not produce or require magnetic fields. They are relatively small in size and weight, and can be mounted in any orientation. They have come into prominence so recently, however, that their potential has not yet been fully explored. The characteristics discussed in the previous sections suggest that they can be used as the only high vacuum pump on systems that do not require pressures much below the 10^{-4} Pa (10^{-6} Torr) range and that are cycled to atmospheric pressure rather frequently. They can also be used as the primary high vacuum pump on systems that are intended to maintain a pressure in the high or ultra-high vacuum range for long periods of time if another pump of modest size is added to handle the inert gases that remain from the initial pumpdown process or are subsequently introduced by small leaks. Bulk getter pumps can also be used in conjunction with oil-free roughing pumps to speed up the pumpdown process. Danielson Associates produce several Tribosorb Modules, which consist of a Sorbodyn pump with an isolation valve and connections for attachment to the roughing line, specifically designed for this purpose. These modules are claimed to reduce pumpdown time by a factor of two, and to decrease the attainable ultimate vacuum by a factor of 10, when used with the Tribodyn pumps described in Section 6.2.3b (p. 268). Obviously, the addition of a bulk getter pump to most vacuum systems that operate below the 10^{-4} Pa (10^{-6} Torr) range would be an inexpensive way to increase the speed and capacity for pumping reactive gases. Some rather unique applications of bulk getters have also been suggested, based on their unusual response to the inert gases and hydrogen. On an apparatus which requires an argon atmosphere, such as an ion mill or a sputter coater, a bulk getter pump can reduce the flow of argon needed to maintain the desired atmosphere by removing reactive gases released by desorption processes and the reactions carried out in the apparatus without removing any argon. Because of their ability to absorb hydrogen and its isotopes near room temperature, and to release them when subsequently heated to higher temperatures, bulk getters can be used to trap, store, and release hydrogen for chemical process applications. The unusual combination of characteristics provided by bulk getter pumps suggest that they will find many other interesting applications in the future. For example, Barry Scientific have recently introduced an accessory package for adding a bulk getter pump to assist in the evacuation of electron gun chambers to improve lanthanum hexaboride filament life. Also, bulk getter strips are being installed inside segments of the storage ring of the Advanced Photon Source, an 1100-metre high-energy positron accelerator which is under construction at the Argonne National Laboratory, and which will use these antimatter particles to produce high intensity beams of X-rays by the synchrotron radiation process, to assist in maintaining the ultra-high vacuum needed to prevent the annihilation of the positrons through collisions with gas molecules.

7.5 The operation of vacuum systems with ion pumps

Fig. 7.12 shows schematically the arrangement of components for an oil-free vacuum system designed to use a sputter-ion pump and a titanium sublimation pump to attain pressures in the ultra-high vacuum range. As noted above (Section 7.1.9) it is not necessary to have a valve, trap, or baffle between the ion pump and the vacuum chamber. To ensure complete freedom from oil contamination it is customary to use sorption pumps to rough pump the system, and it is generally recommended that two such pumps be used sequentially (Section 4.2.1a, p. 158) to obtain a roughing

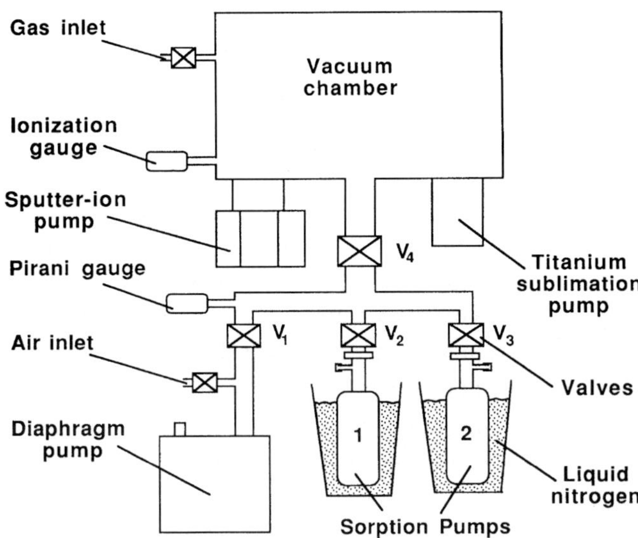

Fig. 7.12 An arrangement of components for an oil-free vacuum system with sputter-ion and titanium sublimation high vacuum pumps, and sorption and diaphragm roughing pumps.

pressure low enough to prevent serious overheating of the ion pump during startup. The system shown also includes a diaphragm pump, which is used for the first stage of the evacuation process. To use multi-staged rough pumping in this way it is necessary to have a roughing manifold with valves V_1, V_2, V_3, and V_4 for isolating the manifold and the individual pumps from the vacuum system. These valves are kept closed except when the pumps attached to them are being used to rough pump the system. It is also necessary to have an accurately-calibrated Pirani or thermocouple gauge to monitor the pressure in this roughing manifold, and it is desirable to have an ion gauge to measure the pressure in the vacuum chamber, although the ion pump current can be used as an indicator of the pressure there. It is also highly desirable to have a titanium sublimation pump in the system to help handle high gas loads, particularly during system startup.

To begin evacuating the system, the sorption pumps are at room temperature, valves V_1, V_2, and V_3 are opened, valve V4 and the inlet valve to the diaphragm pump are closed, and the diaphragm pump is turned on. When the pressure falls to about 10^4 Pa (10^2 Torr) valves V_2 and V_3 above the sorption pumps are closed, and insulated containers are placed around the sorption pumps and filled with liquid nitrogen. It usually requires from 15 to 30 min for the sorption pumps to cool to the operating temperature. During this time valve V_4 is opened, after closing the gas inlet valve for the vacuum chamber, and the vacuum chamber is pumped down to the lowest pressure attainable with the diaphragm pump. At the end of this time valve V_1 above the diaphragm pump is closed, the air inlet valve for the diaphragm pump is opened, and this pump is turned off. Valve V_2 above the first sorption pump is then opened and the pressure is carefully monitored with the Pirani gauge in the manifold. As the pressure approaches 500 Pa (5 Torr) and enters the lower end of the range of viscous flow, valve V_2 is quickly closed so that the inert gases that have been carried into this sorption pump by the viscous flow of other gases are trapped there and thus removed from the system (Section 4.2.1a, p. 158). Valve V_3 above the second sorption pump is then opened and this pump continues the rough pumping operation. When used in this way the sorption pumps should bring the pressure below 10^{-1} Pa (10^{-3} Torr) in 30 min to a few hours, depending on the size and condition of the system. If the system contains a titanium sublimation pump, this is turned on and operated in the continuous sublimation mode when the pressure drops below 5×10^{-1} Pa (5×10^{-3} Torr).

The function selector switch on the power supply for the sputter-ion pump is turned to the position used during startup, frequently the position for monitoring the high voltage, so that the current overload circuit is inactivated. When the pressure drops below 10^{-1} Pa (10^{-3} Torr) the power to this pump is turned on. As discussed above (Section 7.1.6), a low impedance discharge usually develops in an ion pump during startup at such high pressures and the pump becomes unable to sustain more than a few hundred volts potential, but will draw an abnormally high current. If it is operated under these conditions for more than about 5 min the pump may overheat and be damaged. Because the pump power supply is operated during startup in a mode that inactivates the overload safety cutout circuit, it must be watched carefully during this period. If the high voltage does not begin to rise steadily within 3–5 min, the power supply should be turned off for 5 or 10 min to allow it and the pump to cool down, whereupon the startup procedure can be tried again. If the system includes a titanium sublimation pump, it can be used at this time to assist the sorption pump in removing active gases released by the ion pump, and this will be a great help in getting the ion pump started quickly and safely. When the ion pump begins to pump effectively the voltage will rise above 2000 V rather quickly, and then gradually increase. During this time, the current drawn by the pump will decrease steadily, and power dissipation in the pump will drop to a safe level. The function switch on the power supply should then be turned to the normal operating position so that the overload safety cutoff circuit is put into operation to switch off the pump if the ion current subsequently rises to an unsafe level. As the pressure continues to drop the ion pump voltage will continue to rise until it reaches the maximum output of the power supply, and the ion current will decrease steadily (Fig. 7.7). Power supplies usually have a meter that can be used to monitor these current and voltage values, and also to obtain a pressure reading.

As soon as the ion pump begins to pump effectively, valve V_4 is closed to isolate the roughing manifold from the high vacuum chamber, and valve V_3 leading to the second sorption pump is closed. The containers of liquid nitrogen are then removed and the sorption pumps are allowed to warm up to room temperature. The gas released from the molecular sieve material inside these pumps as they warm up must escape through their pressure release valves, and so *great care must always be exercised* to ensure that these valves are in good operating condition. If desired, the sorption pumps can be fully regenerated by heating them at about 250°C for 4–6 h to drive out adsorbed water, as described in Section 4.2.1 (p. 155).

If the system includes a titanium sublimation pump, it is switched over to intermittent operation, as described in Section 7.2.3a (p. 303), when the pressure in the system drops below about 5×10^{-5} Pa (5×10^{-7} Torr). With a system designed to attain pressures in the ultra-high vacuum range, a bakeout procedure is started as soon as the ion pump starts operating effectively. To reach pressures in the 10^{-7} Pa (10^{-9} Torr) range and lower, systems are typically baked out at a temperature of at least 150°C for 10–15 h (after which it usually requires 4–6 h for them to cool to room temperature before being put into operation). On such a system, valve V_4 must be a high-quality valve capable of sealing effectively at pressures in the ultra-high vacuum range, and also capable of withstanding the temperature of the bakeout treatment.

To shut down a system such as this it is only necessary to turn off the power supplies for the ion pump, the vacuum gauges, and the titanium sublimation pump, if one is present, and then to admit dry nitrogen slowly through the gas inlet valve to bring the system up to atmospheric pressure.

A major problem involved in operating well-designed systems with sputter-ion pumps involves rough pumping them to a pressure low enough so that the ion pump will start quickly and smoothly. Although oil-sealed rotary-vane pumps (Section 4.1, p. 134) perform this function very well, they are not recommended for this purpose because of the danger of contaminating the system with pump oil. Sorption pumps, as described above, eliminate all problems with oil contamination, and so this approach is very widely used. Several modifications to the system shown in Fig. 7.12 are frequently encountered. It is possible to replace the diaphragm pump shown in this system with a carbon-vane, piston, or Venturi pump, as suggested in Section 4.2 (p. 155). In many systems this first-stage pump is omitted, and the first sorption pump is used to begin the evacuation at atmospheric pressure, as described in Section 4.2.1a (p. 158). Likewise, a bulk getter pump could be substituted for the titanium sublimation pump, and on many systems this pump is omitted entirely. One very promising recent modification involves replacing the diaphragm and sorption pumps of Fig. 7.12 with a combination diaphragm-molecular drag pump unit, such as the Drytel or Tribodyn units described in Section 6.2.3b (p. 268). These units are capable of starting an evacuation process at atmospheric pressure and reaching an ultimate vacuum below 10^{-4} Pa (10^{-6} Torr), and so are ideally suited for rough pumping a system before a sputter-ion pump is put into operation. One other modification sometimes encountered is the addition of a high-conductance valve between the sputter-ion pump and the vacuum chamber. Although such a valve is not strictly needed, it can make the startup process easier by keeping the ion pump from adsorbing large quantities of gas while the system is open to the atmosphere.

7.6 Concluding remarks

Sputter-ion pumps are widely used on present models of analytical and high-resolution electron microscopes, and it appears certain that they will continue to play a prominent role in the next few generations of electron microscopes. Units of the type shown in Fig. 7.4 which combine an ion pump, a titanium sublimation pump, and a liquid nitrogen trap, are also widely used on photoelectron spectrometers, Auger electron spectrometers, and similar surface analysis instruments, and such instruments are now found with increasing frequency in and around electron microscopy laboratories. Electron microscopists can therefore expect to have to encounter the pumps discussed in this chapter more and more often in the future. Whereas oil diffusion pumps can be considered to be the dominant high vacuum pumps electron microscopists had to deal with in the past, the getter pumps described here may very well fulfill this role in the near future.

8 Cryogenic pumps

Cryogenic pumps remove gas molecules from vacuum systems by collecting them on surfaces cooled to very low temperatures. The condensation of moisture from the air onto the outside of a glass of ice water and the use of traps and anticontamination devices cooled with liquid nitrogen to capture water and oil molecules are examples of the underlying phenomenon involved in cryogenic pumping that are familiar to most of us. Because the gas molecules removed from the vacuum system are retained on cold surfaces inside the cryogenic pumps, these pumps, like the getter pumps described in the previous chapter, are classified as 'entrainment pumps'. However, unlike the getter pumps, which make use of chemical reactions for their pumping action, cryogenic pumps rely only on physical processes in capturing gas molecules. As is true for other types of entrainment pump, cryopumps accumulate so much gas after long periods of operation that their effectiveness decreases to an unacceptable level. At this point they must be *regenerated* by warming them up to room temperature so that the accumulated gases are released and removed.

Cold traps, which are the simplest kind of cryogenic pump, have been in use since the very beginning of vacuum work. In electron microscopy we encounter cold traps mostly in the form of the liquid nitrogen traps that collect and control oil and water vapours in high vacuum systems. However, traps in many other forms, and employing a variety of other cooling agents, have been used for many years and for many other purposes in other types of vacuum systems, particularly those in which gaseous chemical reactions are carried out. As noted in Section 4.2.1 (p. 155), the sorption pump, another common type of cryogenic pump, was introduced in the early 1900s. Most other types of cryogenic pumps came into prominence in the late 1950s and early 1960s in response to the need for high-speed devices to produce a clean ultra-high vacuum for large-scale applications such as space simulators and particle accelerators. Interestingly, the specimen anticontamination devices that are now so widely used on electron microscopes are actually highly specialized cryogenic pumps, and were also introduced at about this time. Cryogenic pumps are now used in a variety of manufacturing processes, particularly in the solid-state electronics industry, and also in many kinds of laboratory-scale scientific instruments. Van Atta (Section 9.9) gives an interesting account of the early work on cryogenic pumps.

The references cited in this chapter are listed in Appendix 1. The suppliers and manufacturers of vacuum equipment referred to here are listed in Appendices 2 and 3.

8.1 Cryogenic pumping mechanisms

In cryogenic vacuum pumps the process of capturing gas molecules on cold surfaces is exploited to the limits of practical feasibility by using the phenomena of cryocondensation, cryosorption, and cryotrapping in a synergistic manner.

Cryocondensation, in this context, refers to the accumulation of gas molecules on a surface which is cooled to a temperature that is low enough to cause the gas to condense to form a solid or liquid phase. When such a temperature is established in a closed system there is an exchange of molecules between the gaseous phase and the condensed phase. If all parts of the system are at the same temperature a condition is ultimately attained where equal numbers of molecules enter and leave the condensed phase each second, and the pressure of the gas above the condensed phase becomes constant. This pressure, attained under conditions of thermal and physical equilibrium, is called the 'equilibrium vapour pressure', the 'saturated vapour pressure', or simply the 'vapour pressure' of the condensed phase. As the temperature is decreased this vapour pressure decreases, because the thermally-activated vibrational energy of the molecules in the condensed phase decreases and they escape into the gas phase at a lower rate. Ice, for example, has an equilibrium vapour pressure of about

665 Pa (5 Torr)	at	0°C (273 K)
10^{-3} Pa (10^{-5} Torr)	at	–100°C (173 K)
10^{-12} Pa (10^{-14} Torr)	at	–150°C (123 K)
10^{-18} Pa (10^{-20} Torr)	at	–196°C (77 K).

0°C is, of course, the normal freezing temperature of water, –100°C and –150°C are temperatures in the range typically attained by specimen anticontamination devices in electron microscopes, while –196°C is the boiling temperature of liquid nitrogen, the coolant most frequently used in vacuum traps. As shown in Fig. 5.6 (Section 5.4, p. 182), the vapour pressures of vacuum pump oils, which have very large organic molecules, are many orders of magnitude lower than the vapour pressure of water at all temperatures, which accounts for the fact that anticontamination devices and liquid nitrogen traps are so effective in removing these materials from vacuum systems. It is also interesting to note that a well-designed anticontamination device capable of reaching temperatures near –150°C, would be much more effective than one that only reaches a temperature near –100°C.

At the boiling temperature of liquid hydrogen (–253°C, 20 K) all gases except hydrogen, helium, and neon have vapour pressures below 10^{-8} Pa (10^{-10} Torr). At the boiling temperature of liquid helium (–269°C, 4 K) the vapour pressure of neon falls below 10^{-8} Pa (10^{-10} Torr), the vapour pressure of hydrogen drops below 10^{-5} Pa (10^{-7} Torr), and only helium remains in the completely gaseous state. These figures suggest that simply by allowing gases to condense onto a surface cooled with liquid helium it might be possible to reduce the pressure in a vacuum system into the ultra-high vacuum range. In fact, many large vacuum systems, such as those used for space simulators and large particle accelerators, are pumped by arrays of large panels cooled by liquid nitrogen and liquid helium. However, when this method of cryocondensation on 'cryopanels' is the only pumping mechanism used, the ultimate pressure is generally limited by the residual amounts of gaseous hydrogen, helium, and neon that remain in the system. These are frequently called the 'permanent gases' because they are so difficult to condense to the liquid or solid state.

Cryosorption is the principal phenomenon that makes it possible to reduce the pressure of these permanent gases into the ultra-high vacuum range by cryogenic means. This possibility arises because there are weak Van der Waals forces of attraction between gas molecules adsorbed on the surface of a solid and the atoms in the surface of the solid itself. Although these forces are small, they can be sufficient to overcome the kinetic energy of gas molecules, particularly if the temperature is low enough. In fact, the pressure of a gas in equilibrium with its molecules adsorbed onto the cooled surface of a reactive solid can be less than the equilibrium vapour pressure for its own solid phase by as much as a factor of 10^{10}. Generally, this *adsorbed pressure* is lowest when only a small fraction of the solid surface is covered with gas molecules, and it usually increases rapidly as surface coverage increases. In the limit, when the adsorbent surface becomes completely covered (about 10^{13} air molecules/mm^2) it is replaced by a surface which consists of adsorbed gas molecules. This is essentially the equivalent of the surface of the condensed phase of the gas itself, and so the process then reverts to cryocondensation and the pressure rises to the equilibrium vapour pressure of the gas for its own condensed phase. Cryosorption is the phenomenon that makes it possible for the sorption pumps described in Section 4.2.1 (p. 155) to use liquid nitrogen cooling to pump large quantities of gas at pressures in the rough vacuum range, and, as we will see shortly, it is the critical mechanism that enables the more sophisticated cryopumps of concern here to reach pressures in the ultra-high vacuum range.

Cryotrapping involves the burial of molecules of one type of gas within a growing deposit of the condensed phase of another. Thus, for example, in a situation where water vapour is rapidly condensing to form ice on a cooled surface, molecules of nitrogen and oxygen that land on the surface of the ice might well be covered over and effectively removed from the gas phase. Although this phenomenon has been used in some special applications, it is not prominently involved in the pumping action of the cryogenic pumps used in most high vacuum applications, because they usually do not operate under conditions that give rise to the continuous formation of large amounts of a condensed phase.

8.2 The design of cryogenic pumps

As suggested above, several different types of cryogenic pumps have been developed. *Liquid-cooled pumps*, which are used mainly on very large vacuum systems, are cooled directly with liquid nitrogen and liquid helium, and rely primarily on the phenomenon of cryocondensation for their pumping action. Some of these, which are essentially highly sophisticated arrangements of cold traps and cryopanels, are referred to as 'liquid-pool pumps' because they contain significant quantities of the liquid cryogens. Others function by pumping the liquid cryogens through coils

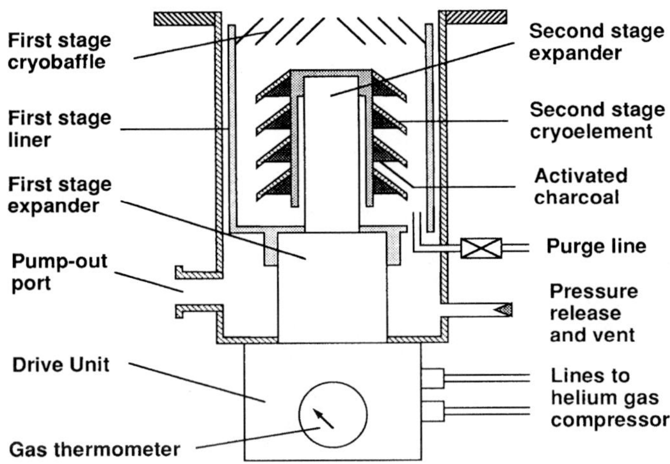

Fig. 8.1 A schematic diagram of the pumping head of a cryogenic pump.

or other types of heat exchangers to cool the pumping surfaces, and are called 'continuous-flow pumps'. Both of these types of pump are often of the 'open-cycle' design, in which the cryogens are allowed to evaporate into the atmosphere. When this is done, the liquid cryogens are frequently produced at gas liquefaction plants located some distance from the site of the pumps. These liquid-cooled pumps and traps, and the principles underlying their design and functioning, are described in some detail by Roth (Section 5.6), Weston (Section 3.5), and Van Atta (Section 9.9). In contrast, the cryogenic pumps most frequently used on electron microscopes and similar laboratory instruments are 'closed-cycle' devices that are cooled by the expansion of helium gas which is recirculated continuously through them by a compressor. Since this is the same general arrangement as is used in ordinary refrigerators, such pumps are also called 'refrigerator cryopumps'. These pumps work by both cryocondensation and cryosorption. It is this type of pump that is usually brought to mind when the term 'cryopump' is used at the present time, and only this type will be considered here. Harris (Chapter 9) and O'Hanlon (Section 9.3.3) give interesting descriptions of these pumps.

The heart of these refrigerator cryogenic pumps is the pumping head, which is built around a two-stage helium gas expansion refrigeration unit of very special design, and which is constructed somewhat as shown in Fig. 8.1. The first stage expander usually reaches temperatures in the range

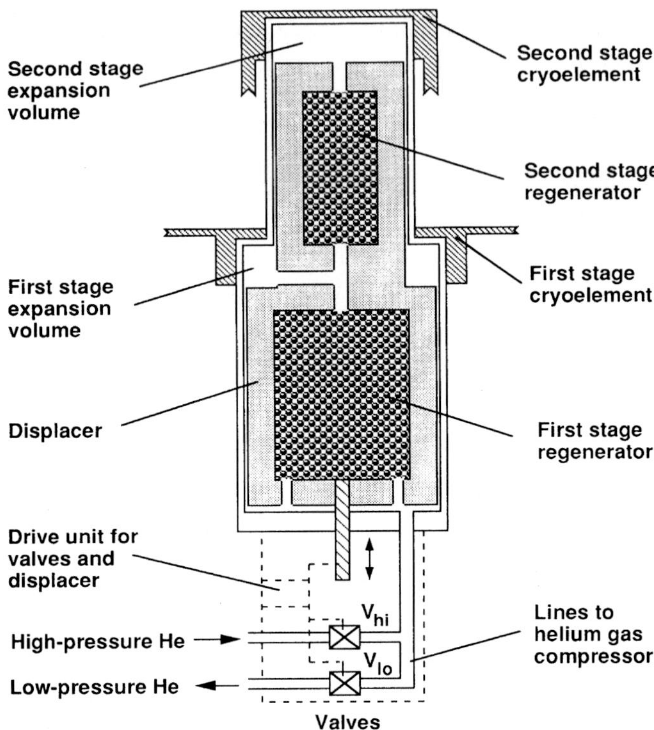

Fig. 8.2 A schematic diagram of the refrigerator mechanism used in most cryopumps.

from 50 to 75 K. This stage cools a liner for the pump housing and a set of baffles across the inlet port which serve as radiation shields to protect the colder inner elements of the second stage from radiative heating by surrounding warmer structures. The second stage expander usually reaches temperatures in the range from 10 to 20 K. It cools one or more cryoelements that are coated with activated charcoal on the inner (under) side. The temperature of this stage is monitored by a hydrogen vapour-pressure thermometer or a semiconductor sensor. The overall design is such that most gas molecules reaching the pump strike the first stage cryobaffle, and such that those that pass through the cryobaffle are directed onto the inner surface of the first stage liner. Virtually all water molecules are pumped on these surfaces by the cryocondensation mechanism, while molecules of other gases give up a significant fraction of their thermal energy here. Gas molecules that are not captured on these first stage surfaces have a high probability of striking the outer surfaces of the second stage cryoelements before reaching the activated charcoal. Oxygen, nitrogen, carbon dioxide, and all other gases except hydrogen, helium, and neon are captured on these surfaces, also by cryocondensation. These last three gases are then left to be pumped by cryosorption on the activated charcoal. Although activated charcoal has a slightly lower sorptive capacity than the molecular sieves which are used in sorption pumps (Section 4.2.1, p. 155), it is used as the adsorbent in this application because it desorbs all gases completely at room temperature, and so there is no need to heat the delicate refrigeration unit to high temperatures during regeneration.

The pumping head is cooled by *gas expansion*. Anyone who has ever used a simple hand-held pump to inflate a high pressure bicycle tyre is painfully aware that rapidly compressing a gas can cause it to heat up, because the pump invariably becomes too hot to hold comfortably long before the tyre is fully inflated. Conversely, when a compressed gas is caused to expand rapidly, its temperature can decrease significantly. This occurs because the molecules in a highly compressed gas are close enough together that the effects of the Van der Waals forces of attraction between them become significant in comparison with their thermally-activated kinetic energy. When the gas is allowed to expand suddenly the energy needed to overcome these attractive forces is obtained by a reduction in the kinetic energy of the molecules, and this is manifested as a decrease in the temperature of the gas.

The basic features of the mechanism used in most current cryopumps to produce this cooling by gas expansion are depicted in Fig. 8.2. As shown, the expander cylinder of the pumping head contains a moveable cylindrical device called a 'displacer', which is made of an insulating material, and which is driven up and down in the expander. In the Gifford–McMahon design the displacer is driven by an electric motor, while in the Longsworth design it is driven by a piston, which is an integral part of the displacer, and which is actuated by helium gas controlled by a set of motor-driven valves. The displacer contains two cavities filled with tightly-packed metallic spheres, gauze, or mesh to provide a large surface area and a high heat capacity. These packed cavities, which are highly efficient heat exchangers, are called 'regenerators'. Channels are bored in the displacer in such a way that all the helium involved in producing the cooling effect passes through the regenerators as it enters and leaves the unit. The flow of helium is controlled by valves in the high-pressure (supply) line and the low-pressure (return) line whose operation is synchronized with the movement of the displacer by the drive unit.

To understand how this device functions to produce the very low temperatures required for effective cryogenic pumping, it is convenient to assume that the displacer has been driven to the bottom of the expander cylinder, as shown in Fig. 8.2, that valve V_{lo} in the low-pressure return line to the compressor is closed, that the valve V_{hi} in the high-pressure supply line from the compressor is open, and that the first and second stage regenerators and the first and second stage expansion volumes are all filled with compressed helium gas.

Then, valve V_{hi} in the supply line is closed, valve V_{lo} in the return line is opened, and the compressed gas is rapidly pumped out of the unit by the compressor. This causes the gas to expand suddenly and to cool the first and second stage expansion chambers and the cryoelements attached to them, and also the packing material in both regenerators. Additional cooling of the regenerators is achieved by driving the displacer to the top of the expander so that the cold gas remaining in the expansion chambers passes through them. From the regenerators, this gas flows into the space that was opened up between the bottom of the displacer and the bottom of the expander cylinder by the upward movement of the displacer, and then it returns to the compressor.

Next, the valve V_{lo} in the return line is closed, the valve V_{hi} in the supply line is opened, and compressed gas is pumped into the space at the bottom of the expander and into the regenerators. The gas that enters the regenerators is cooled by the cold packing material. This cooled gas is then forced into the expansion chambers by driving the displacer down to the bottom of the expander cylinder.

This brings the system back to the configuration assumed at the start, whereupon this process, which classically is known as the 'Solvay refrigeration cycle', is repeated successively. With each repetition slightly lower temperatures are achieved in the expansion volumes, until quasi-equilibrium operating values are attained. In general, the colder a gas is before it is expanded, the lower the temperature it will reach upon expansion. Therefore, the second stage reaches a lower temperature than the first stage, because the compressed gas that enters the second stage expansion volume is cooled by passing through two regenerators, whereas that which enters the first stage is cooled by only one. The highly refined units used in modern cryopumps are capable of maintaining the second stage cryo-elements at temperatures in the range from 10 to 20 K, while temperatures in the range from 35 to 75 K are typical for the first stage. Helium is not one of the most effective refrigeration gases, because the Van der Waals forces between its monoatomic 'molecules' are very small, and so this rather remarkable performance is a tribute to the extent to which the design and production of these units has been refined by their manufacturers.

The high-pressure helium needed by the two stage expander is provided by a compressor unit which is connected to the pumping head by flexible tubing to minimize the transmission of vibrations between the compressor and the pumping head. These compressor units usually contain four major components: the compressor itself, a water-cooled heat exchanger, an oil separator, and an oil adsorber. The compressors are usually oil-sealed devices similar to those used in air conditioners and household refrigerators, although manufacturers now use designs that are modified and refined to make them especially well suited for compressing helium. Ordinary industrial and laboratory grades of helium usually contain small amounts of water, nitrogen, and neon which impair performance by condensing in the pumping head and impeding the flow of helium. Therefore, only ultra-high purity helium (99.995 or 99.9995 percent) is used in these systems. The heat exchanger system serves to keep the compressor from overheating and to cool the helium gas as it leaves the compressor so that it enters the pumping head at room temperature. The helium gas picks up oil as it passes through the compressor, and this must be prevented from reaching the pumping head, where it would solidify and prevent the displacer from moving. The oil separator mechanically collects most of the oil from the helium and returns it to the compressor. Remaining traces of oil are then removed by adsorption on activated charcoal in the oil adsorber. Since the charcoal ultimately becomes saturated with oil, the adsorbers in most compressor units must be replaced after about 10 000 h of service. Manufacturers provide compressor units that are specifically designed for use with their pumping heads. Usually two or three different sizes of compressor units are available, with the smaller units being suitable for running from one to three small pumping heads or a single larger head, while the large units have the capacity to run several small pumps or one of the largest ones.

8.3 Heat flux

One problem unique to cryogenic pumps is the need for concern about the rate of heat input to the pumping head. The operating temperatures attained by the first and second stage cryoelements are determined by the ratio of the refrigerating capacity of the expander units to the level of heat flux they experience. This heat load has three major components that are of concern in understanding the operation of cryopumps: (i) heat conducted from the body of the pump to the cryoelements by free gas molecules inside the pump, (ii) heat radiated to the cryoelements by warmer surrounding objects, and (iii) the heat of condensation released by the gas molecules as they are cooled from room temperature and condensed or adsorbed on the cryoelements. The total heat load a pump can absorb from these sources while maintaining its operating temperature depends on the refrigeration capacity of its expander unit. For pumps with inlet diameters of 100, 200, and 300 mm, for example, this capacity is usually about 15 W (J/s) for the first stage, and about 3 W for the second stage. Pumps are designed so that these capacities are sufficient to accommodate the heat influx under 'normal' operating conditions, and to provide a modest excess capacity to handle minor overload conditions. If serious overloading occurs, however, the refrigeration unit may be unable to handle the extra heat influx, and the temperature of the cryoelements can rise to a level where effective pumping ceases.

The conduction of heat from the outer wall of the pump to the cryoelements by free gas molecules inside the pump head is one of the most important factors contributing to the thermal overloading of cryogenic pumps. As described in Section 3.1.1 (p. 82), the thermal conductivity of gases at pressures below 10^{-1} Pa (10^{-3} Torr), where gas flow is molecular in character, is so low that conductive heat transfer by gas molecules is negligible. Pressures in this range are commonly said to provide an 'insulating vacuum'. As pressure increases through the range of transitional flow, gas conductivity increases by a factor of about 10^3, approximately linearly with pressure, up to about 10^2 Pa (1 Torr). At higher pressures gas flow becomes fully viscous in character and the thermal conductivity remains essentially constant, because it is controlled by gas viscosity which is also constant in this pressure range (see Section 1.11, p. 25). Cryopumps normally operate at pressures below 10^{-1} Pa (10^{-3} Torr) where conductive heat transfer is insignificant. If, however, the pressure in a cryopump rises appreciably above this maximum safe operating level, the rate at which heat is conducted from the wall of the pump to the first and second stage cryoelements can rise to the point where it represents a significant fraction of the refrigeration capacity of the pumping head. Under such conditions, the temperature of the cryoelements will begin to increase. This, in turn, will cause gases to desorb from the cryoelements, exacerbating the situation. If the gas pressure rises high enough, or if the condition persists long enough, the situation can degrade to the point where the refrigeration unit can no longer handle this abnormal heat load and the pumping action will cease.

The rate at which thermal energy is radiated to a cold body at temperature T_{lo} from a surrounding warmer body at temperature T_{hi} is proportional to the difference between the fourth power of their temperatures; that is, $H \propto (T_{hi}^4 - T_{lo}^4)$. At pressures below 10^{-1} Pa (10^{-3} Torr), where conductive heat transfer is negligible, heat radiated from surrounding warmer objects becomes the dominant heat load for the cryoelements in a cryopump. Cryopumps are carefully designed to accommodate this load under normal operating conditions. For example, the second stage cryoelements, which must achieve and maintain the lowest temperature, are shielded from radiation from the warmer outside wall of the pump by the first stage liner, and from warmer objects in the vacuum chamber by the cryobaffle in the inlet port of the pump. The capacity of the first stage refrigeration unit is designed to handle the radiative heat loads to the liner and cryobaffle when these surrounding objects are at temperatures below about 27°C (300 K). Radiative heat loads are therefore unlikely to be a problem when cryopumps are used on instruments such as electron microscopes and surface analysis units which do not have components that run at high temperatures. However, problems can arise with vacuum evaporators and various manufacturing apparatus in which heated filaments, furnace units, and high temperature gas discharges can act as sources of high radiant heat loads. For example, turning on such a device that is openly exposed to the inlet of a cryopump and heating it to 2000 K would increase the radiant heat load to the cryobaffle (at a temperature of 70 K) by about $(2000^4 - 70^4)/(300^4 - 70^4) \approx 2000$ times. The magnitude of the increase in the rate of heat input (in watts) will, of course, depend in a complex way on the size and location of the hot object relative to the inlet of the pump.

Nonetheless, it is clear that under some circumstances the radiant heat load could become quite large. Although cryopumps have a limited reserve capacity that will handle exposures to such abnormal loads for a short time, prolonged exposure can cause the temperature of the cryoelements to rise to the point where captured gas molecules are released from them. This can then introduce problems from increased conductive heat input and ultimately cause the pump to cease functioning. In apparatus that contain sources of high radiative heat loads, the inlet to the cryopump must be shielded by reflective baffles, and these may need to be water cooled under extreme conditions.

The third source of heat input, the release of heat from the condensation of gas molecules onto the cryoelements, is also well within the capacity of cryopumps under normal operating conditions. For example, the amount of heat that must be absorbed to condense oxygen, nitrogen, and most other common atmospheric gases is roughly 15 kJ/mol, or 6×10^{-3} J/Pa-l. When operating at an inlet pressure of 10^{-2} Pa (10^{-4} Torr), a pump with a diameter of 200 mm, and a typical pumping speed of 1500 l/s for nitrogen, condenses $1500\times10^{-2} = 15$ Pa-l of this gas per second. To do this it must absorb $15\times6\times10^{-3} = 0.09$ J of heat per second. This requires a refrigeration capacity of only 0.09 W, and the requirement would be even less at lower pressures. As noted above, the second stage expanders in pumps of this size usually have refrigeration capacities of about 3 W, which is certainly adequate to handle this heat load at these pressures. Even during operation at 10^{-1} Pa (10^{-3} Torr), the upper limit of the normal operating range of cryogenic pumps, the heat load from this source would be only about 0.9 W, which is within the capacity provided. However, if the pressure rises too high the increased heat input from condensing gases will combine with the effects of increased gas conductivity in contributing to the possibility of overloading the refrigeration unit.

Manufacturers indicate the capacity of cryopumps to handle these condensable gases by specifying a throughput value for argon. This is usually given as the rate, in Torr-litres per second, at which argon can be pumped without causing the temperature of the second stage cryoelements to rise appreciably above 20 K, so that hydrogen will not be released from the charcoal bed. For example, the pump just described would typically have a pumping speed for argon of about 1200 l/s and an argon throughput capacity of about 10 Torr-l/s. According to eqn. 2.9 (Section 2.5, p. 39) the maximum operating pressure for this pump would be about $P_p = Q/S_p = 10/1200 = 8\times10^{-3}$ Torr or about 1 Pa. The operation of a pump at pressures approaching this level is not generally recommended, because it is so close to the condition where a thermal overload can develop that a small fluctuation in operating conditions can cause the pump to cease pumping. Operation under such conditions also increases the rate at which gases accumulate inside the pump, making it necessary to shut the pump down more frequently for regeneration.

Interestingly enough, however, cryopumps are widely used in the semiconductor manufacturing industry on apparatus that require argon pressures in the range from 5 to 500 Pa, such as sputtering, reactive ion etching, plasma etching, and low-pressure chemical vapour deposition. This is such a large market that nearly all manufacturers of cryopumps offer models especially designed for such applications. These models have either a fixed or variable aperture in the inlet port. This aperture is designed so that the flow of gas into the second stage of the pump remains at an acceptable level, while the required operating pressure is maintained in the vacuum chamber by controlling the rate at which the argon is admitted to it. O'Hanlon (Chapter 12) discusses the design and operation of systems of this kind in some detail.

8.4 Impulsive gas loads and the crossover pressure

As noted above, cryopumps have a certain reserve heat capacity, determined by the design of their cryoelements and refrigeration systems, that allows them to accommodate a short exposure to an abnormally high heat load without warming up enough to seriously impair the pumping action. One very important function of this reserve capacity is to provide a margin of safety in preventing thermal overloading during minor deviations from normal operating conditions. If, for example, a sudden burst of gas is accidentally admitted to a system, the cryopump will not fail as long as the extra heat released by this *impulsive gas load* does not exceed the pump's reserve capacity. Likewise, if power failure causes the supply of refrigerant to be interrupted for a short time this reserve capacity can prevent a thermal overload from occurring. Pumps will usually recover if the supply of refrigerant is not interrupted for more than 5 or 10 min.

The capability of cryopumps to tolerate impulsive gas loads is particularly important during the pumpdown stage of evacuation, because it is usually not feasible to rough pump a system below the maximum steady-state operating pressure of the cryopump, and so the pump will experience an abnormally high influx of gas whenever the valve between it and the vacuum chamber is first opened. The quantity of gas a pump can handle in this manner without having the temperature of the second stage cryo-elements rise appreciably above 20 K is usually called its 'crossover capacity', C_x. Crossover capacities are almost universally given in units of Torr-litres, and typically have values of about 50, 150, and 300 Torr-l for pumps with inlet diameters of 100, 200, and 320 mm, respectively, although there are large variations among models of a given size. In general, the larger the value of C_x for a given size pump, the better.

The crossover capacity of a pump determines the maximum pressure at which it can be opened to the vacuum system during the pumpdown process. This is commonly referred to as the 'crossover pressure', that is, the pressure at which the pumping is 'crossed over' from the rough vacuum phase to the high vacuum phase. This crossover pressure P_x is determined by both the crossover capacity of the pump and the volume of the vacuum system V_s on which it is being used. For example, the crossover pressure for a pump with a diameter of 100 mm and a typical crossover capacity of 50 Torr-l would be about $P_x = C_x/V_s = 50/25 = 2$ Torr or about 270 Pa on a system with a volume of $V_s = 25$ l, but about $50/100 = 0.5$ Torr or 70 Pa on a system with a volume of 100 l. If a pump with a diameter of 200 mm and a crossover capacity of 150 Torr-l were used on this latter system the crossover pressure would be about $150/100 = 1.5$ Torr or 200 Pa. Thus, crossover capacity is an important characteristic of cryogenic pumps that must be given careful consideration in both the design and operation of vacuum systems on which they are used.

In designing vacuum systems it is desirable to choose a cryopump that is large enough so that the time required for the rough-pumping operation is not excessively long. The cryopump must also be compatible with other characteristics of the rough-pumping method chosen. If the vacuum system is rough pumped with an oil-sealed rotary-vane pump, the rough-pumping operation should be stopped well before the pressure in the roughing line drops below the range of viscous flow, in order to avoid contaminating the system with oil from the rotary-vane pump. This is a very important requirement, because it takes very little oil vapour to contaminate the charcoal bed in a cryopump so seriously that it will no longer adsorb the permanent gases effectively. If such contamination occurs, the pump must be shut down and the charcoal bed in the second stage cryoelement must be replaced. To prevent this, manufacturers usually specify that the rough-pumping operation should be stopped before the pressure drops below about 5 Pa (0.05 Torr) when an oil-sealed rotary-vane pump is used on roughing lines with diameters of 25–50 mm. If we allow a margin of safety and use 10 Pa (0.1 Torr) as the crossover pressure (see Fig. 2.1 in Section 2.2, p. 31), the crossover capacity of the cryopump should be $C_x \geq P_x V_s \approx 0.1 V_s$. This is not a very difficult condition to meet, except for very large systems. For example, the largest glass bell jars commonly in use, which are about 450 mm in diameter and 750 mm high (18 inch×30 inch), have a volume of about 115 l and would require a cryopump with $C_x \geq 0.1 \times 115 \approx 12$ Torr-l. The smallest cryopumps available from most manufacturers, which are only about 100 mm in diameter, have crossover capacities near 50 Torr-l. These small pumps could easily handle the pumpdown load from these large bell jars, and might even be used on systems with volumes up to as much as $V_s \approx C_x / P_x \approx 50/0.1 \approx 500$ l. While we are not directly concerned with the design of vacuum systems, simple calculations such as these do help to illustrate an unusual characteristic of cryogenic pumps; namely, that the proper crossover pressure for systems with cryopumps is determined by *the quantity of gas* in the system rather than by the actual gas pressure.

Another aspect of this is that vacuum systems can be designed for operation at considerably higher crossover pressures with cryopumps than with most other types of high vacuum pumps. As noted in previous chapters, the highest acceptable crossover pressure is about 1 Pa (10^{-2} Torr) for sputter-ion and titanium sublimation pumps, about 30 Pa (0.2 Torr) for oil diffusion pumps, and about 100 Pa (1 Torr) for turbomolecular pumps, regardless of the size of the vacuum system or the size of the pump involved. On the other hand, the crossover pressure can be made significantly higher than any of these values when a cryogenic pump is used, simply by choosing a pump with a sufficiently large crossover capacity. For example, with typical cryopumps the crossover pressure for a system with a volume of 50 l would be $P_x = C_x / V_s = 50/50 = 1$ Torr or 133 Pa with a 100 mm pump, but could be increased to 150/50 = 3 Torr or 400 Pa with a 200 mm pump, or to 300/50 = 6 Torr or 800 Pa with a 300 mm pump. This characteristic can be very useful in designing vacuum systems. For example, it makes it possible to use a multi-stage diaphragm pump (Section 4.2.3, p. 161), which can reach pressures near 250 Pa (2 Torr), for the rough-pumping operation to obtain a completely oil-free system. With this approach, a typical 200 mm cryopump can be used on systems as large as $V_s \approx C_x/P_x \approx 150/2 \approx 75$ l, for example.

It may be difficult to calculate the crossover pressure when it comes to the practical problem of operating a vacuum system that is already equipped with a cryopump. Whereas the crossover capacity of the pump may be available in specifications furnished by the manufacturer, it may be very difficult to estimate the volume of the system with any accuracy. This would certainly be true for most electron microscopes, and for many other pieces of scientific equipment as well. In this situation it is helpful to remember that the crossover capacity of the pump is determined by the impulsive gas load it can accommodate without causing the temperature of the second stage cryoelements to rise significantly above 20 K. On this basis, a compromise can be made among such factors as avoiding oil contamination, reducing the time required for pumpdown, and minimizing the total amount of gas deposited in the pump during the crossover operation, provided the temperature sensors on the cryopump do not show a significant increase in the temperature of the second stage during crossover. As a first try the system can be rough pumped to the lowest pressure that will certainly not permit oil backstreaming, if the system has an oil-sealed roughing pump, or the lowest pressure the roughing pump can reach, if it has an oil-free roughing pump. If the system is properly designed, it should be perfectly safe and acceptable to cross over at this pressure. The only disadvantage of this approach is that it will give a maximum time for the rough-pumping operation. This should not be a problem for the many types of scientific instruments which are seldom cycled to atmospheric pressure. However, a short rough-pumping time may be highly advantageous for apparatus, such as vacuum evaporators and many manufacturing units, which are cycled from atmospheric pressure to high vacuum quite frequently. The approach here is to start with the pressure produced by the longest 'reasonable' rough-pumping time. If the temperature of the second stage does not increase when this pressure is used, then a slightly shorter rough-pumping time can be tried; otherwise, a slightly longer one must be accepted. The use of short rough-pumping times, and the higher crossover pressures they require, has the disadvantage of depositing larger amounts of gas in the cryopump during each cycle, which will make it necessary to regenerate the pump more frequently.

8.5 Ultimate pressure

It is interesting that most manufacturers do not advertise ultimate pressures for the cryopumps they manufacture, and most of the texts cited above do not discuss ultimate pressures in any detail. However, consideration of the characteristics of the pumping phenomena involved can give considerable insight into this matter. As noted above, most atmospheric gases are pumped by the cryocondensation phenomenon, and have very low saturation vapour pressures at the temperature of the cryoelements on which they condense. Most water molecules are pumped on the cryobaffle which is cooled by the first stage expander. At 77 K, the upper end of the operating range of this device, the equilibrium vapour pressure of water is below 10^{-17} Pa (10^{-19} Torr). Nitrogen, oxygen, carbon dioxide, and similar gases are pumped by cryocondensation on the second stage cryoelements. At 15 K, about the middle of the operating temperature range of these elements, the saturated vapour pressure of nitrogen is near 10^{-14} Pa (10^{-16} Torr), and the vapour pressures of all other gases except hydrogen, helium, and neon are even lower. All of these pressures are deep in the ultra-high vacuum range.

Strictly speaking, however, these saturation vapour pressures are attained only when the molecules in the gas phase are at the same temperature as the pumping surface. O'Hanlon (p. 226), Roth (p. 248), and Weston (p. 94) all present analyses based on the kinetic molecular model which indicate that when the gas molecules and surrounding surfaces are at a temperature T_w which is warmer than the temperature of the pumping surface T_p, the ultimate pressure attainable P_{ult} is given by a formula of the form:

$$P_{ult} \approx P_{eq}\left(\frac{fe}{fw}\right)\sqrt{\frac{T_w}{T_p}} \tag{8.1}$$

In this equation f_e is the sticking coefficient for gas molecules in thermal equilibrium with the pumping surface, f_w is the sticking coefficient for the warmer gas molecules, and P_{eq} is the equilibrium vapour pressure of the gas at the temperature of the pumping surface. The sticking coefficients for the cryocondensation processes that occur in refrigerator cryopumps are close to one for the conditions that prevail there, and so the ratio (f_e/f_w) probably does not have a major effect on P_{ult}. At worst, it should not exceed a value of 10. The quantity $\sqrt{T_n/T_p}$ is commonly called the 'thermal transpiration ratio'. This ratio has a maximum value of about $\sqrt{70/10} \approx 2.6$ for the second stage cryoelement, which is surrounded by the first stage liner at 70–80 K, and would have a value of only about $\sqrt{300/10} \approx 5.5$ even if this element were surrounded by surfaces at room temperature. For the first stage cryobaffle, the maximum value for this ratio would be about $\sqrt{300/35} \approx 3.0$. On this basis, the product of the (f_e/f_w) and $\sqrt{T_w/T_p}$ terms probably does not exceed a value of 10 under normal operating conditions, and 100 under extreme conditions. Therefore, the cryocondensation processes should be capable of reducing the pressures of the condensable gases to values which are within the range of their equilibrium vapour pressures at the temperatures of the pumping surfaces, and these expected pressures are all deep into the ultra-high vacuum range.

The effect of the cryosorption phenomenon, which is responsible for pumping hydrogen, helium, and neon, on the ultimate attainable pressure is somewhat less favourable, however. The adsorption pressure of these gases in thermal equilibrium with fresh charcoal at temperatures below 20 K is usually reported to be below 10^{-15} Pa (10^{-17} Torr). When the gases and surrounding fixtures are warmer than the pumping surface, the ultimate pressure attainable is again described by an equation of the form of eqn. 8.1. Now, however, two phenomena make the attainable pressure increase above this optimal value considerably more than for the cryocondensation process discussed above; namely, both the ratio of sticking coefficients (f_e/f_w) and the equilibrium adsorption pressure P_{eq} increase rapidly as the fraction of the adsorbent surface covered with gas molecules increases. Data presented by APD Cryogenics in advertising literature for their cryopumps indicate that the pressure of hydrogen in equilibrium with activated charcoal at 15 K, is about:

10^{-20} Pa (10^{-22} Torr)	for	fresh charcoal
10^{-12} Pa (10^{-14} Torr)	for	25 percent saturation
10^{-7} Pa (10^{-9} Torr)	for	50 percent saturation

10^{-3} Pa (10^{-5} Torr)	for	75 percent saturation
10^{4} Pa (10^{2} Torr)	for	100 percent saturation.

The value for full saturation is, of course, the equilibrium vapour pressure for hydrogen at 15 K. Temperature also has a strong effect. For 50 percent saturation, the pressure of hydrogen in equilibrium with the activated charcoal is about:

10^{-14} Pa (10^{-16} Torr	at	10 K
10^{-7} Pa (10^{-9} Torr)	at	15 K
10^{-3} Pa (10^{-5} Torr)	at	20 K
10 Pa (10^{-1} Torr)	at	30 K.

The adsorption pressures of helium and neon show similar variations with surface coverage and temperature. These relationships suggest that the cryosorption of hydrogen, helium, and neon is the factor most likely to limit the ultimate pressure attainable with a cryopump.

If cryopumps are not under a heavy heat load, so that the second stage temperature is low in the operating range, and if the charcoal is not heavily loaded with adsorbed gases, then ultimate pressures below 10^{-8} Pa (10^{-10} Torr) should be readily attainable. This is consistent with the experience of one of my colleagues who routinely obtains pressures in the middle of the 10^{-8} Pa (10^{-10} Torr) range in a large system that is evacuated only with cryopumps. This system was, of course, constructed of stainless steel by fabrication techniques developed specifically for ultra-high vacuum applications, and it is routinely subjected to extended bakeout treatments above 200°C. As is true for other types of high vacuum pumps, such low ultimate pressures cannot be expected if the vacuum system cannot be baked out (Section 2.10.3, p. 62), or if it was constructed with 'ordinary' methods and materials and has rubber O-rings. For such systems, ultimate pressures in the low 10^{-5} Pa (10^{-7} Torr) or upper 10^{-6} Pa (10^{-8} Torr) range are about the best that can be expected.

8.6 *Pumping speed*

The rigorous definition, calculation, and measurement of the pumping speeds of cryogenic pumps is a complex matter which is discussed in some detail by Roth (p. 161 and p. 250), Weston (p. 94), and Van Atta (Section 9.9). Equations given for the pumping speed of a cryogenically-cooled panel S_{cryo} with an area of A (mm^2) that is openly exposed to the gas molecules in a vacuum chamber at a pressure P_c (Pa) are usually of the form:

$$S_{cryo} \approx 0.0364 \ A f_w \sqrt{\frac{T_w}{M}} \left(1 - \frac{P_{eq}}{P_c} \sqrt{\frac{T_w}{T_p}} \right) \ \text{l/s} \qquad (8.2a)$$

with the variables having the same meanings as given above for eqn. 8.1. For refrigerator cryopumps this equation applies directly only to the cryobaffle in the inlet port of the pump, because this is the only pumping surface that is directly accessible to the gas molecules in the vacuum chamber. This baffle has an area nearly equal to the inlet area of the pump. If the size of the inlet is expressed in terms of its diameter D (mm), this equation can be written as:

$$S_{cryo} \approx 0.0286 \ D^2 f_w \sqrt{\frac{T_w}{M}} \left(1 - \frac{P_{eq}}{P_c} \sqrt{\frac{T_w}{T_p}} \right) \ \text{l/s} \qquad (8.2b)$$

since $A = \pi D^2/4$. This baffle is usually cooled below 77 K where $f_w \approx 1$ for water molecules. Furthermore, at this temperature P_{eq} for water is several orders of magnitude smaller than P_c, and so the term $[1-(P_{eq}/P_c)\sqrt{T_w/T_p}]$ also has a value close to one. Therefore, when applied to the pumping of water this equation reduces to:

$$(S_{cryo})_{water} \approx 2.86 \times 10^{-2} \ D^2 \sqrt{T/M} \ \ \text{l/s} \qquad (8.2c)$$

It is very interesting to note that this is the same as eqn. 2.13d (Section 2.5.1, p. 47) for the theoretical maximum pumping speed S_{tm}, a result which accounts for the remarkable fact that cryogenic pumps *achieve nearly the theoretical maximum pumping speed for water vapour.*

For water at room temperature $M = 18$ and $T = 293$ K, and so this equation simplifies to:

$$(S_{cryo})_{water} \approx 0.115 \, D^2 \ \ \text{l/s} \qquad (8.2d)$$

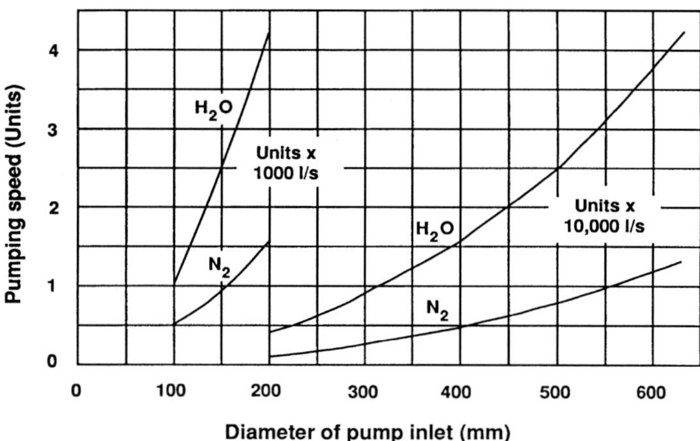

Fig. 8.3 The variation with inlet diameter of the pumping speed of cryogenic pumps for water and nitrogen.

For pumps with inlet diameters of 100, 200 and 400 mm, for example, this gives values of about 1150, 4600, and 18 400 l/s, respectively. Fig. 8.3 is a plot, based on data presented in catalogues of five manufacturers, showing typical pumping speeds for present models of cryopumps. The values shown in this plot for the pumping speed for water are 85–90 percent of these maximum possible values. Such exceptionally high speeds are attained because water molecules in the vacuum system have unobstructed access to the cryobaffle, and because the water molecules are captured on this baffle with great efficiency by the cryocondensation phenomenon. These unusually high pumping speeds for water are very advantageous in most applications because, as was discussed in Section 2.10.3 (p. 60), such large quantities of water desorb from the walls of most vacuum systems that water vapour becomes the principal gas to be handled by the vacuum pumps once most of the free gas molecules have been removed during the initial stages of pumpdown.

The pumping of gases such as nitrogen, oxygen, carbon monoxide, carbon dioxide, and argon, which are captured by cryocondensation on surfaces inside the pumping head, is relatively less efficient than the pumping of water, largely because the cryobaffle significantly obstructs the access of the gas molecules to these surfaces. Even though manufacturers design the cryobaffles and other internal features of their pumps to give gas molecules the best possible access to these internal pumping surfaces consistent with other requirements, the effect is quite significant. For example, for nitrogen the theoretical maximum pumping speed is $S_{tm} = 0.093\ D^2$ l/s, or about 925, 3700, and 14 800 l/s for pumps with inlet diameters of 100, 200, and 400 mm, respectively. The pumping speeds shown for nitrogen in Fig. 8.3 are less than half these theoretical maximum values. Nonetheless, cryopumps are very competitive with other high vacuum pumps in handling these gases. The numerical values for the pumping speeds for nitrogen shown in Fig. 8.3 are nearly as great as those for oil diffusion pumps (see Fig. 5.3 in Section 5.2, p. 175), and they are considerably greater than the pumping speeds provided by turbomolecular pumps (see Fig. 6.5 in Section 6.1.5, p. 239) and sputter-ion pumps (Section 7.1.3, p. 281), when pumps with similar inlet diameters are compared. If water-cooled baffles or liquid nitrogen cold traps are used in conjunction with the oil diffusion pumps, their effective pumping speeds for air are reduced by nearly a factor of two and become less than the speeds of cryopumps of comparable sizes. However, such liquid nitrogen cold traps would provide nearly theoretical pumping speeds for water vapour.

The pumping of hydrogen, helium, and neon is even less efficient, because these gases are pumped by cryosorption on the charcoal which is attached to the undersides of the second stage cryoelements, and access to these areas is even more restricted. On a relative basis the pumping speeds for these gases are only about 15–20 percent of the theoretical maximum possible values. However, these theoretical maximum values are so much larger than those for the other gases (e.g. $S_{tm} = 0.346\ D^2$, and $0.245\ D^2$ for hydrogen and helium, respectively, compared with $0.093\ D^2$ l/s for nitrogen) that the numerical values of the speeds for hydrogen and helium are usually about 25 percent greater than those for oxygen and nitrogen, as shown in Fig. 8.4. Physically, this reflects the fact that the light molecules of hydrogen and helium have higher speeds than the heavier molecules of the other gases. This allows them to travel into the interior of the cryopumps faster and partially compensates for the inaccessibility of the charcoal surfaces. Of course, the pumping speeds for hydrogen and helium decrease as the amount of gas adsorbed on the surface of the charcoal increases.

Cryogenic pumps are widely used in the electronics industry to evacuate apparatus for processes that require the pumping of large quantities of argon, such as sputtering and ion etching. For this reason manufacturers

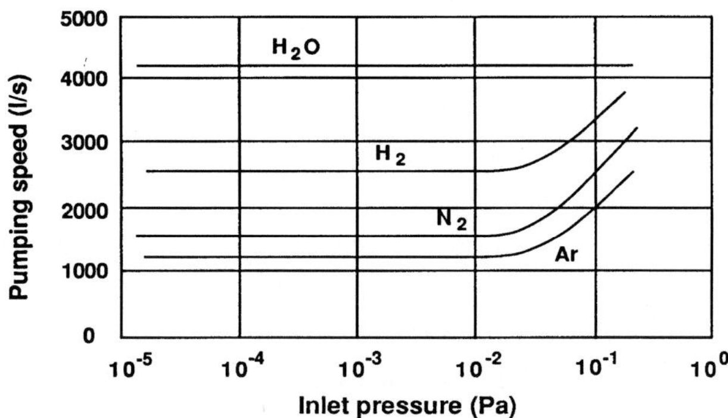

Fig. 8.4 Typical pumping speeds for different gases and their variation with inlet pressure for a cryogenic pump with an inlet diameter of 200 mm.

include the pumping speed for this gas in the specifications for their pumps. Typically, this speed is about 20 percent lower than the speed for nitrogen, as shown in Fig. 8.4 and as would be expected on the basis of eqn. 8.2 (i.e. $S_{Ar}/S_{Nit} \propto \sqrt{M_{Nit}/M_{Ar}} = \sqrt{28/40} = 0.8$).

The pumping speeds for water and nitrogen shown in Fig. 8.3, and the speeds discussed above for other gases, are the speeds cryogenic pumps exhibit when operating at inlet pressures below about 10^{-2} Pa (10^{-4} Torr). Here in the range of molecular flow, pumping speed remains independent of inlet pressure, as shown in Fig. 8.4. These are the speeds manufacturers usually give in specifying the performance characteristics of cryopumps, and these specifications usually include values for water vapour, hydrogen, nitrogen, and argon. At inlet pressures above about 10^{-2} Pa (10^{-4} Torr) the pumping speeds for all gases except water vapour increase. This behaviour is distinctly different from that of most other types of high vacuum pumps, which exhibit a marked decrease in pumping speed in this pressure range, as was described in the discussions relating to Figs. 5.4 (Section 5.2, p. 177), 6.4 (Section 6.1.5, p. 239), and 7.5 (Section 7.1.3, p. 284). The reason for this difference lies in the primitive nature of the cryogenic pumping process, which requires only that the gas molecules have an opportunity to come in contact with the cold pumping surfaces. As was shown in Section 2.5.1 (p. 46) where the equation for the maximum theoretically possible pumping speed S_{tm} was developed, the number of litres of gas that strike a square millimetre of an unobstructed surface per second is independent of pressure, and so the potential pumping speed of the surfaces in cryopumps is essentially independent of pressure. Water molecules are adsorbed on the cryobaffle in the inlet port of the pump. Since access to this baffle is unrestricted, the pumping speed is independent of the pressure of the water molecules in the vacuum system. However, molecules of other gases must pass through this cryobaffle to reach the cold surfaces inside the pump head on which they are adsorbed, and so their pumping speeds are strongly influenced by the conductance of this baffle and by the conductance of the regions inside the pump surrounding the pumping surfaces. Because the conductances of all fixtures increase as pressure increases in this range where gas flow changes from molecular to transitional to viscous in character, the rate at which these gases flow through the baffle into the interior of the pump increases as inlet pressure increases. This makes a greater volume of these gases available each second to come in contact with the pumping surfaces inside the pump head, and this in turn results in the increase in the pumping speeds for these gases shown in Fig. 8.4. Although cryogenic pumps cannot be operated above about 10^{-1} Pa (10^{-3} Torr) for more than a few minutes, for reasons that will be discussed below, these characteristics allow them to bring about the initial stage of a pumpdown process somewhat more rapidly than other types of high vacuum pumps.

8.7 Pumping capacity and regeneration

As noted at the beginning of this chapter, cryogenic pumps are entrainment pumps, because the gas molecules that are removed from a vacuum system by a cryopump are accumulated on the cold surfaces inside its pumping head. After a pump has been in use for some time, the amount of accumulated gas can become so large that the pump no longer functions effectively. Provided the compressor is operating properly, as indicated by proper readings on pressure gauges in the helium supply and return lines, the onset of this condition is usually indicated by one of the following symptoms:

i. There may be a noticeable increase in pumpdown time.
ii. The pump may become unable to handle the normal gas load without an increase in pressure.

iii. It may become impossible to bring the system down to the normal operating pressure, even though the temperature of the pumping head is in the proper operating range.
iv. The hydrogen vapour pressure thermometer or the silicon diode temperature sensor may indicate that the temperature of the pumping head remains above the normal operating temperature.

At this point the pump must be regenerated by removing the gases that have accumulated in it so that it can function normally again.

8.7.1 *Regeneration procedures*

Regeneration involves shutting off the refrigeration unit, warming the pump to room temperature, and allowing the accumulated molecules to return to the gaseous state and escape from the pump through the pressure-release venting valve. It is desirable, although not absolutely necessary, to have a high vacuum valve between the cryopump and the vacuum system that can be closed during this process to prevent the gases released by the pump from re-entering the system. The regeneration operation can be carried out in several ways, which differ in speed, effectiveness, and complexity, as follows:

i. *Unassisted regeneration*: In this procedure the compressor is turned off and the pump is allowed to warm up to room temperature unassisted. This method is the most time-consuming, typically requiring from 5 to 10 h, depending on the size of the pump and the amount of gas it has accumulated. The time required can be reduced somewhat by slowly admitting clean air or nitrogen at atmospheric pressure into the pump at the start of the procedure. This quickly increases the pressure of gas in the pump to a level where its thermal conductivity contributes to warming the cryoelements by transferring heat to them from the outside wall of the pump.
ii. *Regeneration using gas purging*: For this procedure the compressor is turned off and clean dry gas at a pressure of 30–40 kPa (about 5 psi) is flowed through the pump at a rate of 20–30 l/min. Inlet lines for introducing this purge gas are standard features of most cryopumps. In addition to warming up the pump faster, this flow of purge gas sweeps the gases released from the pumping surfaces out of the pump,

giving more complete regeneration. Nitrogen and argon are the gases most often used for this purpose. The use of helium is not recommended because doing so loads the charcoal bed with helium, which then becomes difficult to remove during the pumpdown process that must follow the regeneration treatment. Care must be exercised to be certain that the purge gas used is totally oil-free, otherwise the charcoal will become covered with oil molecules and will have to be replaced because its capacity to adsorb gas molecules will be reduced to an unacceptable level.

iii. *Heat-assisted regeneration*: The purge gas can be made to function even more effectively by heating it before it enters the pump. This will increase the rate at which the cryoelements warm up, thereby decreasing the total time for completion of the regeneration treatment. As the pump approaches room temperature the warm gas is somewhat more effective in removing water vapour. Manufacturers offer accessory heaters that can be attached to the inlet of the purge line for this purpose.

The method of unassisted regeneration is not widely recommended because it is so ineffective in removing water from the pump. When this method is used, the ice on the first stage cryoelements melts and collects as liquid water in the bottom of the pump. Most of it must then be removed when the pump is evacuated again at the end of the regeneration treatment. Both of the methods involving the use of a purge gas are better, because the flowing gas helps markedly in removing water from the pump. The heat-assisted method is, of course, the best because it is the most effective in removing water from the pump, and because it warms the pump up to room temperature most quickly. The time needed to carry out a regeneration treatment by any of these methods can be decreased considerably by heating the body of the pump, and some manufacturers offer accessory heating tapes or mantles for this purpose. However, considerable care must be exercised in heating the pump and the purge gas, because overheating can damage the indium seals, the diode temperature sensor, the hydrogen gas thermometer, and parts of the displacer on most cryopumps. Most manufacturers specify that *neither* the purge gas nor the pump body *should be heated above about 65°*C, and that heating should be stopped before the second stage cryoelements reach *a temperature of 30°C.*

Specifications on temperature limits for each individual pump should be checked carefully and adhered to faithfully. Most manufacturers offer control systems for their pumps that will carry out a regeneration treatment automatically and safely.

The time required for completing a regeneration treatment varies with the size of the pump and the amount and kind of gas accumulated in it, but typically is from 2 to 5 h when the preferred method involving a heated purge gas is used. Once a regeneration treatment has been started it is very important to continue the purging until all accumulated gases have been removed from the pump. Incomplete purging will leave some liquid water in the pump. This will then have to be removed when the pump is later evacuated or it will seriously impair subsequent performance by collecting on the charcoal and second stage cryoelements when the pump is cooled down again.

Before the compressor is turned on to begin the cool-down process, the cryopump must be evacuated to a pressure where the thermal conductivity of the gas in the pump is low enough so that it does not conduct significant heat from the wall of the pump to the cryoelements. An insufficiently low pressure inside the pump will usually be indicated by the formation of frost on the external wall of the pump when the compressor is turned on. This stage of evacuation also removes most gases from the charcoal bed, thereby reducing residual contamination of it by nitrogen, oxygen, and water and giving it maximum capacity to perform its critical function of adsorbing hydrogen, helium, and neon. It has been common practice to use an oil-sealed rotary-vane pump for this pump-out operation. In recognition of this most manufacturers design their pumps so that evacuation to a pressure of 5–15 Pa (0.05–0.1 Torr) is adequate, because this is at the upper end of the range of transitional flow for most rough-pumping lines (see Fig. 2.1 in Section 2.2, p. 31) where there is little danger of oil backstreaming from the pumping line into the cryopump. Nonetheless, great care must be exercised to be sure the charcoal bed does not become contaminated with oil. In particular, it is important to close the valve in the pump-out line before the pressure drops below 5 Pa (0.05 Torr), and this valve should not be opened at any time while the pressure in the cryopump is in the range of molecular flow. It is also advisable to provide an additional margin of safety by installing an oil trap in the pump-out line, and by servicing it regularly so that it always functions effectively (see Section 4.1.5a, p. 147).

Although, as just noted, rotary-vane pumps are widely used in this application, there are two advantages that can be gained by using an oil-free roughing pump instead. First, the danger of contaminating the charcoal bed with oil can be completely eliminated. Second, it becomes permissible to evacuate to a lower pressure, thereby effecting a more complete removal of gases from the charcoal bed. This is very beneficial, because it makes it possible initially to attain significantly lower ultimate pressures, and it will increase the time before the next regeneration treatment becomes necessary. A single sorption pump can evacuate a cryopump to a pressure below 10 Pa (0.1 Torr), while two sorption pumps used in sequence or a sorption pump used in sequence with a diaphragm pump, as described in Section 4.2.1 (p. 159), can reach a pressure below 1 Pa (10^{-2} Torr). An even more effective approach is to use a combination diaphragm-molecular drag pump (DP/MDP; see Section 6.2.3b, p. 268) for this pump-out operation. These pumps are totally oil-free, they can begin operation at atmospheric pressure and reach ultimate pressures below 10^{-2} Pa (10^{-4} Torr), and they have unusually high pumping speeds at pressures below 10 Pa (0.1 Torr) as shown in Fig. 6.16 (Section 6.2.3b, p. 270). They also have sufficient speed and capacity to serve as the main roughing pump for most vacuum systems, thereby giving totally oil-free pumping.

As noted above, the charcoal bed becomes saturated with water during regeneration. Before the displacer and compressor are turned on again to cool the pump down to operating temperature it is very important to be certain that as much as possible of this adsorbed water has been removed so that the charcoal has full capacity for adsorbing helium, hydrogen, and neon. To test for this, manufacturers usually recommend performing a *rate-of-rise* test. This is done by closing the valve in the pump-out line and noting the rate at which the pressure in the pump increases. If the rate of rise is less than 2 Pa (0.015 Torr) per minute [i.e. a total pressure increase of less than 10 Pa (0.075 Torr) over a 5 min period] it is usually considered acceptable to start cooling the pump down. If the rate of rise is greater than this, the pump should be purged again with heated gas until the temperature of the second stage reaches 25°C, and then evacuated again. If one or two repetitions of this purge-and-pumpdown cycle do not lead to a satisfactory rate of rise it is likely that there is a leak in one of the vacuum seals in the pump head assembly. When a satisfactory rate of rise is attained, the displacer and compressor are turned on to cool the pump down to operating temperature. This approach to determining when a regeneration treatment is complete is not very sensitive or accurate and is somewhat time-consuming. To overcome this problem, US Thin Films have recently developed a water sensor (sold under the Gen-A-Torr trade name) that is installed in the outlet of the purge gas line. This device will signal when water is no longer detected in the purge gas, thereby providing a reliable indication that the regeneration process is finished. The time needed for cryopumps to reach operating temperature after the completion of a regeneration treatment varies from about 1 h for the smallest pumps to 5 h for the largest ones.

There are certain precautions that must be followed in carrying out regeneration treatments. It is of the *utmost importance* to be sure that the pressure release valve is always in good working order, because it usually provides the route by which gases escape from the pump when they are released as the pumping surfaces warm up during regeneration. If the pressure release valve is not functioning properly, and if no other means of escape is provided, the pressure inside the pump can become great enough to cause the pump to explode. The pressure release valve also serves the critical safety function of preventing the pump from exploding if a power failure, malfunctioning of the compressor or displacer, or some similar unexpected event allows the pumping elements to warm up during normal operation. In some manufacturing applications large quantities of toxic, corrosive, or flammable gases can accumulate in a cryopump. When there is a possibility that this has occurred the pump should be flushed with a generous flow of nitrogen during regeneration to ensure that these gases are completely removed from the pump; furthermore, the gas mixture that emerges from the vent valve must be collected and disposed of in accordance with the procedures recommended for the handling of these hazardous materials. Manufacturers produce couplings for attaching tubes to the pressure release valve to facilitate this. As its cryoelements warm up, a cryopump that has been used to pump appreciable amounts of hydrogen or other flammable gases can release explosive mixtures of these gases with air or oxygen. Therefore, it is *always important* to be sure that all vacuum gauges that generate a gas discharge or that have heated filaments (even thermocouple and Pirani gauges) are turned off whenever the pump is likely to warm up so that they will not ignite such mixtures and cause an explosion.

8.7.2 Operating time between regeneration treatments

The period of time a pump can be operated between regenerative treatments t_{br} depends on its capacity to accumulate gases and the conditions under which it is used. Current models have capacities for several hundred grams of the gases which are pumped by cryocondensation. The limit for water occurs when the deposits of ice become so thick that the openings in the cryobaffle are reduced enough to interfere with the flow of other gases into the interior of the pump. For gases such as oxygen, nitrogen, and argon, which are pumped by cryocondensation on the second stage cryoelements, it occurs when the deposits of solidified gases become so thick that they act as an insulating layer, effectively increasing the temperature of the pumping surfaces. Very thick deposits may also reduce the access of hydrogen, helium, and neon to the charcoal bed. Because cryopumps are so widely used on apparatus for processes such as sputtering and ion etching, where large amounts of argon must be pumped, manufacturers specifications usually indicate the total amount of argon their pumps can accumulate without significant degradation in performance. Capacities for other gases that are also pumped by cryocondensation on the second stage elements, such as oxygen, nitrogen, and carbon dioxide, are roughly the same as the capacity for argon. This specification is almost universally given in units of standard litres (see Table 1.2 and Section 1.3, p. 16), and ranges from about 100 l(STP) for the smallest pumps to more than 3000 l(STP) for the largest ones. Typically, the argon capacity $C_A \approx 500$ l(STP) for pumps with inlet diameters of 200 mm. As noted above, pumps of this size have a maximum throughput for argon of about $Q_A \approx 10$ Torr- l/s. If argon were pumped steadily at this rate it would take about $t_{br} \approx C_A/Q_A \approx (500\times815)/10 = 4.0\times10^4$ s or 12 h to load a pump up to the point where regeneration would be necessary. This is an extreme example, because cryopumps are unlikely to be used for continuous service at maximum throughput. Usually the argon throughput in sputtering and etching processes is less than one-fifth of this level, and t_{br} is correspondingly longer. When used to evacuate systems during operation in the high and ultra-high vacuum ranges, cryopumps have sufficient capacity for condensable gases to give them very long operating lifetimes. For example, at 10^{-4} Pa (10^{-6} Torr) pumps 200 mm in diameter typically have pumping speeds of about 1500 l/s for atmospheric gases. At this pressure $Q \approx 1500\times10^{-6} \approx 1.5\times10^{-3}$ Torr- l/s, and $t_{br} \approx C_A/Q \approx (500\times815)/1.5\times10^{-3}$

$\approx 2.7\times10^8$ s $\approx$ 75 000 h or about 8 years. On most laboratory apparatus cryopumps probably experience the greatest gas load during pumpdown. Even here, however, their capacity is high enough to be very adequate for most practical purposes. As noted above, for example, pumps 200 mm in diameter have crossover capacities of about $C_x \approx 150$ Torr-l, and so should be able to perform $N_x \approx C_A/C_x \approx (500\times815)/150 \approx 2700$ pumpdown cycles before accumulating enough condensable gases to require regeneration.

The capacity of cryopumps for the gases that are pumped by cryocondensation on the cryobaffle and the second stage cryoelements is not the only factor of importance in determining the time between regeneration treatments. In many applications the need for regeneration arises because the charcoal bed becomes so loaded with adsorbed gases that the equilibrium adsorption pressure becomes high enough to limit the ultimate pressure attainable. The primary function of the charcoal bed is to pump hydrogen, helium, and neon by cryosorption. Most manufacturers indicate the capacity of the charcoal beds in their pumps by giving the amount of hydrogen, in standard litres, that must be adsorbed to cause its equilibrium adsorption pressure to rise to about 10^{-4} Pa (10^{-6} Torr) when the temperature of the second stage is maintained below 15 K. Sorptive capacities for helium and neon are roughly the same as that for hydrogen. Typical capacities for pumps with diameters of 100, 200 and 320 mm are 3, 10, and 30 l (STP), respectively. There are, however, large variations among pumps of a given size, and several manufacturers produce pumps with extra-large charcoal beds for use in applications where large amounts of the permanent gases must be handled. For example, ion implantation is frequently carried out at total pressures near 3×10^{-3} Pa (3×10^{-5} Torr) using gas mixtures in which hydrogen is the major component. If a cryopump 200 mm in diameter, with a speed of about 2500 l/s for hydrogen, were used for such a process the hydrogen throughput would be of the order of $Q_H = 2500\times(3\times10^{-5}) = 0.075$ Torr-l/s. The time needed to saturate the charcoal bed in a standard pump with a hydrogen capacity of $C_H \approx 10$ l (STP) would then be about $t_{br} = C_H/Q_H \approx (10\times815)/0.075 = 1.1\times10^5$ s or 30 h, and about 90 h for a pump with an extra-large charcoal bed and a hydrogen capacity of 30 l (STP). This represents one of the most demanding applications for cryopumps, and probably yields about the shortest regeneration time likely to be encountered.

In contrast, pumps on vacuum systems that operate continuously in the high and ultra-high vacuum ranges accumulate hydrogen, helium, and neon at a much lower rate. Even in such applications, however, the rate may not be entirely negligible. Whereas hydrogen, helium, and neon represent only about 2×10^{-3} percent of the atmospheric gas mixture, hydrogen is usually a significant component of the gas in vacuum systems at pressures in the high vacuum range and below. It diffuses readily through O-rings and surprisingly well through stainless steel and some other metals, it is evolved from many materials, and it is formed by the dissociation of water and organic compounds. In fact, it is often the dominant gas in ultra-high vacuum systems that have been well baked out to remove water vapour. It is, of course, difficult to know the amount of permanent gases being pumped in most apparatus; however, cryopumps have very adequate capacities for most applications. As an illustration, we might assume that a standard 200 mm diameter cryopump, with $S_p \approx 1500$ l/s and $Q_H = 10$ l(STP), is operating at $P = 10^{-4}$ Pa (10^{-6} Torr) under conditions where one-quarter of the gas is hydrogen. Then the hydrogen throughput is $Q_H \approx 0.25 S_p P = 3.8\times10^{-4}$ Torr- l/s, and $t_{br} = C_H/Q_H \approx (10\times815)/3.8\times10^{-4} \approx 2.2\times10^{7}$ s, or about 8 months.

Calculations of the kind carried out in the previous paragraphs are of benefit mainly in showing the extent to which different factors can contribute to determining the time between full regeneration treatments, and the nature of these factors. Quite obviously, the useful pumping life in practice can vary widely, depending on the conditions under which pumps are used. Daily, weekly, or monthly regeneration may be needed on apparatus that are cycled to atmospheric pressure frequently or that handle a high throughput of gases, while yearly regeneration may be sufficient for instruments that are not opened frequently but operate continuously in the high or ultra-high vacuum range where the throughput of gas is very low.

It can readily be appreciated that stopping a manufacturing unit for the several hours needed to fully regenerate a cryogenic pump can be a rather costly event. Therefore, various methods have been devised for overcoming this problem, most of which take advantage of the fact that the need for regeneration usually arises first from saturation of the charcoal bed, as discussed above. For example, Danielson Associates suggest that on systems that use a DP/MDP combination pump (Section 6.2.3b, p. 268) to perform the pump-out function it is possible to increase the time between full regeneration treatments by taking advantage of the unusually high pumping speed of these pumps at pressures below 10 Pa (0.1 Torr) to carry out a *partial regeneration* or *cold regeneration*. To do this the DP/MDP pump is first turned on and allowed to reach full operating speed. Then the high vacuum valve between the cryopump and the vacuum system is closed, the valve in the pump-out line leading from the cryopump to the DP/MDP pump is opened, and the compressor and displacer drive unit on the cryopump are turned off. This allows the charcoal bed to warm up slowly and release most of the gases that have been adsorbed on it. Because of its high pumping speed at low pressures, the DP/MDP pump can remove these gases as fast as they are released, thereby maintaining an insulating vacuum in the pumping head so that gas conduction does not cause the pump to become thermally overloaded. When the temperature of the second stage rises to between 25 and 30 K the compressor and the displacer drive are turned on again, the valve in the pump-out line is closed, and the DP/MDP pump is turned off. After the second stage temperature falls into the normal operating range the high vacuum valve to the system is opened and normal operation resumed. A partial regeneration treatment of this kind, which requires only 15–30 min to carry out, can be very useful in industrial processes, where saturation of the charcoal bed occurs long before the capacity of the pump for condensable gases is reached and a full regeneration becomes necessary. Similar results can be achieved with a patented low pressure electrical regeneration system recently introduced by Leybold, which involves the installation of computer-monitored and -controlled heaters on the first and second stage cryoelements. With this system, a partial regeneration treatment, in which only the second stage element is heated, can be carried out in less than 45 min. This drives off both the adsorbed and condensed gasses that have accumulated on the second stage, and restores full capacity for the pumping of all gases except water vapour. A full regeneration treatment, in which both stages are heated, also drives the condensed water off the first stage and requires less than 3 h with this system.

8.8 Design features

Manufacturers presently offer cryopumps in a greater range of sizes than is available for any other type of high vacuum pump except oil diffusion

Fig. 8.5 A photograph of several different sizes and types of cryogenic pumps. (Courtesy of APD Cryogenics Inc.)

pumps. Pumps with pumping heads of the general design shown in Fig. 8.1 are available from most manufacturers with standard ASA and ISO inlet flanges ranging in diameter from 100 mm up to 630 mm and with the pumping speeds shown in Fig. 8.3. Fig. 8.5 is a photograph of some pumps produced by one manufacturer. In addition, pumps with flanges up to 1600 mm (48 in ASA) and pumping speeds for water and air as great as 180 000 and 60 000 l/s, respectively, are also available from a few manufacturers. In these large pumps the first stage elements are usually cooled with liquid nitrogen, because the full capacity of the refrigeration unit is needed to cool the large second stage cryoelement.

In addition to coming in a wide range of sizes, cryopumps are available with a variety of special features. Most manufacturers produce 'low-profile' pumps, in which the expander and its drive unit extend outwards at right angles to the pump head, as shown in Fig. 8.5, for use in situations where space is limited. Pumps with integral high vacuum valves are available for ultra-high vacuum applications. This design eliminates one vacuum joint that is otherwise needed for installation of a high vacuum valve. As noted above, pumps with fixed and variable apertures, with specially designed second stage elements, and with fast regeneration systems are produced for semiconductor manufacturing and similar applications where high operating pressures and large quantities of gas must be handled. Recently, considerable emphasis has been placed on reducing the vibrations produced by cryopumps, and so special low-vibration models are available from some manufacturers. One of the most recent developments is the introduction of microprocessor-based control units that provide a wide range of automatic control functions. These units continuously monitor critical operating parameters, such as the temperature of the second stage elements, the helium pressure in the inlet and outlet lines from the compressor, readings of system vacuum gauges, and the temperature of the water in the heat exchanger. This information can be recorded for a permanent record of operating conditions, and can be transmitted to a remotely-located monitoring station. Some control units also keep a record of operating time and conditions and use this to issue a warning when the time for a regeneration treatment approaches, and some are even capable of carrying out such a treatment automatically. Obviously, such control units can be of great value in many manufacturing operations.

8.9 The operation of systems with cryopumps

The arrangement of components recommended by most manufacturers for systems that are evacuated by cryogenic pumps is shown in Fig. 8.6. Although not absolutely essential in all systems, a high vacuum valve provides many advantages and is almost universally recommended. It is essential in systems that are cycled to atmospheric pressure frequently because it allows the cryopump to be isolated from the vacuum chamber so that it does not have to be warmed up to room temperature and then cooled down again, both of which are time-consuming operations, during each cycle. During regeneration treatments this valve can be closed to prevent the gases released by the cryopump from flowing back into the vacuum chamber. When the pump is being cooled down after a regeneration treatment the charcoal bed becomes cold enough to actively adsorb gases, particularly water vapour, before the outer cryoelements become cold enough

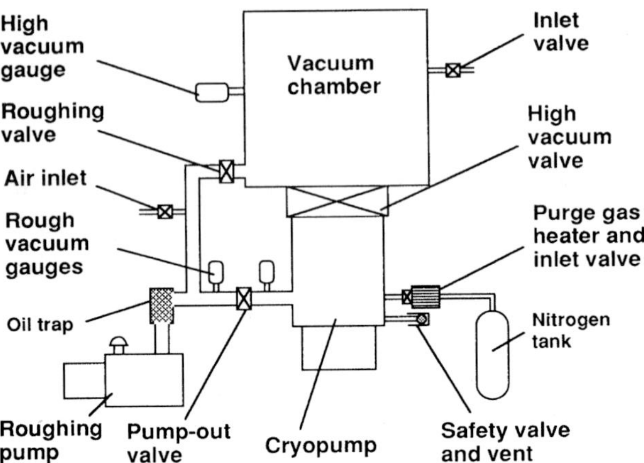

Fig. 8.6 A vacuum system with a cryogenic pump and a standard complement of three vacuum valves.

to capture gases effectively by cryocondensation. Closing the high vacuum valve while the pump is cooling down greatly reduces the quantity of gas that can reach the second stage cryoelements. This significantly reduces the amount of oxygen, nitrogen, and water that can adsorb on the charcoal as the pump cools down, and leaves the charcoal bed with an appreciably greater capacity for performing its critical function of pumping the permanent gases. The high vacuum valve also makes it possible to isolate the chamber from the pump while performing the rate-of-rise test described above (Section 8.7.1, p. 356), and in carrying out other repair and test functions. It is advantageous if this valve is of a type that will automatically close in the event of power failure so that gases released as the pump warms up will not return to the vacuum chamber. The system shown has two rough vacuum gauges, which may seem somewhat redundant. However, the one between the pump-out valve and the cryopump is useful for performing the rate-of-rise test, while the other is essential for monitoring the pressure in the system during the rough-pumping operation, when the pump-out valve is closed, to determine when the proper crossover pressure is reached. If an oil-sealed rotory-vane pump is used for rough-pumping functions it is highly advisable to have an oil trap in the vacuum line leading from this pump to help prevent oil from migrating into the cryopump and contaminating the charcoal bed. This trap must be serviced carefully and frequently to ensure that it is always in good condition (Section 4.1.5a, p. 147). It is also *essential* to maintain the pressure-release safety valve on the cryopump in good operating condition; otherwise, the pressure in the pump can build up enough to cause an explosion in the event of a power failure.

To carry out a regeneration treatment the pump-out, high vacuum, and roughing valves are all closed, and the pump is flushed thoroughly with warm, oil-free, dry nitrogen. This gas should be admitted through the purge gas heater and inlet valve at the temperature, pressure, and flow rate recommended by the manufacturer of the pump, and is allowed to escape from the pump through the pressure-release safety valve and vent. A cryopump that has been open to the atmosphere should also be flushed with heated dry nitrogen in this manner. When the temperature sensors of the cryopump show that the temperature of the second stage has risen slightly above 25°C, the purge gas inlet valve is closed and the gas heater is turned off. Next, the air inlet in the roughing line is closed and the roughing pump is started. The pump-out valve is then opened and the cryopump is evacuated to a pressure between 5 and 15 Pa (0.05 and 0.1 Torr). When this pressure is reached, the pump-out valve is closed and the rate-of-rise test described above is performed by watching the pressure reading on the rough vacuum gauge between the pump-out valve and the cryopump. As soon as possible after the pump-out valve is closed air should be admitted to the roughing pump via the air inlet valve in the roughing line and this pump should be turned off to minimize the opportunity for oil to migrate from it into the roughing line. If the pressure rises less than 10 Pa (0.075 Torr) over a period of 5 min it is safe to turn on the compressor and cool the cryopump down to operating temperature. This will require from 1 to 3 h, depending on the size of the pump. When the temperature sensors in the cryopump indicate that the temperature of the second stage has dropped below 20 K, the cryopump is ready to be used to evacuate the vacuum chamber. However, in this configuration, with the pump-out, high vacuum, and purge gas inlet valves all closed, the cryopump is fully isolated, and so the system can safely be left standing while work is done in the vacuum chamber, or while the system is not actively in use. This then is the standby configuration for this system, and is functionally equivalent to the configuration shown in Fig. 5.12a (Section 5.7.1, p. 205) for the system with an oil diffusion pump.

When it is desired to evacuate the vacuum chamber, the inlet valves to the chamber and the roughing line are closed, the roughing pump is turned on, the roughing valve is opened, and the vacuum chamber is pumped down to the crossover pressure selected according to the procedures discussed in Section 8.4 (p. 344). As soon as this crossover pressure is reached the roughing valve is closed, the roughing pump is vented and shut off, and the high vacuum valve is opened. Typically, the pressure in the system will drop rapidly to near 10^{-2} Pa (10^{-4} Torr) as soon as the the high vacuum valve is opened, because cryopumps provide such high pumping speeds for the water, nitrogen, and oxygen that comprise most of the free gas in the system. The high vacuum gauge can now be turned on to monitor the pressure in the vacuum chamber. The time required to reach lower pressures will then depend on the rate of gas evolution within the system. For reasons discussed in Section 2.10.3 (p. 62), pressures much below 10^{-5} Pa (10^{-7} Torr) usually cannot be reached unless the vacuum chamber is baked out at a temperature well above 100°C.

If it is desired to bring the vacuum chamber up to atmospheric pressure without warming the cryopump up to room temperature, the system is put into the standby configuration described above, with the high vacuum, pump-out, roughing, and purge gas inlet valves all closed. The high vacuum gauge is turned off and dry air or dry nitrogen is admitted through the inlet valve on the vacuum chamber. An operating vacuum can be attained again by repeating the rough-pumping and crossover operations as described in the previous paragraph. Normally, once a cryopump is cooled down it is not allowed to warm up again until regeneration is required. Whenever the rotary-vane roughing pump is not actually being used it should be vented and shut off to minimize the opportunity for contaminating the roughing line with oil vapours.

As mentioned at the start of this section, systems are sometimes constructed without a high vacuum valve between the cryopump and the vacuum chamber. The main reason for doing this is to eliminate the relatively high cost of a large valve of this kind, and so to reduce the initial cost of constructing the system. For such a system there is no particular advantage in using the pump-out port on the cryopump, and so this port is usually blocked off and the roughing pump is connected to the system only through the roughing line for the vacuum chamber, as shown in Fig. 8.7. While a system of this design can be operated with an oil-sealed rotary-

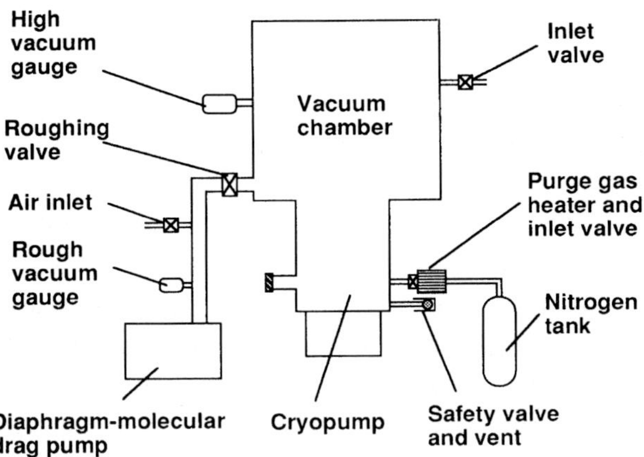

Fig. 8.7 A vacuum system with a cryogenic pump, but with no high vacuum valve.

vane roughing pump, it is highly advantageous to employ a DP/MDP to perform the rough-pumping functions instead. With an oil-sealed pump it is not safe to rough pump the system below about 5 Pa (0.05 Torr), and this leaves a considerable amount of gas in the system to be taken up by the charcoal bed of the cryopump as it cools down. With a DP/MDP pump, pressures near 10^{-2} Pa (10^{-4} Torr) can be reached, leaving a much smaller amount of gas in the system and providing totally oil-free pumping, as discussed in Section 8.7.1 (p. 356). In the following discussion of operating procedures it will be assumed that the system has a DP/MDP pump. It will become quite obvious that operation without a high vacuum valve is useful only on systems that do not need to be cycled between atmospheric pressure and an operating vacuum frequently or rapidly.

To open such a system to the atmosphere, the roughing valve and inlet valve to the vacuum chamber are left closed, and heated dry nitrogen is flushed through the cryopump via the purge gas heater and inlet valve, and allowed to escape through the pressure-release safety valve, as described above. When the second stage of the cryopump warms up to about 20°C the inlet valve for the vacuum chamber is opened and the vacuum chamber is also flushed with the warm dry nitrogen. When the temperature of the cryopump rises to 25°C the flow of nitrogen is turned off. The vacuum chamber can now be opened. Overall, this may require several hours.

To establish a high vacuum in such a system, the vacuum chamber is closed, but the inlet valve on it is left open to act as a gas outlet, and the entire system is flushed thoroughly with warm dry nitrogen. If the system has been open to the atmosphere for some time, and particularly if the humidity has been high, it is desirable to pump the system out and then flush it again with dry nitrogen. To do this the inlet valves to the vacuum chamber and the roughing line are closed, the roughing pump is turned on, the roughing valve is opened, and the system is pumped down to about 10^{-1} Pa (10^{-3} Torr). The roughing valve is then closed and nitrogen is admitted slowly through the purge gas heater and valve. When gas begins to escape through the pressure-release safety valve, indicating that the system is at atmospheric pressure, the inlet valve on the vacuum chamber is opened and the entire system is flushed with warm dry nitrogen until the temperature of the second stage of the cryopump rises slightly above 25°C. The flow of nitrogen is then stopped, the inlet valve on the chamber is closed, the roughing valve is opened, and the system is pumped down to the lowest pressure that can be reached by the DP/MDP pump. Since rapid cycling between atmospheric pressure and an operating vacuum is not to be expected with a system of this kind, patience is a critical part of the operating protocol, and so it might well be rough pumped overnight or over a weekend. At the end of the rough-pumping operation the roughing valve is closed, the roughing pump is vented and turned off, and the compressor is turned on. As the cryopump cools down to operating temperature it will, of course, establish a high vacuum in the vacuum chamber.

8.10 Concluding remarks

Cryogenic pumps offer a number of important advantages. They are totally oil-free and thus cannot contribute to contaminating a vacuum system with oil. Because they can tolerate unusually high crossover pressures they can be combined with various oil-free roughing pumps to produce totally oil-free systems. They have no moving parts exposed in the vacuum system, and do not require a backing pump. They are available in a wide range of sizes and with a variety of features, and so are suitable for many different applications. Compared to other types of high vacuum pumps of comparable size, they provide unusually high pumping speeds for water and competitive speeds for all other gases. They eliminate the need for liquid nitrogen traps to handle water vapour. They can produce ultimate pressures well into the ultra-high vacuum range, and they are relatively trouble-free and inexpensive to operate. Their main disadvantage is the need for periodic regenerative treatments. In spite of their obvious advantages, cryopumps have found only limited use on electron microscopes, probably mainly because of the frequency with which problems from mechanical vibration, noise, and failure of components in the refrigeration system were encountered in early models. Cryopumps have become widely used in many other applications, particularly in the electronics industry, however, and this has led to many improvements in design and manufacture so that current models are relatively reliable and trouble-free. Special models have even been developed that are sufficiently free of vibrations to have been used successfully on high-resolution electron microscopes. These have not been widely adopted by electron microscope manufacturers in their efforts to develop instruments with oil-free vacuum systems, probably because satisfactory results have already been achieved with ion pumps and turbomolecular pumps. Whether or not they are ultimately utilized by electron microscope manufacturers, they are very widely used to great advantage in a large number of other applications.

9 Practical vacuum systems

Having reviewed the characteristics of the various types of vacuum pumps and gauges, it now becomes possible to consider the ways in which these devices are incorporated into some practical vacuum systems, especially those found on electron microscopes. As was true for the previous chapters in this book, the emphasis here will not be on designing vacuum systems, because most electron microscopists are unlikely to become involved in this activity. For the few who do, such matters are discussed in a general way by Harris (Chapters 12 and 13), O'Hanlon (Chapters 10, 11, 12, and 13), and Weston (Chapter 7). Instead, most electron microscopists are more likely to be concerned with understanding the vacuum systems on the instruments in their laboratory in a way that makes it possible to use them effectively without causing vacuum problems, and so this is the general point of view that will be emphasized in this chapter.

It is not necessary to discuss most of the simpler vacuum apparatus here, however, because the characteristics and operation of the vacuum systems found on these units have already been described in considerable detail in preceding chapters. For example, the operation of the vacuum systems on film prepump chambers, sputter coaters, carbon fibre vaporizers, plasma reactors, freeze-drying units, and similar apparatus that have only a rotary-vane roughing pump is covered thoroughly in Section 4.1.8 (p. 152). The vacuum system shown in Fig. 5.11 is identical to those on the standard vacuum evaporators which are evacuated by an oil diffusion pump, and which have been in use for so many years. The operation of this type of system is discussed in great detail in Section 5.7.1 (p. 202). Likewise, the systems shown in Figs. 6.7 and 6.8 are the same as those on vacuum evaporators with turbomolecular pumps, and the operation of such systems is described in detail in Section 6.1.8 (p. 249). Moreover, the principles and methods described in these previous chapters are fundamental to vacuum practice, and it will be assumed that the readers of this chapter are familiar with this information.

Before proceeding with this discussion, however, it is probably wise to recognize a fact that will enter the minds of most electron microscopists as soon as they start to read this chapter; namely, that the operation of the vacuum systems on most electron microscopes produced since the early 1960s is carried out by automated valve-operating systems. All the operators of these instruments have to do is push a correspondingly labelled button to put the vacuum system into the configuration for a particular purpose, such as admitting air to the gun, camera, or column, or establishing an operating vacuum, or starting the instrument up, or shutting it down. Questions may then be raised as to the value of examining these operations in detail, inasmuch as they are totally out of the control of the operator. A careful examination of these matters can be justified, however, because blind reliance on automated systems usually gives rise to feelings of helplessness and uncertainty, and in effect puts the operator at the mercy of the instrument. If it becomes necessary to decide whether or not a system is performing properly and then to select a course of action when a problem is detected, uncertainty often escalates to panic. When working with a vacuum system as complex as those on current electron microscopes it is very beneficial, and quite comforting, to have a good understanding of the sequence in which the components of the vacuum system are being manipulated by the automatic valve-operating system, and to understand the reason this sequence is being used. This will lead to a much more comfortable relationship between the operator and the automated instrument, and will often prevent the occurrence of serious problems. In addition, many instruments have a set of switches that can be operated manually to over-ride the automatic valve-operating system. Those who understand the operational protocol for a vacuum system can use these switches to great advantage in carrying out many service and maintenance operations, and particularly in preventing damage to the system during emergencies. The goal of this chapter, then, is to review several rather complex vacuum systems in some detail, showing how to analyse such systems and to understand their general operating characteristics and procedures.

The references cited in this chapter are listed in Appendix 1. Manufacturers and suppliers of vacuum equipment referred to here are listed in Appendices 2 and 3. Owners of this book have permission to copy the figures presented in this chapter so that they can refer to them conveniently to follow the operational procedures described in the text.

9.1 Freeze-fracture and freeze-etching apparatus

It is convenient to start this discussion of more complex vacuum systems by considering a system of the type that is presently being used on apparatus which are designed to produce high-resolution replicas from biological specimens by the freeze-fracture method. This is a very versatile technique which has proved particularly useful for studying the structure of membranes, and which has been used to study a variety of types of biological materials, including solid tissue, cell suspensions, tissue cultures, and even food products such as ice cream and salad dressings. Bozzola and Russell (Chapter 14) give a concise, general description of this method, while Robards and Sleytr (Chapter 6) and Willison and Rowe (Chapter 7) discuss it in detail. The apparatus used in the *freeze-fracture and replication* processes requires a vacuum system with rather unusual and interesting characteristics, such as the one shown schematically in Fig. 9.1, and so it will be useful to outline the method in a general way to provide an understanding of the basis for these requirements.

Briefly, the frozen specimen is quickly transferred into the apparatus and then fractured by one of several methods, the simplest of which involves simply pressing a knife blade against it. Both the specimen stage and the fracturing device are cooled with liquid nitrogen, usually to –150 to –170°C. After fracturing, the specimen may be *freeze etched* by warming it for a few minutes to a temperature near –100°C where water readily sublimes from its surface onto the surrounding colder cryoshields. The fracture surface is next shadowed by evaporating a layer of platinum a few nanometres thick onto it at an angle of 45–75°, in the manner commonly used to enhance contrast in replication procedures (Hall, Section 10.6; Willison and Rowe, Chapters 2 and 3). Finally, a carbon film 10–20 nm thick is evaporated onto it from directly overhead to produce a replica of the surface (Willison and Rowe, Chapters 3 and 6). The specimen is then removed from the freeze-fracture apparatus, warmed slowly to room temperature, and treated with chemical reagents to dissolve the tissue off the replica. The cleaned replica is mounted on a specimen grid for examination in a transmission electron microscope.

It is not difficult to understand why the condensation of water, vacuum pump oil, and similar volatile materials onto the fracture surface constitutes a major problem in the freeze-fracture and replication processes.

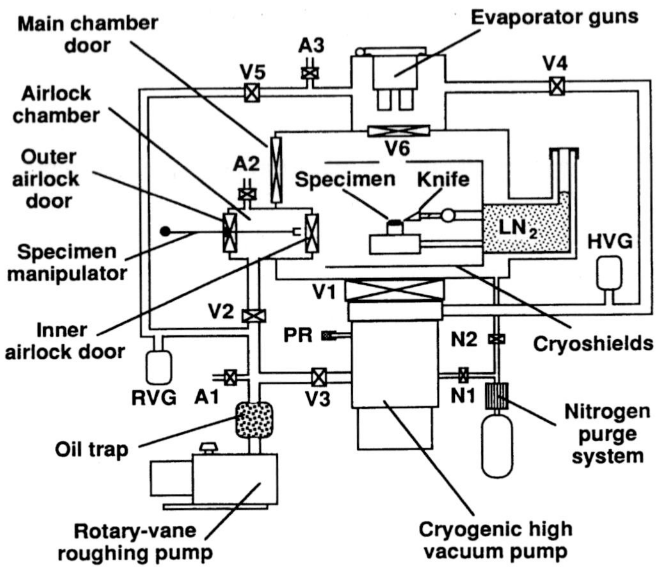

Fig. 9.1 A schematic diagram of a vacuum system for a freeze-fracture apparatus. A1 to A3, inlet valves; V1 to V6, vacuum valves; LN_2, liquid nitrogen; N1 and N2, nitrogen inlet valves; PR, pressure release valve; RVG, rough vacuum gauge; HVG, high vacuum gauge.

Control of this problem requires a vacuum system which can be maintained at a pressure near or below 10^{-5} Pa (10^{-7} Torr), plus the use of a set of cryopanels cooled to near –170°C that completely surround the specimen and the knife in such a way that there is no line-of-sight path by which water molecules can reach these objects from any part of the apparatus that is not cooled by liquid nitrogen. It is also advantageous if specimens and knives can be exchanged, and the evaporator guns serviced, without bringing the entire system up to atmospheric pressure, so that a large amount of water vapour is not introduced into the system each time one of these operations has to be carried out. Interestingly, there is an even more subtle source of water vapour; chips of the specimen produced during the fracturing process can warm up and release enough water to contaminate the fracture surface. Control of this phenomenon requires cooling of the fracture device, the specimen stage, and all other surrounding objects onto which these chips might fall, to near –170°C. Rash and Yasumura (*Journal of Microscopy Research and Technique*, 1992, 20, 187–204) describe methods for modifying freeze-fracture units manufactured by JEOL and Balzers to meet these requirements.

The vacuum system shown in Fig. 9.1 is similar, although not identical, to the one on the JEOL Model 9010 freeze-fracture apparatus (marketed in North America by RMC), which incorporates many of the features recommended and developed by Rash and Yasumura. Apart from its merits for use on a freeze-fracture apparatus, this vacuum system was chosen for discussion here because it can be used to illustrate the operating protocol for a system with a cryogenic pump and to introduce the organization and operation of airlock chambers. Robards and Sleytr (Section 6.3.2) review a variety of other designs for vacuum systems on freeze-fracture units.

9.1.1 *Pumpdown from atmospheric pressure*

Although a system of this kind will be kept under vacuum as much of the time as possible to minimize the introduction of water vapour into it from the atmosphere, it will occasionally be necessary to open it for cleaning, for servicing some internal part, or for regenerating the cryogenic pump (Section 8.7.1, p. 353). The procedure for starting up again after such an event, when the cryogenic pump has been turned off and warmed up to room temperature, is as follows:

1. *Cool the cryogenic pump down to operating temperature*: The cryogenic pump is first flushed, rough pumped, and cooled down to operating temperature, as described in Section 8.9 (p. 365). To initiate this process close valves V3, V4, and V5, the main chamber door, and the inner door to the airlock chamber, and open valves V1 and V6, and air inlet A3. Now open nitrogen inlet N1, and use the nitrogen purge system to flush the pump and the main vacuum chamber with warm, oil-free, dry nitrogen. When the temperature of the second stage of the refrigeration unit on the cryopump rises above 25°C, close the nitrogen inlet valve N1, close valves V1 and V2 and air inlet A1, open valve V3, turn on the rotary-vane roughing pump, and evacuate the cryopump with the roughing pump. When the rough vacuum gauge RVG indicates that the pressure has dropped to 10 Pa (0.1 Torr), promptly close valve V3 to prevent the pressure from dropping into the range of molecular flow where oil can backstream into the cryopump from the rough vacuum line (Section 2.3, p. 32). A rate-of-rise test should now be performed in the manner described in Section 8.7.1 (p. 356) to be sure water has been satisfactorily removed from the cryopump. If the

results of this test are satisfactory, turn on the compressor and cool the cryopump down to operating temperature, a process that will require several hours. During this time, open nitrogen inlet valve N2 and continue flushing the main vacuum chamber with warm dry nitrogen to remove as much adsorbed water from surfaces inside it as possible.

2. *Rough pump the main vacuum chamber*: When the sensors in the cryopump indicate that the temperature of the second stage has dropped below 20 K, the pump is ready for use and it is time to rough pump the main vacuum chamber. Start by closing the outer airlock door, air inlets A2 and A3, and nitrogen inlet N2. Then open the inner airlock door and valve V2, and allow the rotary-vane pump to evacuate the main chamber. When the pressure in it drops to 10 Pa (0.1 Torr), close the inner airlock door and valve V2 immediately.
3. *Establish an operating vacuum in the main chamber*: Finally, open main vacuum valve V1 and allow the cryopump to produce an operating vacuum in the main vacuum chamber.

Typically, the cryopump used on a system such as this will have an inlet diameter of 200 mm and will provide a pumping speed greater than 3000 l/s for water vapour, and greater than 1000 l/s for air. With such high pumping speeds, the free atmospheric gases are pumped out so quickly that the pressure inside the chamber will usually drop below 10^{-2} Pa (10^{-4} Torr) almost immediately after valve V1 is opened. However, it is generally considered desirable to reach a pressure in the lower end of the 10^{-5} Pa (10^{-7} Torr) range before performing the freeze-fracture and replication processes, to reduce the amount of the water adsorbed on surfaces inside the chamber to a level suitable for obtaining high quality replicas. This will usually require an additional 24–48 h of pumping. The desorption of water can be speeded up by baking the unit out, if it is designed to tolerate such treatment, or by making use of ultraviolet light to dislodge the water molecules, as described in Section 2.10.3 (p. 63). It is best not to cool the specimen stage, the fracture device, and the cryoshields with liquid nitrogen during this pumpout period, so that they do not become covered with a heavy layer of condensed water that could impair their performance later on. After the pressure reaches the lower end of the 10^{-5} Pa (10^{-7} Torr) range, and at least an hour before a specimen is introduced and the fracture, etching, and replication processes are undertaken, the liquid nitrogen reservoir should be filled with liquid nitrogen, and it should be faithfully kept filled while these processes are in progress.

9.1.2 Exchanging specimens

The diagram in Fig. 9.1 is intended to represent a freeze-fracture apparatus that is designed to permit specimens and knives to be exchanged through an airlock. This arrangement makes it possible to perform these operations quickly and conveniently without bringing the main vacuum chamber up to atmospheric pressure and, if well designed, can allow several samples to be processed per hour. Although airlock systems are widely used, some readers may not have encountered them previously. Since they were not included in any of the vacuum systems described in the previous chapters it is appropriate to introduce them at this point.

The most common type of airlock system includes an airlock chamber, an outer door between the airlock chamber and the atmosphere, and an inner door between the airlock chamber and main vacuum chamber, as shown in Fig. 9.1. The system also includes an inlet valve, A2, which is used to bring the airlock chamber up to atmospheric pressure so that the outer door can be opened, and a valve, V2, which allows the airlock chamber to be pumped down to a pressure in the rough vacuum range with the roughing pump, so that only a very small amount of air remains in the airlock to be admitted to the main vacuum chamber when the inner airlock door is opened. The airlock system is used for exchanging specimens in the following manner:

1. *Evacuate the airlock chamber*: Close the outer airlock door and air inlet valve A2, open valve V2 in the rough vacuum line, and allow the roughing pump to evacuate the airlock chamber. When the rough vacuum gauge RVG (which may be either a thermocouple gauge or a Pirani gauge) indicates that the pressure has dropped to 10 Pa (0.1 Torr), close valve V2 to minimize the diffusion of oil molecules from the roughing pump into the airlock chamber.
2. *Remove the old specimen from the specimen stage*: Open the inner airlock door and use the specimen manipulator device, which is usually attached to the outer airlock door, to remove the specimen holder from its place on the specimen stage and to withdraw it into the airlock chamber.
3. *Exchange specimens*: Close the inner airlock door and admit air to the airlock chamber by opening air inlet valve A2. Open the outer door of the

airlock chamber and remove the specimen holder from the specimen manipulator.

Attach a new specimen holder, carrying a new specimen, to the specimen manipulator and insert it into the airlock chamber, closing the outer airlock door. Close air inlet valve A2, open valve V2, and pump the airlock chamber down to a pressure of about 10 Pa (0.1 Torr) with the roughing pump. Promptly close valve V2 again.

4. *Transfer the new specimen to the specimen stage*: Open the inner airlock door and use the manipulator to set the specimen holder in place on the specimen stage. Then withdraw the manipulator into the airlock chamber, close the inner airlock door, and re-establish an operating vacuum in the main vacuum chamber with the high vacuum pump.

The above steps describe the manipulation of an airlock and specimen exchange system in fairly general terms. This procedure must be carried out skilfully to insert a frozen specimen into a freeze-fracture apparatus and still maintain it at cryogenic temperatures. Usually the specimen holder, carrying a new frozen specimen, is attached to the manipulator device under liquid nitrogen. The operations of transferring the specimen into the airlock, evacuating the airlock chamber, and then moving the specimen holder onto the cold specimen stage inside the main vacuum chamber must then be carried out in such a way that the specimen neither thaws nor becomes heavily coated with frost. The apparatus described by Rash and Yasumura has an airlock chamber with a volume less than 25 ml that can be evacuated in less than 15 s. When combined with airlock doors and a specimen manipulator that are very easy to operate, this makes it possible for a skilled operator to transfer a specimen into the unit in very satisfactory condition in less than 30 s. Other specialized methods are used with other models of freeze-fracture apparatus.

9.1.3 Servicing the evaporator guns

Most freeze-fracture units use evaporator guns to deposit the platinum and carbon films needed in the shadowing and replication processes. The way these guns are designed and function is described by Willison and Rowe (Section 2.3.4). Periodically it becomes necessary to replace the filaments or evaporant electrodes in the evaporator guns, or to clean the gun chamber. The vacuum system shown in Fig. 9.1 has a set of valves (V4, V5, V6 and inlet valve A3) that are used in a manner similar to those of the airlock system just described to make it possible to carry out these operations without disturbing the vacuum in the main vacuum chamber. Functionally, the gun chamber is equivalent to the airlock chamber, the demountable gun assembly and valve V6 correspond to the outer and inner airlock doors, respectively, while valve V5 and inlet valve A3 play the same roles as valve V2 and inlet valve A2 in the airlock system. However, the high vacuum pumpout line containing valve V4, that allows the gun chamber to be evacuated into the high vacuum range before it is opened to the main chamber, is a feature not found on the simple airlock system discussed above. This arrangement of valves and the demountable vacuum seal is used to service the evaporator guns as follows:

1. *Open the gun chamber*: Isolate the gun chamber by closing valves V4, V5 and V6, and admit air to it by opening inlet valve A3.
2. *Service the guns*: Expose the gun assembly for servicing by releasing the demountable joint that seals it to the gun chamber and lifting it out of the chamber.
3. *Rough pump the gun chamber*: After the service operation is completed install the gun assembly in the gun chamber again, close air inlet A3, open valve V5, evacuate the gun chamber to about 10 Pa (0.1 Torr) with the roughing pump, and then promptly close valve V5.
4. *Establish a high vacuum in the gun chamber*: Close valve V1 and pump out the gun chamber with the cryopump by opening valve V4 in the high vacuum pumpout line that leads from it to the gun chamber. When the pressure in the gun chamber falls into the 10^{-4} Pa (10^{-6} Torr) range, turn on the electron guns and heat them to operating temperature to thoroughly outgas them and the gun chamber. To resume normal pumping close valve V4 and open valves V1 and V6.
5. *Using the evaporator guns*: After the specimen has been fractured and is ready for shadowing with platinum, tilt it to the desired shadowing angle. Close valve V6 and open valve V4 in the high vacuum pumpout line, leaving valve V1 open. In this way, both the gun chamber and the main chamber are evacuated by the cryopump, but there is no direct line-of-sight path for water molecules or radiation to travel from the gun to the specimen while the gun is being warmed up. The gun to be used for the evaporation of platinum is then turned on. As soon as a stable rate of

evaporation is established, open valve V6 and the aperture in the cryoshield surrounding the specimen. When the desired amount of platinum has been deposited onto the specimen immediately close valve V6 and the aperture again, and then turn off the gun. This minimizes the time the specimen, and objects immediately surrounding it, are exposed to water molecules and radiant energy from the hot gun.

Then tilt the specimen for deposition of carbon at a 90° angle of incidence and follow a similar sequence of steps to evaporate the carbon replica film onto it from the second evaporator gun.

9.1.4 *Servicing devices in the main chamber*

The arrangement of valves on this system also makes it possible to service devices inside the main vacuum chamber without turning off the cryopump or disturbing the vacuum in the gun chamber. This is a particularly useful feature, because it avoids the long delays involved in warming a cryopump up to room temperature when it is turned off, and in cooling it down to operating temperature when it is started up again. Starting with the system in the normal operating configuration, proceed as follows:

1. *Warm the cryoshields up to room temperature*: Blow the liquid nitrogen out of the reservoir and dry the reservoir with a gentle stream of air or nitrogen, and then allow the apparatus to stand over night so that all surfaces inside the main vacuum chamber reach room temperature.
2. *Isolate the main chamber*: Close valve V1 to isolate the cryopump from the main vacuum chamber. Close valve V6 to isolate the gun chamber from the main vacuum chamber. Open valve V4 so that the cryopump continues to evacuate the gun chamber through the high vacuum pumpout line.
3. *Bring the main chamber up to atmospheric pressure*: Admit dry nitrogen through inlet valve N2 to bring the main chamber up to atmospheric pressure. Open the main chamber door and perform the necessary service procedures. Maintaining a slow flow of dry nitrogen through the chamber will help prevent atmospheric moisture from entering it during this time.
4. *Re-evacuate the main vacuum chamber*: Close the main chamber door, the outer airlock door, and inlet valves N2 and A2. Open the inner airlock door and valve V2, and allow the roughing pump to evacuate the main chamber. When the pressure reaches 10 Pa (0.1 Torr) promptly close valve V2 and the inner airlock door. Close valve V4 and evacuate the chamber

with the cryopump by opening valve V1. When the pressure drops to about 10^{-4} Pa (10^{-6} Torr) open valve V6 and resume normal pumping. As noted in Section 9.1.1 above, it will usually require 24–48 h to reach an operating vacuum again.

9.1.5 *Turning the cryogenic pump off*

Occasionally the cryogenic pump must be turned off while it is regenerated, or while some other major service operation is performed on the system. The procedure for doing this is discussed in Section 8.7.1 (p. 353) and outlined in Section 8.9 (p. 365). The procedure for turning the cryopump on again is described in Section 8.9 and Section 9.1.1.

9.2 The basic diffusion pump system for transmission electron microscopes

A typical arrangement of components for the vacuum systems on many transmission electron microscopes manufactured during the 1960s and early 1970s is shown in Fig. 9.2. There is only one high vacuum pump in this system, an oil diffusion pump (Section 5.1, p. 171) that evacuates the entire instrument through a high vacuum manifold which has branches leading into the gun, specimen, and viewing chambers, and into two cavities which contain aperture manipulators. There is both a water-cooled baffle and a liquid nitrogen trap above the diffusion pump to reduce the backstreaming of pump oil into the microscope column (Section 5.5, p. 190). The trap is located in such a position that it also acts as a very effective cryogenic pump for the water that is given off by the photographic film in the camera chamber. This speeds up the pumpdown process after exchanging film and reduces the amount of water that reaches the specimen and gun chambers. Because the trap is used in this dual role, the main high vacuum valve V1, which isolates the diffusion pump from the column when it is necessary to admit air to the column, actually has to be two valves that operate simultaneously.

In this system there are two rotary-vane rough vacuum pumps (Section 4.1, p. 134). Roughing pump RVP1 serves only to back up the oil diffusion pump, while roughing pump RVP2 performs all other rough

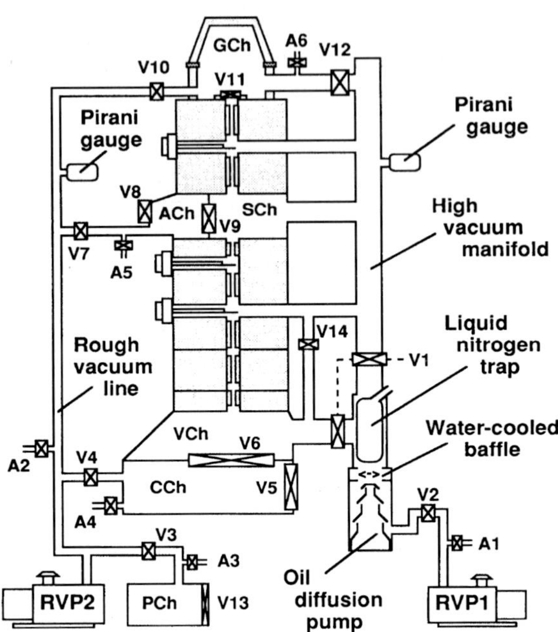

Fig. 9.2 A typical vacuum system for transmission electron microscopes on which the only high vacuum pump is an oil diffusion pump. A1 to A6, air inlet valves; V1 to V14, vacuum valves; ACh, airlock chamber; CCh, camera chamber; GCh, gun chamber; PCh, pre-pump chamber for photographic film; SCh, specimen chamber; VCh, viewing chamber; RVP1 and RVP2, rotary-vane roughing pumps.

pumping functions through the rough vacuum line that enters the column through the three branches which contain valves V4, V7, and V10. The various evacuation processes are performed by manipulating the six inlet valves A1 to A6 and 14 vacuum valves V1 to V14 that are incorporated into the pumping lines at strategic locations. As described below, these valves are arranged so that it is possible to replace a filament, and to exchange specimens and photographic film, without admitting air to the electron optical column.

In discussing operating procedures in the following subsections it will be assumed that valve V6 is kept open except when photographic film is being exchanged, and that valves V11 and V12 are kept open except when air must be admitted to the electron gun chamber. Likewise, valve V9 is ordinarily closed except when a specimen is being inserted into or removed from the specimen stage, and valve V3 in the line to the film prepump chamber is closed before any other roughing valve (V4, V7, or V10) is opened.

9.2.1 *Pumpdown from atmospheric pressure*

It is sometimes necessary or convenient to turn off the vacuum pumps and cut all electrical power to the instrument. The procedure for doing this is described in the next section, while the procedure for evacuating the system again is described in the following paragraphs. As noted in the introduction to this chapter, these procedures are usually carried out by an automatic valve-operating system (AVOS). When the instrument is shut down the AVOS usually closes valves V1, V2, V3, V4, V7, and V10, so that the column and the diffusion pump are both sealed under vacuum, and then opens air inlets A1 and A2 as the rotary-vane pumps are turned off, so that these pumps are left standing under atmospheric pressure (Section 4.1.8, p. 154). A typical procedure for an AVOS to follow in evacuating the vacuum system shown in Fig. 9.2 after being shut down in this way is as follows:

1. *The oil diffusion pump is heated to operating temperature*: When the operator turns on the main power switch the AVOS usually closes inlet valves A1 and A2 and turns on both roughing pumps. Next, valve V2 is opened, the supply of cooling water for the baffle and diffusion pump is turned on, and then the electrical power for the diffusion pump turned on. The liquid nitrogen trap must now be filled, and it should faithfully be kept filled as long as the system is under vacuum.
2. *The column is rough pumped*: After allowing about 20 min for the oil diffusion pump to reach operating temperature the AVOS opens valve V14, to provide a high-conductance path between the viewing chamber and the column, and then opens valves V4 and V10, so that roughing pump RVP2 can establish a good rough vacuum in the entire system. To minimize the opportunity for oil molecules to diffuse from the rough vacuum line into the electron optical column, this rough pumping operation should be terminated by closing valves V4, V10, and V14, when the Pirani gauges show that the pressure is approaching 10 Pa (10^{-1} Torr), the bottom of the pressure range for viscous flow (Section 2.2, p. 30).
3. *An operating vacuum is established in the column*: The main valve V1 is now opened so that the oil diffusion pump can produce an operating

vacuum in the instrument. The AVOS now uses the output of the Pirani gauge in the high vacuum manifold to determine when to allow the high voltage and filament to be turned on. Valve V3 is now opened to evacuate the film stored in the prepump chamber, PCh

9.2.2 Shutdown procedures

Normally it is advisable to maintain an operating vacuum in an electron microscope as much of the time as possible, even over weekends and short holidays. Under some circumstances, however, it may be desirable to turn the pumps off but to leave the column sealed under vacuum. A procedure that typically might be used by an automatic valve-operating system for doing this, starting with the vacuum valves in the operating configuration that was established above, is as follows:

1. *The oil diffusion pump is first turned off*: Valve V1 is closed, the electric power to the diffusion pump heater is turned off, and the oil in the diffusion pump is allowed to cool well below its boiling point. This requires from 20 to 30 min, depending on the size of the pump. The cooling water for the diffusion pump is turned off at the end of this time.

2. *The roughing pumps are now turned off*: To do this, valve V2 is closed so that a vacuum is retained in the oil diffusion pump, inlet valve A1 is opened, and roughing pump RVP1 is immediately turned off. Valve V3 is then closed, inlet valve A2 is opened, and roughing pump RVP2 is turned off. In this configuration all vacuum pumps are turned off, the electron optical column and the film prepump chamber are sealed under vacuum, and the roughing pumps are both at atmospheric pressure.

3. *Admitting air to the column*: Obviously, the electron optical column can now be brought up to atmospheric pressure, if this should be necessary (for a major service operation, for example) simply by opening valve V14, to establish a high-conductance path between the viewing chamber and the column, and then admitting air through valve A4. Whether or not this is done, the procedure outlined in Section 9.2.1 above is followed to evacuate the system again.

Caution: if an electron microscope is equipped with thin-film apertures it is very important to retract these apertures fully, so that they do not extend into the centre of the electron optical column, before air is admitted. If this is not done, these delicate apertures will be severely damaged by strong currents of air that rush through the column. As an additional safety measure it is also advisable to remove the specimen holder from the specimen stage, to be sure that pieces of a specimen are not blown around inside the column.

Many electron microscopists are firmly devoted to the practice of admitting only dry air or dry nitrogen to electron microscopes. The question of whether or not this is likely to be beneficial, thereby justifying the extra cost and effort involved, needs to be examined from a realistic, practical point of view. If an instrument of the general type that we are dealing with here does not have a liquid nitrogen trap, water molecules given off by the photographic film can move freely from the camera chamber into all other parts of the vacuum system, and so there is little to be gained by admitting dry air or dry nitrogen. Even if the system does have a liquid nitrogen trap there would be little point in admitting dry gas if the procedure described in Step 3 above is followed. This is because the gas is admitted through inlet valve A4 directly into the camera chamber where it will pick up water from the film and carry it into the rest of the vacuum system.

If, however, Step 3 were to be carried out by opening valve V14 and then admitting gas through inlet valve A6, there would be some benefit in using dry air or dry nitrogen, because now the entering gas will flow from the column into the camera chamber and will not carry water from the photographic film into other parts of the instrument. It will, of course, also be helpful to close valves V6 and V14 after the flow of gas stops, to minimize the diffusion of water from the camera chamber into the column while the system is standing at atmospheric pressure. When the system is subsequently evacuated the liquid nitrogen trap will prevent most water molecules from diffusing from the camera chamber into the optical column. However, because there is an open path of high conductance through the trap from the camera chamber into the column, the pressure in the column will be the same as in the camera chamber, which is usually determined by the pressure of water vapour existing there.

9.2.3 Exchanging photographic film

Most instruments of this type have an auxiliary vacuum chamber for pumping water out of the photographic film before it is inserted into the camera chamber. The automatic valve-operating systems usually close valve V3 in the line leading to this prepump chamber, PCh, whenever roughing pump RVP2 is being used for another rough pumping function. The system is also designed so that valve V3 closes and inlet valve A3 opens as the prepump chamber door, V13, is opened. Most instruments also have a set of valves that are used in a manner analogous to the valves of the airlock system described above in Section 9.1.2 to isolate the camera chamber from the rest of the vacuum system so that film can be exchanged without admitting air to the electron optical column. These are valves V4, V5, V6, and A4 in Fig. 9.2, and they are manipulated by the AVOS as follows:

1. *Air is admitted to the camera chamber*: With valves V4 and the camera chamber door, V5, closed, as they are during normal operation, valve V6 is closed to isolate the camera chamber from the optical column, and air is admitted by opening inlet valve A4. Obviously, the photographic film gives off so much water that there is nothing to be gained by using dry air or dry nitrogen here. The door to the camera chamber, V5, is then opened and the exposed photographic film is removed.
2. *Transfer new film into the camera chamber*: The operator can now open the door to the prepump chamber, V13, remove the unexposed, 'pre-pumped' film from this chamber, and transfer it directly into the camera chamber. It is, of course, important to make this transfer promptly so that the film does not have time to absorb water from the atmosphere again.

   It is advisable to turn the filament and high voltage off before proceeding to the next step, because the volume of the camera chamber is necessarily so large that enough air will be released into the rest of the vacuum system when valve V6 is eventually opened to seriously damage the filament if it is left on. Usually the pressure will increase enough to cause the AVOS to turn both the high voltage and filament off anyway.
3. *The camera chamber is rough pumped*: After the camera chamber door, V5, is closed by the operator, the AVOS closes inlet valve A4 and opens valve V4, so that roughing pump RVP2 can evacuate the camera chamber. The AVOS now monitors the output of the Pirani gauge in the rough

vacuum line, and when a pressure of about 10 Pa (0.1 Torr) is reached it closes valve V4 again.

4. *An operating vacuum is established*: Valve V6 is then opened, and the oil diffusion pump is allowed to produce an operating vacuum in the entire instrument. The AVOS now monitors the output of the Pirani gauge in the high vacuum manifold to determine when to allow the high voltage and filament to be turned on again.

9.2.4 Exchanging specimens

Many weird and wonderful devices have been designed to allow specimens to be inserted into, and removed from, the specimen stages of electron microscopes without significantly disturbing the vacuum in the column. Valves A5, V7, V8, and V9 and the associated specimen airlock chamber ACh shown schematically in Fig. 9.2, are usually integral parts of these specimen manipulators. Functionally, valves V8 and V9 are the counterparts of the outer and inner doors of the airlock chamber, respectively, while A5 corresponds to the inlet valve, and V7 to the pumpout valve, of the airlock system described above in Section 9.1.2. Most microscopes produced prior to the mid-1970s have top-entry specimen stages. For these stages, the specimen is held in the bottom of a tapered cylindrical *specimen holder* which is inserted into a matching tapered hole in the top of the stage, hence the 'top-entry' designation. These specimen holders are usually about 15 mm in diameter and 30 mm long. To handle such large objects the specimen manipulators for these stages are usually large complex devices which require large airlock chambers. When used to exchange specimens, starting with inlet valve A5 and valves V7, V8, and V9 already closed in the normal operating configuration, these specimen manipulators usually perform the following operations:

1. *The old specimen is removed from the specimen stage*: In this phase of its operating cycle the specimen manipulator first opens valve V9 (the inner door of the airlock chamber). Then it enters the specimen chamber, engages the specimen holder and lifts it out of the specimen stage, and then withdraws into the airlock chamber and closes valve V9 again.
2. *Specimens are exchanged*: In the next stage of its operating cycle the specimen manipulator opens inlet valve A5 and the outer door of the airlock,

V8, exposing the specimen holder, which is then removed from the manipulator. A specimen holder carrying a new specimen is then loaded into the specimen manipulator.

3. *The specimen airlock chamber is rough pumped*: The specimen manipulator is then moved into the airlock chamber, inlet valve A5 and valve V8 are closed, valve V7 is opened, and roughing pump RVP2 is allowed to evacuate the airlock chamber to a pressure near 10 Pa (0.1 Torr).

For the system shown in Fig. 9.2 this evacuation process can be monitored by the Pirani gauge in the rough vacuum line. Many instruments do not have a vacuum gauge in the rough vacuum line, however, and so this operation is then terminated after a time specified by the manufacturer. This is not a good procedure. If the specified time is too short (which it may be, to make it appear that the specimen exchange process is not very time-consuming) then the amount of gas subsequently admitted to the column will be greater than necessary, as will the time then needed to pump down to an operating vacuum. On the other hand, if the specified time is too long there is a danger that the pressure in the pumping line will drop into the range of molecular flow, increasing the likelihood of contaminating the airlock with pump oil (Section 9.2.7).

4. *The specimen is transferred to the specimen stage*: Further operation of the specimen manipulator then closes valve V7, opens valve V9, and inserts the specimen holder into the specimen stage. Usually the manipulator then disengages from the specimen holder and withdraws into the airlock chamber, and valve V9 is closed while the specimen is being examined.

The procedure just described applies to a basic specimen manipulator for a top-entry specimen stage. When exchanging specimens with this type of manipulator it is necessary to turn the filament off, because the volume of the airlock chamber needed to accommodate these complicated devices is so large that opening valve V9 introduces enough air into the column to seriously damage a hot filament. However, many instruments with top-entry stages have more complex manipulators that allow several specimen holders to be loaded into the airlock chamber. The filament must be turned off when the first of these holders is inserted into the specimen stage. However, all the other specimen holders that were loaded in the airlock can later be moved into the specimen stage for examination without turning the filament off. This is possible because an operating vacuum is established in the airlock chamber when valve V9 is opened to transfer the first specimen holder into the stage. It is only necessary to open valve V9 to move the other specimen holders between the airlock chamber and the specimen stage, and doing this will not now change the pressure in the column or gun significantly.

Most electron microscopes manufactured since the late 1970s have *side-entry* specimen stages. For these stages the specimen carrier is a long rod, and the specimen is held by a spring or screw clip in a hole that is bored through a flattened portion of the end of this rod. This specimen rod fits snugly into a hole that is bored horizontally through the side of the specimen stage (hence, the 'side-entry' name), and the airlock chamber consists only of the small space between the rod and the wall of this hole. The interior end of this hole is sealed by valve V9. Valve V8 is an O-ring that is held in a groove at the back end of the rod. As the rod is first inserted into the hole in the specimen stage this O-ring makes contact with the wall of the hole in the manner shown in Fig. 10.12 (Section 10.11, p. 455), effectively sealing the airlock chamber from the atmosphere. After this seal is established, further insertion of the rod into the stage moves a small pin on the rod into a groove in the wall of the stage where it contacts a switch that closes inlet valve A5 and opens valve V7. This contact also prevents the rod from moving into the stage far enough to open valve V9. After the airlock is rough pumped, the specimen rod is turned so that the pin releases the switch, closing valve V7. The pin then enters another channel that allows the specimen rod to be inserted fully into the specimen stage. As this is done valve V9 is opened and the specimen is moved into the viewing position. On many instruments, the volume of the airlock space is so small that very little air enters the column when valve V9 is opened, and so it is not necessary to turn a tungsten filament off during this process. Lanthanum hexaboride emitters require a better vacuum, and should be turned off, however.

9.2.5 *Replacing a filament*

Valves V10, V11, and V12 and inlet valve A6 in the vacuum system shown in Fig. 9.2 are functionally analogous to valves V5, V6, V4, and A3, respectively, which are associated with the evaporator gun chamber of the vacuum system for the freeze-fracture apparatus shown in Fig. 9.1. The procedure for using this arrangement of valves to replace a burned-out filament without bringing the rest of the vacuum system up to atmospheric pressure is described above in Section 9.1.3.

9.2.6 *Removing apertures*

To remove the apertures from the instrument for cleaning, or to perform other service operations that require access to the interior of the electron optical column, it is necessary to admit air into the column, but it is not necessary to turn off the oil diffusion pump. To prepare for this, it is advisable to remove specimen holders and apertures from the column, so that they cannot be damaged by the strong currents of air that can flow through the column. The filament and high voltage must, of course, be turned off. Assuming the vacuum valves are in the normal operating configuration when the operator presses the control button that signals a desire to admit air to the column, the AVOS first closes valve V1, so that air cannot reach the hot oil in the diffusion pump. Then valve V6 is closed, to isolate the camera chamber, valve V14 is opened, to provide a path for air to flow freely from the high vacuum manifold into the viewing chamber, and inlet A6 is opened to bring the system up to atmospheric pressure.

If the column is going to be disassembled, so that most of it will be openly exposed to the atmosphere, there is little advantage to admitting dry gas here. If, however, only a small hole is opened in the column, such as by the removal of an aperture manipulator, it will be beneficial to admit dry air or dry nitrogen. Then it will also be beneficial to cover this opening loosely with aluminum foil and to maintain a slow flow of dry gas through the system while the manipulator is being serviced.

When the service operation is completed and the operator presses the control button for evacuating the column, the AVOS closes inlet valve A6 and opens valve V10 to rough pump the column with roughing pump RVP2. When the output of the Pirani gauge in the high vacuum manifold indicates that the pressure is about 10 Pa (0.1 Torr), valves V10 and V14 are closed, valves V1 and V6 are opened, and the oil diffusion pump is allowed to establish an operating vacuum. The AVOS now continues to monitor the output of this Pirani gauge to determine when the pressure becomes low enough to allow the high voltage and filament to be turned on.

9.2.7 Variations in design features

While the vacuum system shown in Fig. 9.2 contains most of the important features found on instruments of this type, there have been nearly as many minor variations in design as there have been models of electron microscopes produced. On some, there is no liquid nitrogen trap, and these require only a single main vacuum valve (V1) located directly above the water-cooled baffle (as in Fig. 9.3). On others there is only a single rotary-vane roughing pump, and a large buffer tank is used to back up the oil diffusion pump while the roughing pump is being used for pumpdown operations as described in Section 5.7.2 (p. 212) and Section 9.3 below. Many early models do not have valves V11 and V12, and so air has to be admitted to the entire electron optical column to change a filament. On some early instruments the vacuum valves were driven by electric motors. These were slow, subject to mechanical problems, and could not follow a safe shutdown procedure in the event of an electrical power failure, and so they were ultimately abandoned in favour of valves driven by magnetic solenoids and air pistons. Solenoid valves allow fail-safe shutdown if electrical power is lost, but must be located carefully so that the magnetic fields they generate do not disturb the electron beam. Systems of air-operated valves are more complex, because they require an air compressor and roughly twice as many high pressure air lines and air valves as there are vacuum valves. They operate almost instantaneously, however, and enough reserve air can be stored in a rather small tank to carry the system through a safe shutdown procedure in almost any emergency.

Thousands of electron microscopes have been built with vacuum systems of this basic type. These instruments are generally reliable, straightforward to operate, and quite adequate for much routine microscopy. However, as electron optical capabilities have been refined and become more elaborate, and as interest in analytical techniques and ultra-high resolution imaging has increased, specimen contamination has become an overriding problem. It can easily be seen that this problem is practically unavoidable in these instruments. First of all, for reasons discussed in Section 5.5.2 (p. 193), even a water-cooled baffle and a liquid nitrogen trap cannot prevent oil molecules from the oil diffusion pump and its backing pump from reaching the electron optical column over the long run. Furthermore, the rough pumping line is a serious potential source of oil contamination. It is evacuated well into the pressure range of molecular flow by the rotary-vane roughing pump, day in and day out, and so low molecular mass fragments of pump oil molecules can freely diffuse from this pump into all parts of it. If the pressure is then allowed to fall into the range of molecular flow during a roughing operation, when valve V4, V7, or V10 is open, these oil molecules can move into the column and ultimately onto apertures, pole pieces, and the specimen. Ideally, these roughing operations are always terminated before the pressure in the roughing line drops below the range of viscous flow, as suggested in Section 4.1.5 (p. 145). However, this is controlled by the AVOS and may not be performed with a high degree of reliability, because this system may not have been properly set at the factory, and, in any event, is likely to drift out of adjustment after a period of use. Foreline traps (Section 4.1.5a, p. 146) located in the roughing lines leading to both roughing pumps would be helpful, but are almost never provided. In addition to these problems with oil contamination, vacuum systems of this kind do not produce an operating vacuum low enough to permit a lanthanum hexaboride (LaB_6) emitter to be used in the electron gun. An LaB_6 emitter is advantageous because it provides illumination of higher brightness and better coherence than a standard tungsten hairpin filament (Spence, p. 250), but requires a vacuum in the bottom of the 10^{-5} Pa (10^{-7} Torr) range. Consequently, the use of vacuum systems of this kind, which are evacuated by only a single oil diffusion pump, has decreased markedly in recent years.

9.3 Transmission electron microscope vacuum systems with ion pumps

As related in Section 1.1 (p. 9), most high-resolution electron microscopes produced since 1980 have vacuum systems in which the electron optical column is evacuated by an oil-free high vacuum pump to reduce the level of oil contamination in the specimen chamber and to attain a vacuum in the gun chamber that is low enough for an LaB_6 emitter. On a majority of these instruments the oil-free pump is a sputter-ion pump (Section 7.1, p. 276) and, although many different designs have been used, the vacuum system shown in Fig. 9.3 has most of the important features found in systems of this kind. Usually an oil diffusion pump (Section 5.1, p. 171) is used to evacuate the viewing and the camera chambers. A valve, V6, is provided to isolate these chambers from the remainder of the system when photographic film is being exchanged. Typically, a vacuum in the 10^{-4} Pa (10^{-6} Torr) range can be maintained in these chambers. During observation of the specimen, when valve V6 must be open, the diffusion of gases from these chambers into the electron optical column is reduced to a negligible level by one or more small apertures in the projector and intermediate lenses. The specimen chamber and the electron gun chamber are pumped by a sputter-ion pump, which typically has a speed of 100–200 l/s. Although this is lower than the speed of the oil diffusion pumps that are commonly used on electron microscopes, it is adequate for two reasons: first, it is not necessary to have valves, traps, and baffles, which reduce the speed of evacuation, between the ion pump and the electron optical column; and second, the ion pump does not evacuate the large viewing and camera chambers and is not required to handle the large amount of water given off by the photographic film. Typically, pressures low in the 10^{-5} Pa (10^{-7} Torr) range can be maintained in the specimen and electron gun chambers. This is low enough to permit the use of an LaB_6 emitter, and to substantially reduce the rate of specimen contamination. Because there is a significant difference between the pressure in these upper chambers and that in the lower chambers of the system, this is often called a 'differential pumping system', and the apertures that make this pressure differential possible by restricting the diffusion of gas between the two sets of chambers are often called 'differential pumping apertures'.

On most instruments two rotary-vane pumps are used to perform the rough pumping functions in substantially the same manner as described above in Section 9.2. In other instruments a buffer tank is substituted for the rotary pump that was dedicated to serving the diffusion pump, as shown in Fig. 9.3. The pressure in this tank is constantly monitored by Pirani vacuum gauge PG1, and whenever this pressure drops towards a level that is unsafe for backing the oil diffusion pump the AVOS closes valve V10, opens valve V2, and allows the roughing pump to reduce the pressure to a safe level again. Another common variation, usually offered as an option at extra cost, involves the substitution of a turbomolecular pump (Section 6.1, p. 228) for the oil diffusion pump. Most manufacturers also offer an oil trap for the roughing line as another option. Both of these options should be selected if funding permits.

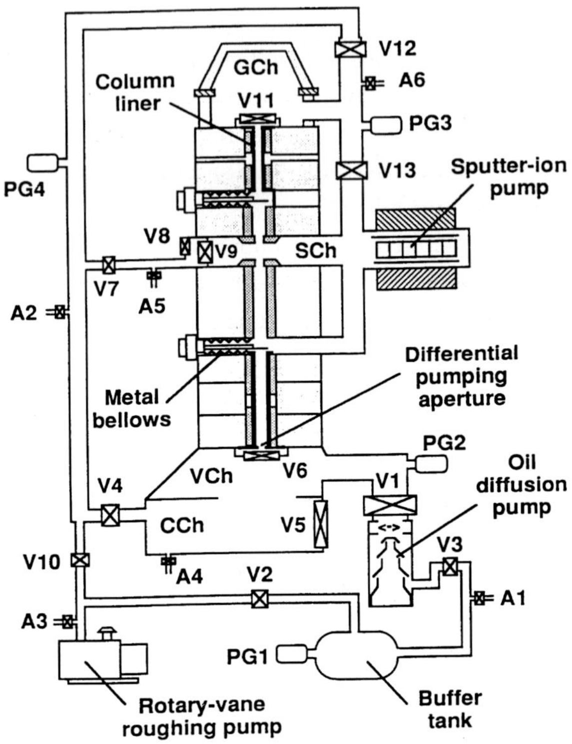

Fig. 9.3 A typical vacuum system for transmission electron microscopes on which the electron optical column in evacuated by a sputter-ion pump. A1 to A6, air inlet valves; V1 to V13, vacuum valves; PG1 to PG4, Pirani gauges; CCh, camera chamber; SCh, specimen chamber; VCh, viewing chamber.

Several other measures are usually taken to enhance the quality of the vacuum in the specimen and gun chambers. The total volume of these chambers is kept as small as possible and the components they contain are designed and manufactured to minimize gas evolution and virtual leaks (Section 2.10.4e, p. 76). The volume and the internal surface area of the electron optical column are minimized by non-magnetic, vacuum-tight, stainless steel tubes that run along the bore of the column from the electron gun to the specimen chamber and from the objective lens to the bottom of the projector lens. Column components that produce high levels of outgassing, such as the joints between lenses and beam deflecting coils, are built outside these *column liners*. The use of lubricated O-rings, which can evolve organic compounds (Section 10.11, p. 451), is minimized by sealing the vacuum joints associated with these column liners and the joints between the gun and the column with gaskets of indium metal, and by making the vacuum seals in the aperture mechanisms with metal bellows. In most instruments a bakeout treatment can be carried out by running the lenses at a high current without the cooling water. Some microscopes have quartz-halogen lamps installed in the specimen chamber. These emit enough radiant energy to raise the temperature in the specimen chamber above 100°C. The ultraviolet radiation they emit is also effective in dislodging water molecules from surfaces by radiative excitation (Section 2.10.3, p. 63). In addition, all instruments now have an efficient specimen anticontamination device which is cooled by liquid nitrogen to a temperature near –150°C when specimens are being observed. At this temperature the vapour pressure of water is near 10^{-8} Pa (10^{-10} Torr), and that of pump oils is even lower (Fig. 5.6 in Section 5.4, p. 182), and so this device plays a major role in controlling the rate of specimen contamination by acting as a small cryogenic pump (Section 8.1, p. 331) which removes these materials from the region immediately surrounding the specimen.

The operating protocols for the vacuum system shown in Fig. 9.3 are similar to those for the system with the oil diffusion pump that was discussed in detail in Section 9.2, with a few important variations. In the following discussion it is assumed that valves V11 and V13 are open except when it is necessary to admit air to the electron gun, and that valve V12 is closed except when the column or the electron gun is being rough pumped. Obviously, valve V6 must be opened while a specimen is being examined; otherwise, it should be closed to minimize the diffusion of gas from the viewing chamber into the electron optical column. The opening and closing of the valves in the vacuum systems of these advanced instruments are, of course, carried out by a microprocessor-based AVOS.

.3.1 Initial pumpdown

When a system of this kind is shut down under vacuum with all pumps turned off, valves V1 and V3 are closed to isolate the diffusion pump, valves V4, V5, and V6 are closed to isolate the camera and viewing chambers, valves V9 and V12 are closed to isolate the column, valves V2 and V10 are closed to isolate the roughing pump, inlet valve A3 is open so that the roughing pump is at atmospheric pressure while it is turned off, and all other inlet valves are closed. Typically, an AVOS will proceed as follows to establish an operating vacuum again:

1. *The oil diffusion pump is heated up to operating temperature*: When the main power switch for the instrument is turned on the AVOS usually closes air inlet A3 and turns on the roughing pump. Valves V2 and V3 are then opened, so that the roughing pump will back up the diffusion pump directly, the cooling water and the heater for the oil diffusion pump are turned on, and 20–30 min are allowed for the diffusion pump to warm up to operating temperature.
2. *The viewing and camera chambers are rough pumped*: When the period allowed for the diffusion pump to warm up has passed the AVOS closes valve V2, so that the buffer tank backs up the diffusion pump, and so that air cannot reach the hot diffusion pump through the rough vacuum line. Valves V4 and V10 are then opened to allow the rotary-vane pump to rough pump the camera and viewing chambers. When the output of Pirani gauge PG2 indicates to the AVOS that the pressure has dropped to about 10 Pa (0.1 Torr) valves V4 and V10 are closed, terminating this rough pumping operation.
3. *A high vacuum is produced in the viewing chamber*: Valve V2 is now opened so that the roughing pump again directly backs up the diffusion pump. Then, valve V1 is opened and the oil diffusion pump is allowed to reduce the pressure in the viewing and camera chambers into the 10^{-3} Pa (10^{-5} Torr) range. Throughout this process the AVOS monitors the backing pressure for the diffusion pump with Pirani gauge PG1, and the pressure in the viewing and camera chambers with Pirani gauge PG2.
4. *The electron optical column is now rough pumped*: To prevent air from entering the oil diffusion pump through its backing line, the AVOS closes valve V2. The buffer tank now backs up the oil diffusion pump. Valves V10 and V12 are opened and the rotary-vane pump is allowed to evacuate the column until the output of Pirani gauge PG3 indicates that the pressure there is about 10 Pa (0.1 Torr). Valve V10 is then closed. Note that valves V11 and V13 are normally left open except when a filament is being changed.
5. *A high vacuum is established in the column*: This process is usually carried out in two stages. Since the sputter-ion pump will start up more quickly, and with fewer problems from gas release due to overheating, if the pressure in it is in the 10^{-2} Pa (10^{-4} Torr) range (see Section 7.1.6, p. 290), the oil diffusion pump is first used to bring the pressure down to this level, and then the ion pump is turned on. To prepare for the first stage of this

process the AVOS opens valve V2, so that the roughing pump again backs up the diffusion pump directly. Then, keeping valve V1 open, valve V4 is opened and the diffusion pump is allowed to evacuate the column through valve V12 and the rough vacuum line, which is usually somewhat enlarged for this purpose.

6. *Evacuate the column with the ion pump*: When the output of Pirani gauge PG3 signals to the AVOS that the pressure in the column and gun has dropped into the lower end of the 10^{-2} Pa (10^{-4} Torr) range, valves V4 and V12 are closed, and the power to the sputter-ion pump is turned on.

The precautions and procedures discussed in Section 7.1.6 (p. 290) for preventing overheating of the sputter-ion pump during startup should now be observed carefully. Once the sputter-ion pump is operating stably, the value of the ion current in it can be monitored and used as an indicator of the vacuum in the electron optical column (see Section 3.2.3, p. 104). If the instrument has provisions for a bakeout treatment, this can be started as soon as the voltage to the sputter-ion pump rises above 2500 V. After a microscope has been shut down for a period of time, it should be pumped out at least over night before turning on the high voltage and the electron gun. Then, when the gun is first turned on it should be brought up to operating temperature very slowly, preferably by increasing the heating current in small increments and allowing a period of several minutes for each, to give the sputter-ion pump ample time to remove the gas released within the electron gun as it heats up.

9.3.2 Other vacuum operations

The process for changing a filament on an instrument with the vacuum system shown in Fig. 9.3 involves basically the same steps as described for servicing the evaporator guns on the freeze-fracture apparatus discussed in Section 9.1.3, with valves V11, V12, V13, and A6 performing functions analogous to valves V6, V5, V4, and A3, respectively, in Fig. 9.1. The sequence of operations for servicing the electron gun is the same as described in Section 9.1.3. After the service operation is completed, the system should be pumped out thoroughly, at least overnight, before the gun is turned on. If the instrument has provisions for a bakeout treatment, this will usually be beneficial too. When the gun is turned on for the first time the heating current should be increased in small increments, and the gun should be allowed to stabilize for a period of several minutes after each incremental increase, to give the sputter-ion pump ample time to remove the gas released within the electron gun as it heats up.

The process of exchanging film follows the same general procedure described above in Section 9.2.3, except that now it is not necessary to disturb the vacuum in the electron optical column, because the oil diffusion pump is used only to evacuate the viewing and camera chambers, and these chambers can be fully isolated from the column by keeping valve V6 closed throughout the procedure.

Likewise, the process of changing specimens is accomplished by operating the specimen manipulator, and involves essentially the same routine as described in Section 9.2.4, except that valve V2 is closed before valve V7 is opened to rough pump the specimen airlock chamber. People who are accustomed to operating microscopes with the type of vacuum system described in Section 9.2, which are evacuated by an oil diffusion pump, usually note that it takes a much longer time for the pressure in vacuum systems with sputter-ion pumps to drop back to an operating level after exchanging specimens. This occurs for two reasons: first, the sputter-ion pumps usually have much lower pumping speeds than the oil diffusion pumps; and, secondly, most of the instruments with ion pumps have an LaB_6 emitter which requires a much lower operating vacuum than the tungsten filaments used on the other instruments. The time to pump down after exchanging specimens can only be reduced by being very careful to operate the specimen airlock mechanism properly so that the amount of air admitted to the specimen chamber when valve V9 is opened is no larger than necessary.

Lanthanum hexaboride emitters are used in most electron microscopes with vacuum systems of this type, because they provide much higher brightness than tungsten hairpin filaments. These emitters are easily damaged, and so the electron beam should be turned off during the specimen exchange process, and any other time when there is a likelihood that the pressure in the gun will rise appreciably. For long emitter life and high electron beam intensity, the pressure in the gun chamber should be below 7×10^{-5} Pa (5×10^{-7} Torr) before an LaB_6 emitter is turned on. When the emitter is turned on, the heating current should be increased very gradually to minimize warping of the supporting wires and damage to the emitter tip. A commonly-used procedure involves estimating the value of the heating current that will be needed for saturation (based on past experience or the manufacturer's recommendations) and dividing this value by 10 to obtain an incremental amount by which the heating current will be increased as the gun is turned on. Starting with the heating power completely off, the current to the gun is increased by this incremental amount, the gun is allowed to stabilize for 20–30 s, and then this process is repeated until saturation is established.

A vacuum system of this general design is quite flexible. As already described, it is possible to admit air to the electron gun, and to the camera chamber, without disturbing the vacuum in the specimen chamber and column. It is, of course, possible to bring the column up to atmospheric pressure, to service apertures or the specimen stage, for example, without disturbing the vacuum in the camera chamber. This is accomplished by turning off the electrical power to the ion pump, closing valve V6 (leaving valve V12 closed, and valves V11 and V13 open, as in normal operation) and admitting dry air or dry nitrogen to the column through inlet valve A6.

If needed, air can now be admitted to the entire vacuum system, by closing valve V1 and opening inlet valve A4. The power to the diffusion pump can then be turned off, if desired. After it has cooled to room temperature, valves V2 and V3 can be closed, inlet valve A3 can be opened, and the roughing pump can also be turned off.

9.3.3 *General operating and design considerations*

In considering the operation of an electron microscope with a vacuum system of this kind it is important to keep in mind the basic reason for incorporating the sputter-ion pump into it in the first place; namely, to minimize the rate of specimen contamination by drastically reducing the amount of oil vapour reaching the specimen chamber from the vacuum pumps. When specimens are being observed, valves V4, V7, V9, and V12 are all closed, and so there is no path open for oil molecules to move into the specimen chamber from the rotary-vane and oil diffusion pumps (except through the differential pumping apertures at the bottom of the column, which are so small that the flow is negligible). During this time, then, the electron optical column, which is evacuated by only the sputter-ion pump, constitutes an essentially oil-free vacuum system. On this basis it should be clear that the danger of introducing oil contamination is greatest during periods when one of these valves between the rough vacuum line and the electron optical column is open; that is, when some part of the electron optical column is being rough pumped. This will normally occur after replacing a filament, during the specimen exchange process, and when the entire instrument is being re-evacuated after having been opened to the atmosphere for a major service operation. It therefore follows that these operations should be carried out in such a way as to minimize the time these roughing valves are open, and particularly to avoid having them open when the pressure in the roughing line is below about 10 Pa (0.1 Torr) and conditions for molecular flow prevail in this line.

It is rather surprising to find that many operators, and even some manufacturers, of these instruments are apparently not aware of this basic operating principle. I recently visited a laboratory where it was standard practice to turn the ion pump off over nights and weekends, during which time the oil diffusion pump was used to evacuate the electron optical column (in effect, by opening valves V4 and V12). The reason given by the laboratory manager for doing this was to "save the ion pump". He was totally unaware that in doing this he was defeating the basic principle underlying the design of the advanced vacuum system on his instrument. In effect, he was steadily converting his instrument into the equivalent of one with the standard type of vacuum system described in Section 9.2. He could have saved a considerable amount of money, and have developed an annoying level of specimen contamination much more quickly, by purchasing a microscope of that type in the first place.

At another laboratory, a heater was placed in the liquid nitrogen container of the specimen anticontamination device to warm it up quickly whenever an operator finished using this device. Then the ion pump was turned off and the oil diffusion pump was brought into use to evacuate the column during the warm-up process (by opening valves V4 and V12). The reason given for using this procedure, which reportedly was recommended by the manufacturer of the instrument, was that it prevented the ion pump from having to handle 'the extra gas load' released as the anticontamination device warmed up. Incredibly, neither the operator nor the manufacturer of the instrument apparently realized the fallacy of this procedure. Unless there is a serious design flaw, such as a leak that opens and admits air as the anticontamination device cools down, pouring liquid nitrogen into a container that is totally isolated from the vacuum system has no effect whatsoever on the amount of gas that enters the system. The use of the anticontamination device does not introduce any 'extra' gas into an electron microscope. All it does is to accumulate gas from the region of the specimen chamber by the cryocondensation process (Section 8.1, p. 330). This gas is either already in the instrument or is introduced into it during specimen exchange operations. If the same specimen exchange and viewing processes are carried out, the same amount of gas will be in the instrument, and will have to be handled by the ion pump, whether or not the anticontamination device is cooled with liquid nitrogen. Thus, the gas that is released from the anticontamination device as it warms up does not represent an extra load of gas for the ion pump to handle; it only involves a delay in the time at which gas that is normally in the instrument must be handled. Any overloading of the ion pump can be attributed directly to using the heater to warm up the anticontamination device quickly, because this may very well release the gas condensed on the anticontamination device so rapidly that the ion pump cannot sustain the vacuum in the column at an acceptable level. This would not occur, however, if the anticontamination device were merely allowed to warm up naturally, without the aid of the heater. Then the gas collected on it would be released slowly, the ion pump would not become overloaded, there would be no incentive for bringing the oil diffusion pump into use for evacuating the electron optical column, and this opportunity for contaminating the column with pump oil would be avoided. When these considerations were called to the attention of one of the manufacturer's service engineers he suggested that the procedure could also be justified because using the oil diffusion pump in this way reduces the total amount of gas the ion pump must handle over the long run, thereby prolonging its life considerably. While this is certainly true, if this were an important objective underlying the design of the vacuum system then the life of the ion pump could be extended indefinitely by redesigning the system slightly so that the column could always be evacuated effectively with the oil diffusion pump, and then never turning the ion pump on at all.

When operating an electron microscope with a vacuum system similar to the one shown in Fig. 9.3, the oil diffusion pump should be used to evacuate the electron optical column only when *absolutely necessary*. In addition, it is highly desirable to have a liquid nitrogen trap or a high-conductance oil trap in the pumping line that runs between the oil diffusion pump and the column (i.e. in the line leading to valve V12), and this trap should be very carefully maintained to prevent oil contamination from occurring over the long run (Section 5.5.3, p. 196).

From a practical, operational standpoint it is difficult to avoid a slow increase in the level of oil contamination in the specimen chamber after an electron microscope of this kind has been in use for a long time, even without following fallacious operating procedures of the type just described. As noted in Section 7.1.6 (p. 291), it is highly desirable not to attempt to start the sputter-ion pump on an electron microscope until the pressure in it has been reduced to a value near 10^{-2} Pa (10^{-4} Torr). Otherwise, the pump may overheat and release materials which apparently are deposited on the ceramic insulator in the electron gun, and this leads to instabilities in the high voltage from microdischarges along the gun insulator. Therefore, it is common practice to use the oil diffusion pump for a considerable time after conditions for molecular flow have been reached in the pumping line that leads from it into the electron optical column each time the electron gun chamber is pumped down following a filament replacement operation, and each time the entire electron optical column is evacuated after it has been let up to atmospheric pressure to carry out a major service operation. During such times oil can backstream from the diffusion pump into the microscope column. Fortunately, these operations are not carried out very often, and so they should not represent a serious source of contamination.

The specimen exchange operation is a much more serious potential source of contamination, however. As pointed out at the end of Section 2.3 (p. 37), and again in Section 9.2.7, it is almost certain that the pressure in the rough vacuum line will drop into the range of transitional or molecular flow during periods between specimen exchange operations. Under these circumstances oil molecules from the roughing pump are free to diffuse up this line and into the airlock chamber, where they can find their way onto the specimen during the time this chamber is being rough pumped in preparation for inserting the specimen holder into the specimen stage. In addition, it is easy to contaminate the specimen holder itself with finger prints and other organic materials (Section 2.10.4b, p. 68), which are then carried into the instrument in close proximity to the specimen. These phenomena effectively defeat the entire purpose of the special vacuum system in the worst possible location, the specimen itself. Since the specimen exchange process is carried out so frequently this is a major potential source of increased specimen contamination. Interestingly, these problems very likely will not be present when a microscope is first installed and the initial performance tests are carried out, because then the rough vacuum line and the specimen holders are new and clean. Consequently, tests of contamination rate performed at the time of installation can give quite favourable results. After an instrument has been in use for a few months, however, the backstreaming of oil up the rough vacuum line and contamination of the specimen holder can lead to very different results.

While it should in theory be possible to keep the specimen holder clean, it is virtually impossible to stop the diffusion of oil molecules up the rough pumping line. Therefore, the incorporation of the oil-sealed rotary-vane pumps and the oil diffusion pump into these vacuum systems can be considered to be basic shortcomings in their design. A minimal approach to overcoming these shortcomings can be achieved by installing a high conductance oil trap in the rough vacuum line. However, this will be effective over the long run only if this trap is designed so that it can be, and in operation actually is, removed from the vacuum line frequently and properly cleaned or replaced, as recommended in Section 4.1.5a (p. 146). An additional improvement can be achieved by using a turbomolecular pump to evacuate the camera chamber and to perform the intermediate pumping functions. Anyone purchasing one of these instruments should most certainly obtain these modifications if it is at all possible to afford to do so. Once an instrument becomes contaminated with oil, it is virtually impossible to clean it up, and its essential characteristics are lost.

Notwithstanding these possible shortcomings and potential difficulties, recent models of electron microscopes with columns evacuated by ion pumps are a tremendous improvement over instruments on which oil diffusion pumps are used for this purpose. Pressures in the mid and low 10^{-5} Pa (10^{-7} Torr) range are routinely attained, and the upper end of the 10^{-6} Pa (10^{-8} Torr) range can be reached in some instruments. This makes it possible to obtain good performance from LaB_6 emitters (Spence, Chapter 7). Manufacturers show mass spectra (obtained on new instruments, of course) that demonstrate partial pressures for hydrocarbons in the specimen chamber in the 10^{-8} Pa (10^{-10} Torr) range. Most important, users report being able to make high-resolution microscopic observations, and to carry out microanalyses, on a given area of a specimen for periods of several minutes without significant carbon contamination.

9.4 A vacuum system for a scanning electron microscope with a cold field-emission electron gun

One of the most recent advances in the design of electron microscopes is the development of instruments with cold field-emission electron guns. The electron source in these guns is a piece of fine tungsten wire (about 0.1 mm in diameter) that has been electrolytically etched to produce a tip with a very small radius. As indicated in Fig. 9.4, most field-emission guns have two anodes, one of which is known as the 'extractor electrode', and the other as the 'accelerator electrode'. The extractor electrode is maintained at a positive direct current potential of a few thousand volts relative to the emitter tip. Assuming that the potential on this electrode is $E_e = 2500$ V and that the radius of the tip is $r = 500$ nm, the electrostatic field strength at the tip has a value of about $F \approx E_e/r \approx 2500/5\times10^{-7} \approx 5\times10^{9}$ V/m. A field strength of this magnitude is great enough to draw electrons from the tip without the need for heating it (hence, the 'cold field-emission' name). The potential on the accelerator electrode is adjusted to give the total desired electron accelerating voltage. These guns produce an electron beam with a smaller focused spot size and much higher brightness than guns with tungsten hairpin filaments or LaB_6 emitters (Reimer, p.88), making it possible for a scanning electron microscope to produce high-intensity, high-resolution images, even when working at very low electron accelerating voltages.

For stable operation, a field-emission electron gun requires a clean vacuum that is well into the ultra-high vacuum range, and so the great challenge in developing the vacuum systems for these instruments involved finding a means to maintain such a vacuum in the electron gun chamber while retaining the flexibility and convenience microscopists are accustomed to in the insertion and manipulation of specimens. This has been achieved very successfully by employing a vacuum system of the type shown in Fig. 9.4, which is very similar to the one on the Hitachi S-800 FEG SEM. Here, the two gun electrodes, together with a differential pumping aperture in the top of the second electron lens, divide the upper

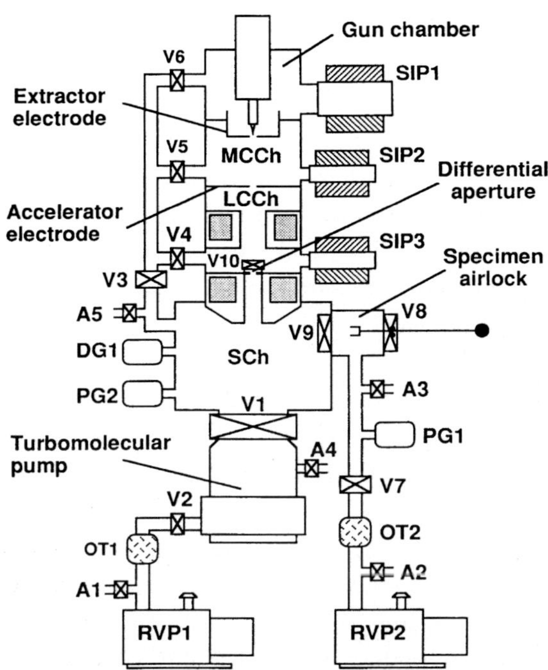

Fig. 9.4 A typical vacuum system for scanning electron microscopes with field-emission electron guns. A1 to A5, inlet valves; V1 to V10, vacuum valves; DG1, cold cathode discharge gauge; PG1 and PG2, Pirani gauges; SP1 to SP3, sputter-ion pumps; MCCh, middle column chamber; LCCh, lower column chamber; SCh, specimen chamber; OT1 and OT2, oil traps; RVP1 and RVP2, rotary-vane pumps.

part of the electron optical column into three chambers, each of which is evacuated by a separate sputter-ion pump. The holes in the two electrodes and in the differential aperture are small enough so that this arrangement of apertures, in combination with the pumps on the lower two column chambers, reduces the amount of air that reaches the electron gun chamber by diffusion up through the electron optical column to a negligible level, even when valve V10 is open. This makes it possible for the ion pump on the gun chamber to maintain the ultra-high vacuum needed for operation of the cold field-emission gun. The attainment of pressures in the ultra-high vacuum range in the upper two column chambers is further facilitated by constructing the upper part of the column so that it can be baked out at temperatures above 200°C. Typically, the pressure can be maintained below 10^{-7} Pa (10^{-9} Torr) in the gun chamber, below 10^{-6} Pa (10^{-8} Torr) in the middle column chamber MCCh, and below 10^{-5} Pa (10^{-7} Torr) in the lower column chamber, LCCh. In order to establish these low pressures, however, the flux of molecules up through the column must be very low, and so it is also necessary to hold the pressure in the specimen chamber, SCh, well down into the 10^{-4} Pa (10^{-6} Torr) range while valve V10 is open. Therefore, these instruments have a specimen airlock to reduce the time needed to pump the specimen chamber down to such a low pressure after exchanging specimens. Because of the several different levels of vacuum that exist in them, these are often called 'multi-stage' or 'multi-level' differential vacuum systems.

Although oil diffusion pumps are used to evacuate the specimen chamber on some instruments of this kind, it is better to use a turbomolecular pump for this purpose, as shown in Fig. 9.4, because this drastically reduces the extent to which problems with oil contamination are likely to arise. It is also common to find a pumping manifold leading from the specimen chamber to the electron gun chamber, with branches into each of the intermediate vacuum chambers of the upper column. This makes it possible to use the high vacuum pump that serves the specimen chamber to evacuate these upper chambers during the initial stages of the pumpdown process. The valves in this manifold, V3, V4, V5, and V6, are usually manually operated, and are kept closed except when these chambers are being pumped down in preparation for starting up the ion pumps. These valves must, of course, be able to withstand the high temperatures generated during the bakeout treatment. Valve V10 is closed whenever a specimen is not being examined, to reduce the total gas load that must be handled by the ion pumps. A brief review of the operation of a system of this kind will provide an opportunity to illustrate procedures for using turbomolecular pumps and field-emission guns, both of which are being used with increasing frequency on newer electron microscopes.

9.4.1 Pumping down from atmospheric pressure

The procedure for establishing a vacuum after the entire system has been brought up to atmospheric pressure is as follows:

1. *The gun and column chambers are rough pumped*: The three chambers in the upper electron optical column are pumped down to a pressure near 10^{-2} Pa (10^{-4} Torr) before the sputter-ion pumps are turned on. To open paths between these chambers and the turbopump and roughing pumps,

valves V3, V4, V5, V6, and V10 are all opened. The evacuation process can then be carried out in either of two ways:

(a) *One way* is to use rotary-vane pump RVP2 to rough pump the specimen and gun chambers while the turbomolecular pump is being started up. To do this, all inlet valves are closed, isolating the system from the atmosphere, and valve V1 is closed to isolate the turbomolecular pump from the system. Valve V2 is then opened, both the turbomolecular pump and roughing pump RVP1 are turned on, and the turbomolecular pump is allowed to come up to operating speed while isolated from the vacuum system. Then valves V7 and V9 are opened, and rotary-vane pump RVP2 is turned on and allowed to establish a rough vacuum while the turbomolecular pump is coming up to operating speed. As soon as Pirani gauge PG2 indicates that the pressure in the vacuum system is approaching 10 Pa (0.1 Torr), the lower end of the pressure range for viscous flow, valves V7 and V9 are closed, valve V1 is opened, and the turbopump is allowed to take over the evacuation process.

(b) *A better way* to carry out this rough pumping phase of the evacuation process, because it minimizes the possibility of introducing oil from the rotary-vane pumps into the specimen chamber, is to take advantage of the fact that a turbomolecular pump can be started up while its inlet pressure is well above the normal operating level, as described at the end of Section 6.1.6 (p. 243) and in Section 6.1.8b (p. 252) To use this approach, valve V9 and all inlet valves are closed, valves V1 and V2 are opened, the turbopump and its backing pump RVP1 are turned on at the same time, and the backing pump is allowed to rough pump the system through the turbopump while the turbopump is running up to operating speed. In this way the turbopump will gradually take over the direct evacuation of the system as its speed increases, and will become fully operational before the pressure drops into the range of molecular flow, thereby protecting against the backstreaming of oil vapours from the rotary-vane pump into the vacuum system.

2. *A high vacuum is now established in the system*: When the turbomolecular pump reaches full operating speed it should bring the vacuum in the instrument down into the high vacuum range within a few hours. Vacuum gauge DG1, which is usually a cold cathode discharge gauge, is used to monitor this process. It is beneficial to continue pumping in this way until

a vacuum well below 10^{-3} Pa (10^{-5} Torr) has been established so that the sputter-ion pumps can be started up with a minimum likelihood of overheating (Section 7.1.6, p. 291).

3. *Establish an ultra-high vacuum in the gun chamber*: With a good high vacuum in the gun chambers, the power supplies for the sputter-ion pumps are turned on. Although these ion pumps should start up quickly and without overheating under these conditions, it is nonetheless advantageous to leave valves V3, V4, V5, and V6 open, and to continue pumping on the ion pumps with the turbomolecular pump so that it can help handle any gas that may be released inside them as they start up. If the ion pumps are equipped with individual heaters it will also be advantageous to turn these on at this time to speed up the removal of the gas molecules that adsorbed on the surfaces inside these pumps while they were at atmospheric pressure.

   After the high voltage to the ion pumps has increased to a value above 2500 V and it has become clear that they are operating in a stable manner, valves V3, V4, V5, V6, and V10 are closed and the ion pumps are allowed to evacuate the electron gun and column chambers for several hours, preferably overnight, whereupon the bakeout procedure that is necessary for obtaining an ultra-high vacuum is started. The way in which this procedure is implemented will depend on the design and construction of the gun chamber and its heating apparatus for the particular instrument involved, and so the manufacturer's instructions must be followed carefully to avoid damage to the system. When the bakeout treatment is completed the heaters are turned off and the instrument is allowed to cool overnight before it is used for viewing specimens.

Ed Birko, a regional service engineer for Hitachi FEG SEMs, reports that his experience with several of these instruments shows that it is highly advantageous to carry out the bakeout treatment for a total heating time of from 75 to 100 h. Although this may seem unreasonably long, the results it produces are impressive. For example, our Hitachi S-800 FEG SEM, which was baked out in this way when it was installed, is now over 5 years old. Throughout its lifetime it has been used daily by a large number of different students and faculty members. During all this time it has never been necessary to open the upper column. The original emitter is still performing to specifications, and the vacuum in the gun chamber and middle column chamber MCCh is still below 10^{-7} Pa (10^{-9} Torr), the lowest value that is indicated on the pressure scale of the ion pump controllers. In the lower column chamber LCCh the pressure routinely remains in the low end of the 10^{-6} Pa (10^{-8} Torr) range during operation when valve V10 is open, and drops into the 10^{-7} Pa (10^{-9} Torr) range when this valve is closed.

Note that if the specimen chamber were evacuated by an oil diffusion pump instead of by a turbomolecular pump, it would be advisable to stop pumping on the gun chamber in Step 2 as soon as the pressure drops below the 10^{-1} Pa (10^{-3} Torr) range, to minimize the opportunity for the gun to become contaminated with pump oil. This would necessitate starting up the ion pumps at a much higher pressure, which, in turn, would make it necessary to watch them more closely to be sure that they do not overheat and cause problems of the kind described in Section 7.1.6 (p. 290).

9.4.2 *Exchanging specimens*

The specimen airlock shown in Fig. 9.4 is functionally identical to the one in the vacuum system for the freeze-fracture apparatus shown in Fig. 9.1, with valves V8 and V9 in Fig. 9.4 corresponding to the inner and outer airlock doors, and with valves V7 and A3 corresponding to valves V2 and A2 in Fig. 9.1, respectively. Therefore, it should not be necessary to describe the use of such a system again. It should be noted, however, that in preparing to exchange specimens in the scanning electrom microscope, it is of great importance to be certain that valve V10 is closed so that there will not be an undesirably high influx of gas into the ultra-high vacuum section of the system when the pressure in the specimen chamber subsequently rises. Most of the valves in these systems are operated manually, and so operators of these scanning electron microscopes must understand the functions of the valves clearly so that they are able to manipulate them correctly, otherwise serious vacuum problems can be generated.

9.4.3 *Servicing the field-emission electron gun*

Replacing the field-emission emitter and performing other work in the ultra-high vacuum section of an instrument of this kind usually requires so much time and effort that the instrument is fully shut down and the entire vacuum system is brought up to atmospheric pressure while these operations are being performed. To do this the electrical power to the turbomolecular pump and to all the ion pumps is first turned off. Valves V1 and V2 are closed, inlet valve A1 is opened, and roughing pump RVP1 is turned off. When the speed of the turbopump has decreased to about 60 percent of its operating value dry nitrogen is admitted slowly into it through inlet A4. Valves V3, V4, V5, V6, and V10 are then opened, valve V9 is closed, and dry nitrogen is slowly admitted to the entire vacuum system through inlet A5. When the system reaches atmospheric pressure the gun chamber can be opened as required. In the meantime, valve V7 is closed, inlet valves A2 and A3 are opened, and roughing pump RVP2 is turned off.

9.4.4 *Opening the specimen chamber*

Parts in the specimen chamber can be serviced without disturbing the vacuum in the electron gun. The specimen chamber is first isolated by closing valves V1 and V10. Valve V7 is closed, valve V9 is opened, and dry nitrogen is slowly admitted through inlet A5. When the service procedure is completed, the outer airlock door V8 and inlet A5 are closed, and valve V9 is opened. Valve V7 is then opened and the specimen chamber is rough pumped with rotary-vane pump RVP2 to a pressure near 10 Pa (0.1 Torr), as indicated by Pirani gauge PG2. Then valves V7 and V9 are closed, Valve V1 is opened, and the turbomolecular pump is allowed to establish an operating vacuum in the specimen chamber. The pressure in the specimen chamber is monitored by cold cathode discharge gauge DG1.

9.5 *A transmission electron microscope vacuum system of advanced design*

With rather strong encouragement from electron microscopists who want instruments with the best possible capabilities, manufacturers are now building new models of electron microscopes with vacuum systems that are even less subject to oil contamination than the one described in Section 9.3. Fig. 9.5 is a diagram showing one approach that has been taken. This diagram is based largely on a Hitachi Model HF-2000 200 kV transmission electron microscope with a cold field-emission electron gun which has a vacuum system that was modified by the manufacturer in accordance with specifications formulated by L. F. Allard and T. A. Nolan of the Oak Ridge National Laboratory. This represents an attempt to come as close

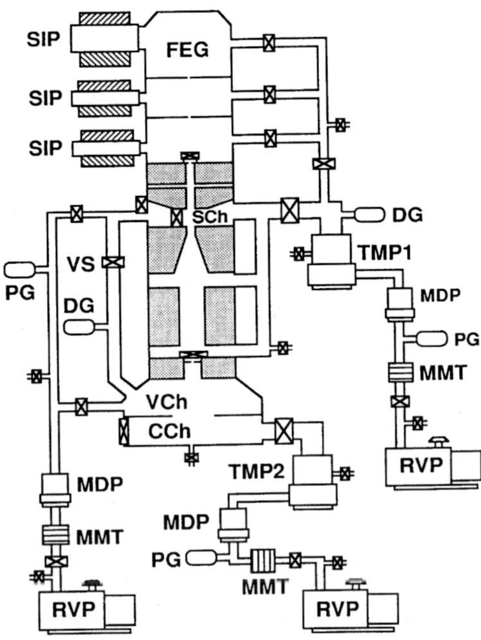

Fig. 9.5 A vacuum system for a transmission electron microscope with a field-emission electron gun which has special features to reduce oil contamination. FEG, field-emission electron gun; MDP, molecular drag pump; SIP, sputter-ion pump; TMP, turbomolecular pump; MMT, Micromaze oil trap; DG, discharge gauge; PG, Pirani gauge; SCh, specimen chamber; CCh, camera chamber; VCh, viewing chamber; VS, high vacuum valve for the specimen airlock.

as possible to achieving a totally oil-free vacuum system, while retaining all of the reliability, flexibility, and versatility that users have come to expect in modern ultra-high-resolution transmission electron microscopes. Although the field-emission electron gun on this instrument is quite large and complex, to allow operation at 200 kV, the ultra-high vacuum subsystem associated with it is generally similar in construction to the one for the scanning electron microscopes described in Section 9.4, and so vacuum manipulations involving this subsystem are generally the same as those previously described. Likewise, the arrangement of the vacuum valves associated with the specimen airlock and the camera chambers are much the same for this instrument as for the other transmission electron microscope described in Section 9.2 and Section 9.3. The sequence of operations involved in manipulating these valves when exchanging photographic film or specimens is the same as already described, and so need not be repeated here. Instead, it will be more interesting and useful to consider some of the special features of this vacuum system and their contributions to enhancing the capabilities of this microscope over the more commonly-encountered ion-pumped system of the type described in Section 9.3.

As shown schematically in Fig. 9.5, this is a rather complex vacuum system — it has 11 vacuum pumps, 17 vacuum valves, nine inlet valves, and involves five stages of differential pumping. However, this vacuum system makes it possible to incorporate a cold field-emission electron gun, capable of operating at 200 kV, into this instrument, and this gives improved electron optical characteristics in nearly all respects over transmission electron microscopes with LaB_6 emitters or hairpin tungsten filaments. The brightness of the beam (Agar, Alderson and Chescoe, Chapter 1 and Hall, Chapter 7) produced by a cold field-emission electron gun is about 100 times greater than can be obtained from an LaB_6 emitter, and nearly 1000 times greater than from a hairpin tungsten filament (Reimer, Chapter 4). In addition, the effective diameter of the electron source is only about 3 nm, compared with about 15 nm for an LaB_6 emitter, and about 30 nm for a hairpin tungsten filament. When combined with the high beam brightness and the advanced capabilities of modern electron optical systems, this small source of electrons makes it possible to produce a beam spot at the specimen that is only 1 nm in diameter, but that carries a current of the order of 1 nA. With such a small, high-intensity beam, X-ray and electron energy loss spectroscopic analyses can be performed on areas of the specimen that are 100 times smaller in diameter than can be analysed when the source is an LaB_6 emitter. Because the electrons are drawn from the filament without heating it, the energy spread in the electron beam from a cold field-emission gun is much smaller than from guns with heated sources; typically, 0.4 to 0.5 eV, at levels of emission current that are practically useful, compared to several electron volts for heated sources. This means that the electron beam has excellent coherence (Hall, Chapter 7 and Spence, Chapter 4) and is therefore particularly well suited for performing experiments that involve electron optical interference phenomena. Manufacturers publish micrographs of holes in carbon films taken with instruments of this kind that show more than 100 Fresnel interference fringes. The most important and exciting consequence of this high coherence, however, is that, when combined with the high quality and stability of modern electron optical systems, it becomes possible to produce electron holograms (Lichte, p. 25), something electron microscopists have been striving to do since the concept of holography was introduced more than 40 years ago. The use of holograms promises to make it possible to correct for the aberrations of the electron lens system, thereby giving a substantial improvement in resolution, and to provide information from phase-contrast phenomena that is not contained in ordinary micrographs.

The bakeable, three-stage, differential, ultra-high vacuum subsystem associated directly with the field-emission electron gun maintains the pressure in the gun chamber in the 10^{-8} Pa (10^{-10} Torr) range during operation, giving a source life which is typically greater than 5000 h. The specimen chamber and all parts of the electron optical column below the electron gun are constructed with column liners, metal gaskets, bellows seals, and similar refinements which facilitate obtaining a clean high vacuum, as described above in Section 9.3. However, these parts of the vacuum system are evacuated by a turbomolecular pump, TMP1, rather than by a sputter-ion pump. The speed of this turbopump is 400 l/s, which is more than twice the speed of the ion pumps used on most of the instruments of the type described in Section 9.3. This makes it possible to maintain an oil-free vacuum in the specimen chamber that is near the bottom of the 10^{-6} Pa (10^{-8} Torr) range. This microscope also has a specimen anticontamination device of advanced design.

No oil diffusion pump is used on this instrument; instead, the viewing and camera chambers are evacuated by a second turbopump, TMP2, with a speed of 570 l/s which easily maintains a vacuum in the 10^{-4} Pa (10^{-6} Torr) range. Since the vacuum in this segment of the system is now oil-free, a pumping line is extended from it to the line leading into the specimen airlock. After the specimen airlock chamber is pumped out with the rough pumping system (which is somewhat unusual, and which is described in more detail below), valve VS in this line is opened for a short time and the airlock is further pumped down into the high vacuum range by turbopump TMP2. This operation, combined with the high speed of evacuation provided in the specimen chamber by turbopump TMP1, gives almost instantaneous recovery of the vacuum in the specimen chamber after a specimen exchange operation. Both turbopumps have magnetic levitation bearings, which effectively eliminate the problems with mechanical vibrations and bearing replacement that previously plagued turbomolecular pumps (Section 6.1.7, p. 248). Recent experience of the Hitachi Company indicates that the design and construction of these pumps has now been refined to the point where they will give many years of continuous trouble-free service.

Perhaps the most difficult part of designing this vacuum system was finding a solution to the problem of oil backstreaming from oil-sealed rotary-vane roughing pumps. Interestingly, these pumps have such a good combination of performance characteristics and such a high level of operational reliability that it is very difficult to find a good substitute for them. This may well be the most challenging problem currently facing designers of high-performance vacuum systems for electron microscopes. Initially it had been intended to eliminate rotary-vane pumps from the system entirely by using a combination unit consisting of a molecular drag pump in series with a set of diaphragm pumps (Section 6.2.3b, p. 268) to back up each of the turbopumps and to perform other rough pumping functions. However, Hitachi engineers had encountered problems with diaphragm pumps that made them reluctant to use these units in a situation of this kind where the pumps have to run continuously for many months, and where any malfunction could have serious effects on a very expensive instrument. In the end, and as a compromise, it was decided to still make use of rotary-vane roughing pumps, but to provide as much protection against the backstreaming of oil from them as is reasonably possible. Therefore, each turbomolecular pump is backed up directly by a molecular drag pump. Each molecular drag pump is, in turn, backed up by an oil-sealed rotary-vane pump, but with a continuously-heated Micromaze oil trap inserted into the pumping line between them. This arrangement provides three levels of protection against the backstreaming of oil from the rotary-vane pumps into the microscope column: first, the turbopumps are themselves totally oil-free, and are generally considered to afford very effective protection against oil backstreaming from other sources (Section 6.1.4, p. 237); second, molecular drag pumps have characteristics similar to turbopumps in this regard (Section 6.2.1, p. 260); and third, the continuously-heated Micromaze traps appear to be one of the most effective types of traps currently available for preventing oil backstreaming from rotary-vane pumps (Section 4.1.5a, p. 147). As shown in Fig. 9.5, a third molecular drag pump, which is also backed up by a rotary-vane pump through a heated Micromaze trap, is used to perform general rough pumping functions, providing two levels of protection in this segment of the system.

The freedom of this vacuum system from oil contamination after 2 years of operation is very good indeed. In recent experiments, L.F. Allard did not observe any carbon contamination, and could not detect the carbon spectral line with an energy dispersive X-ray spectrometer with an ultra-thin window, when the electron beam, with a current of 30 μA, was focused to a diameter of 10 nm and held stationary on a specimen of Si_3N_4 for more than 15 min. After this specimen was removed from the instrument, coated with 5 nm of carbon in a clean vacuum evaporator (evacuated by a turbomolecular pump), and re-examined under the same conditions, the build-up of carbon contamination was easily observed and the carbon spectral line appeared in the X-ray spectrum. The specimen was then taken out of the instrument and the carbon coating was removed from it in a clean ion mill, whereupon it could again be examined without having detectable carbon contamination develop. Furthermore, high-resolution electron images and electron holograms are routinely obtained without encountering any problems with specimen contamination. These results suggest that the vacuum system on this instrument is so clean that it makes no detectable contribution to specimen contamination. Instead, it appears that any contamination that may be observed on specimens examined in this instrument is produced from material that is present in the specimen itself or that is carried into the instrument on the specimen or the specimen holder. These are very encouraging results, because they show that by means of presently-available methods, which are convenient to apply and use, it is possible to construct vacuum systems that are clean enough to meet the requirements of the most advanced electron microscopy techniques presently being used.

From a practical point of view, it may well be that vacuum systems for electron microscopes are now approaching a plateau in performance that is generated by limitations introduced by the design of the specimen stage itself. Nearly all current instruments use side-entry goniometer specimen stages, and these stages all require the use of several well-lubricated O-rings to achieve their wide range of performance capabilities. Since these lubricated O-rings are a potential source of carbonaceous material (Section 10.11, p. 451) that is located in close proximity to the specimen itself, they are likely to be the most significant contributors to the carbon contamination problem in the entire instrument. Therefore, until a comparably versatile specimen stage is designed which does not require the use of O-rings, it is unlikely that it will be of much benefit to make drastic improvements in other parts of the vacuum system.

9.6 A scanning electron microscope vacuum system for wet specimens

The final type of vacuum system that will be discussed here is found on the 'environmental', 'natural', 'variable pressure', or 'wet' scanning electron microscopes that have achieved considerable popularity in recent years. The development of this type of instrument was pioneered by the ElectroScan Corporation. These vacuum systems are interesting, because unlike those previously discussed, which are designed to provide the lowest possible vacuum and a very clean environment for the specimen, these are designed to allow the specimen to be examined in the presence of air, water vapour, and other gases at pressures as high as 5000 Pa (50 Torr). The general features of the vacuum systems for instruments of this kind are shown schematically in Fig. 9.6. The system is, again, a fairly complex one involving five stages of differential pumping. The electron gun chamber is evacuated to a pressure in the 10^{-5} Pa (10^{-7} Torr) range by a sputter-ion pump, making it possible to use an LaB_6 emitter. A differential pumping chamber DPC1 in the upper part of the electron optical column is separated from the gun chamber and from the segment of the column below it by several small apertures, and is maintained at a pressure in the 10^{-4} Pa (10^{-6} Torr) range by oil diffusion pump ODP1 and its backing pump RVP1. Vacuum valve VC can be closed to prevent gas from the lower stages of the system from diffusing into this chamber, and from there into the electron gun, when the electron beam is not on. Differential pumping apertures located below the objective lens form two more differential pumping chambers. The upper one of these chambers, DPC2, is evacuated into the 10^{-2} Pa (10^{-4} Torr) range by oil diffusion pump ODP2 and its backing pump RVP2, while the lower one, DPC3, is held in the 10 Pa (0.1 Torr) range by a large rotary-vane pump RVP3. The specimen chamber is independently evacuated by another large rotary-vane pump RVP4.

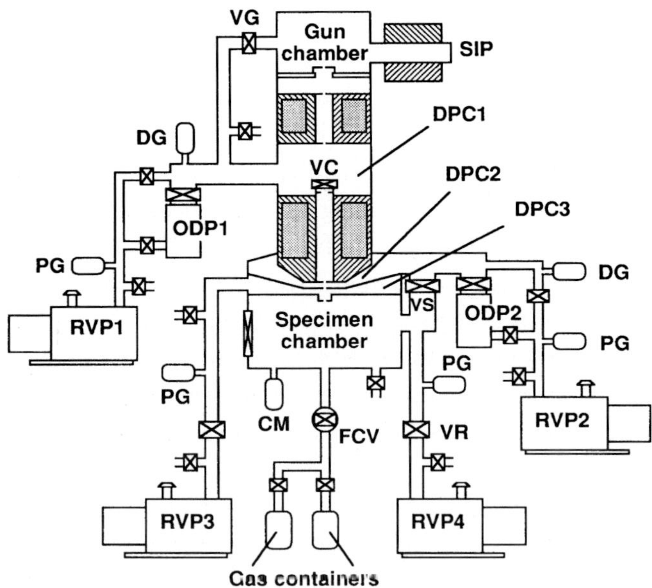

Fig. 9.6 A vacuum system for a scanning electron microscope that allows specimens to be examined in the presence of air, water vapour, and reactive gases. ODP1 and ODP2, oil diffusion pumps; RVP1 to RVP4, rotary-vane pumps; SIP, sputter-ion pump; DPC1 to DPC3, differential pumping chambers; DG, discharge gauge; PG, Pirani gauge; FCV, flow control valve; CM, capacitance manometer; VC, VG, VR, and VS, vacuum valves.

Again, it should not be necessary to describe operating procedures for this system in great detail, and so they will only be outlined briefly. To start up this system, the valves leading into the column from the oil diffusion pumps and their backing pumps are closed, their backing valves are opened, their backing pumps are turned on, cooling water is turned on, the power to their heaters is turned on, and they are allowed to warm up to operating temperature (Section 5.7.1a, p. 205). To evacuate the upper part of the electron optical column, valve VC in chamber DPC1 is closed, valve VG in the manifold to the gun chamber is opened, the upper diffusion pump is then isolated for a few minutes by closing its backing valve, while backing pump RVP1 is used to rough pump the gun and column chambers. Then the diffusion pump is used to evacuate the column and gun chambers into the 10^{-2} Pa (10^{-4} Torr) range, whereupon the gun chamber is isolated (by closing valve VG) and the sputter-ion pump is turned on. When the vacuum in the gun chamber drops into the low end of the

10^{-5} Pa (10^{-7} Torr) range the LaB_6 emitter can be turned on, and the instrument is ready for use in examining specimens.

After a specimen is inserted into the specimen chamber, the three roughing pumps RVP2, RVP3, and RVP4, are used to rough pump the specimen chamber and the lower two differential pumping chambers DPC2 and DPC3. Then differential pumping chamber DPC2 is further evacuated into the high vacuum range by oil diffusion pump ODP2. If it is desired to use the instrument as a conventional scanning electron microscope, the specimen chamber is also pumped down into the high vacuum range with ODP2 by closing roughing valve VR and opening valve VS in the high vacuum manifold. More often, however, it will be desired to take advantage of the special capabilities of the instrument that make it possible to observe a specimen while it is exposed to air, water vapour, or other gases. To expose a specimen to water vapour, for example, a flask containing a small amount of water is connected to the gas inlet manifold, valve VS is kept closed, roughing valve VR in the line to roughing pump RVP4 is also closed, and water vapour is admitted to the specimen chamber by opening the appropriate valves in the manifold. Since the water vapour is slowly pumped out of the specimen chamber through the holes in the differential pumping apertures below the objective lens, the flow control valve FCV is adjusted to admit water vapour to the specimen chamber continuously at a rate sufficient to balance this loss and maintain the desired pressure. Usually, a capacitance manometer, CM, is used to monitor this pressure, because of the very high accuracy and sensitivity this type of gauge can provide (Section 3.5, p. 128), and the output of this manometer is used by a computer as a basis for adjusting the throughput of the flow control valve. It is, of course, possible to expose the specimen to gases, or mixtures of gases, other than water vapour by attaching containers of these materials to the gas intake manifold.

There are several advantages to having a vacuum system of this kind on a scanning electron microscope. Obviously, it is possible to observe the progress of reactions between the specimen and the gaseous environment inside the specimen chamber, and the range of conditions can be extended by using special stages for heating, cooling, and deforming the specimen. Biological specimens can be examined in the natural 'wet' state without having to be dried, frozen, coated, or otherwise processed before examination. Air or water vapour at a pressure of a few pascals will suppress the buildup of surface charges on most insulating materials, making it possible to observe specimens of non-conductive materials such as ceramics, minerals, textiles, paper, and plastics without coating them with metal or carbon.

9.7 Concluding remarks

A surprising variety of vacuum systems are presently being used on electron microscopes and other vacuum apparatus used in and around electron microscopy laboratories. Several relatively simple systems of the type used on such apparatus as vacuum evaporators and sputter coaters were discussed in earlier chapters, and six somewhat more complex systems are discussed in this chapter. Taken together, these contain most of the important features found on the vacuum systems commonly encountered by electron microscopists. I have also described most of the common protocols involved in the operation of these systems, and in doing so have attempted to give illustrations of the kinds of procedure that can give rise to problems, so that readers will be sensitive to such matters and can avoid incurring problems while operating their own vacuum systems. From discussions I have had with a large number of electron microscopists over the past 30 years, it is apparent that many do not have a very high level of confidence in their understanding of the vacuum systems on their instruments. This is a very uncomfortable situation, because it more or less puts the instrument in control of the microscopist. I hope the material presented in this book will help alleviate this situation.

10 Vacuum leaks, seals, and greases

Leaks are the bane of all who work with vacuum systems. Unfortunately, nearly every vacuum system will develop a leak at some time, and so nearly everyone closely involved in operating a vacuum system will eventually have to deal with a leak. Hablanian (Chapter 11) and Harris (Chapter 14) give practical descriptions and illustrations of the numerous methods available for detecting leaks, while Weston (Chapter 8) gives a more theoretical discussion of the physical principles and mathematical relations that are useful for describing the magnitude of leaks and for understanding and applying the various methods of leak detection. In this chapter, only those methods for detecting leaks that are the most convenient for use on electron microscopes and other vacuum apparatus commonly found in electron microscopy laboratories are discussed.

The references cited in this chapter are listed in Appendix 1. Manufacturers and suppliers of vacuum equipment referred to here are listed in Appendices 2 and 3.

10.1 The symptoms of a leak

Fundamentally, the effect of the development of a leak in a vacuum system is to make it impossible to reach an ultimate pressure P_u as low as could be reached before the leak arose. The reason for this can be explained on the basis of eqn. 2.16 (Section 2.6, p. 48), $P_u = q/S_e$. When a system develops a leak the gas influx due to the leak q_l increases the total gas influx q (see Section 2.10, p. 55) while the speed of evacuation S_e remains unchanged, and so P_u must increase. A secondary effect, but the one that is usually noticed in practice before the increase in ultimate pressure becomes evident, is a decrease in the rate at which the system pumps down. This can be accounted for on the basis of eqn. 2.17 (Section 2.6, p. 49), $dP/dt = -(S_e/V_c)(P_c - P_u)$. Since the development of the leak causes the value of P_u to increase, as just noted, the magnitude of the quantity $(P_c - P_u)$ becomes smaller than it was before the leak developed at every value of the chamber pressure P_c. However, the leak does not affect either the volume of the chamber V_c or the speed of evacuation S_e, and so the term (S_e/V_c) remains unchanged. The net result is to cause the rate at which the pressure decreases dP/dt to be smaller than it was without the leak, and this in turn increases the time needed to reach an operating pressure, even though this pressure may be significantly greater than the ultimate pressure attainable by the system.

Operationally, then, there is reason to suspect that a vacuum system has developed a leak if it pumps down noticeably slower than usual, and ultimately does not reach as low a pressure as it should or previously did. Unfortunately, these symptoms are not totally unambiguous. Note in particular that it is the total gas influx q that determines the level of ultimate pressure attainable. As discussed in Section 2.10 (p. 55), the leak rate q_l is only one factor contributing to gas influx; others are gas permeation q_p, gas desorption q_d, and gas evolution q_e, with $q = q_p+q_d+q_e+q_l$. The influx due to gas permeation is determined by the materials used in constructing the vacuum system, and is unlikely to change after an apparatus has been delivered by the manufacturer. However, the rates of gas desorption and gas evolution can increase from a variety of causes, as discussed in Section 2.10, and if this should happen the initial symptoms will be the same as if a leak had developed. Since leak testing procedures can be tedious, uninspiring, and quite time-consuming, it is wise to be as certain as possible that there is indeed a leak present before undertaking the task of finding and stopping it.

10.2 Distinguishing gas evolution from a leak

The question then arises as to what methods can be used to distinguish a leak from an increase in gas influx from these other sources. This can usually be done in a straightforward manner if the instrument has a residual gas analyser (RGA), by comparing a spectrum taken when the leak is suspected to be present with reference spectra recorded previously when the instrument was leak-free, as suggested in Section 3.4.4 (p. 123). The development of a leak usually causes the peaks due to nitrogen and oxygen at $m/z = 28$ and 32 to increase significantly relative to peaks due to water and other gases characteristically found in the system, as illustrated by the RGA spectra of Fig. 3.20 (Section 3.4.4).

In the absence of an RGA, a sensitive high-vacuum gauge, such as a hot cathode or cold cathode ionization gauge (Chapter 3), can often both confirm the presence of a leak and reveal its location. For example, if the gauge shows an increase in pressure when a mechanism that transmits motion through the vacuum wall, such as an aperture control or a stage drive, is operated, the vacuum seals associated with this mechanism are not functioning properly. Since the mechanism allows an easily detectable burst of gas to enter when it is operated, it is very likely to be admitting a steady influx of gas large enough to constitute a significant leak even when not in use. Specimen chamber airlock mechanisms are used so frequently, and often so harshly, that they are very common sites for leaks. This is particularly true for those on side-entry goniometer stages, which are delicate devices that usually involve a complex arrangement of several small O-rings. To provide a basis for recognizing abnormal behaviour that might be indicative of a leak, it is very helpful to have carefully documented the pressure variations associated with the specimen exchange operation when the microscope was known to be performing normally.

When such obvious indications are not present, more indirect methods must be used to distinguish internal gas influx from a true leak. If the instrument has a sensitive total-pressure, high vacuum gauge which provides a numerical display of pressure readings, it may sometimes be possible to gain useful information by performing a rate of pressure rise test (often also called a rate of leak-up test). Hot cathode ionization gauges (Section 3.2.1, p. 92), which have good sensitivity and stability, are well-suited for use in a test of this kind. Cold cathode ionization gauges (Section 3.2.2, p. 99) are less satisfactory because they are likely to be too unstable to yield meaningful data in the upper end of their operating range. To perform this test on the vacuum system shown in Fig. 10.1, for example, the system is first evacuated to as low a pressure as possible, valves V1 and V3 between the pumps and the high vacuum portion of the system are then closed, and the pressure registered by the high vacuum gauge is recorded at intervals of 10 or 15 s until it rises to the top of the operating range of the gauge. The data obtained in this way may involve a pressure variation spanning several orders of magnitude, and so are best plotted on a log–log basis. The blank page of graph paper in Fig. 10.2 can be enlarged by photocopying and used for this purpose. On the vertical axis of this graph note the units in which the pressure is measured and add exponents

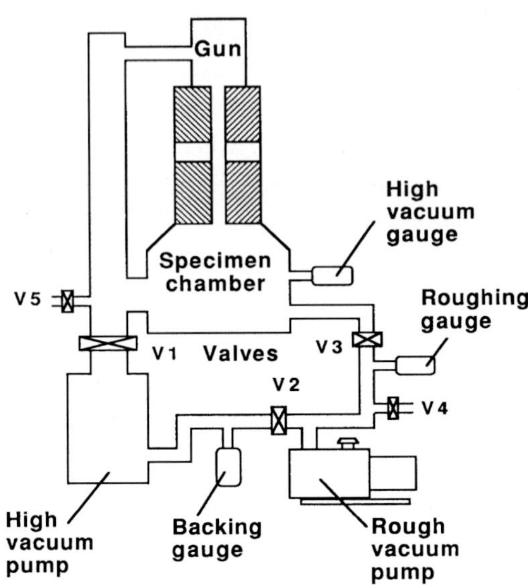

Fig. 10.1 A vacuum system for a scanning electron microscope with vacuum gauges at locations that facilitate testing for leaks

for the 10s so that the pressure range of the graph corresponds to that of the test. Ideally, a leak will cause the pressure to rise at a nearly constant rate over the entire range of the measurement, as shown by line A in Fig. 10.3. If, however, the problem is gas desorption or gas evolution from materials within the vacuum system the pressure should ideally increase linearly at first, but then the rate of rise should decrease as the pressure in the system approaches the equilibrium pressure of these materials, as depicted by curve B in Fig. 10.3. A leak in combination with a significant level of outgassing should produce a result somewhat like curve C in Fig. 10.3. Although this type of test is widely recommended in texts and instruction manuals, it is often difficult to carry out and may not yield results that are entirely unambiguous. In practice, it is most likely to be truly useful for ultra-high vacuum systems, since they have inherently low base leakage rates and are routinely baked out to produce low residual outgassing rates. For such systems, the pressure at the beginning of the test may be near the ultra-high vacuum range, and since it can be allowed to rise to the top of the high vacuum range before the high vacuum gauge has to be turned off, the test range covered may be large enough to reveal

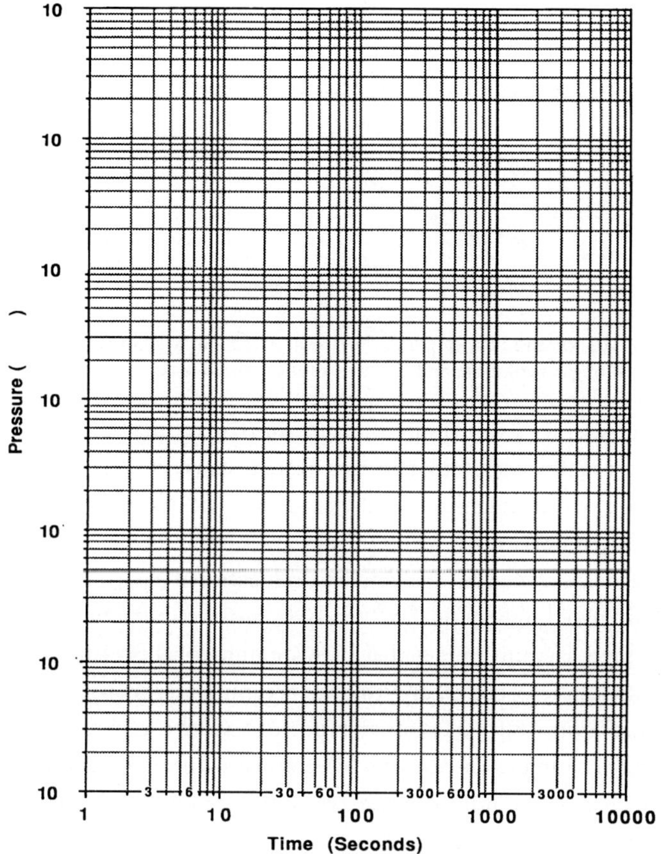

Fig. 10.2 A page of graph paper that can be copied for plotting pumpdown and leakup data. On the vertical axis, indicate the units used to measure the pressure and add exponents to the 10s to cover the range over which pressure is measured in the test.

definitive features of the type shown in Fig. 10.3. For most vacuum instruments found in electron microscopy laboratories, however, the results are likely to be less satisfactory. For one thing, such instruments cannot be baked out, and so they are certain to have a relatively high inherent rate of outgassing. Furthermore, if there is a significant problem it is unlikely that such systems can be evacuated to a pressure much below 10^{-4} Pa (10^{-6} Torr). Since high vacuum gauges cannot be operated above about 10^{-1} Pa (10^{-3} Torr), this will limit the range over which pressures can be measured to only two or three orders of magnitude, and this may not be enough to show the effects of a leak in the presence of the inherently high

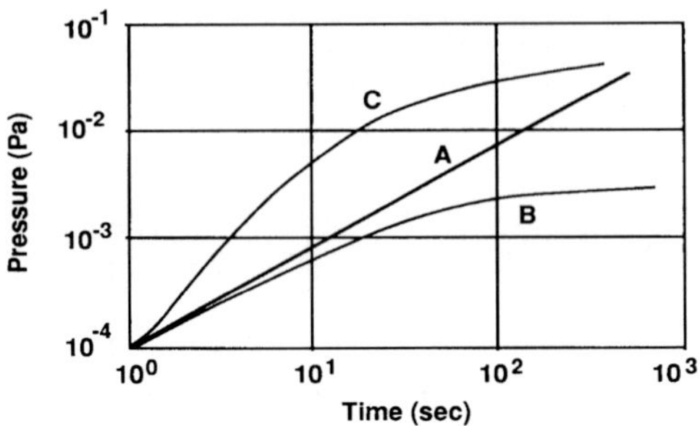

Fig. 10.3 Idealized curves for the rate of pressure rise in a vacuum system with: A, a true leak; B, a high level of gas evolution; and C, gas evolution plus a leak.

rate of outgassing. Even though ideal results may not be obtained, however, it is usually useful to run this test, because it may yield valuable information on the magnitude and character of the problem, particularly if test data taken previously when the apparatus was known to be performing properly are available for comparison (Section 3.3.3, p. 112).

As discussed in Section 3.3.2 (p. 108), most manufacturers do not put sensitive high vacuum gauges that provide numerical pressure readings on their electron microscopes, and this is becoming increasingly true for vacuum evaporators and many other types of vacuum apparatus now being manufactured for use in research laboratories. Instead, such instruments usually have only an indicator light, or some similar 'go-no go' device, that is activated by the control circuits of the automatic vacuum valve operating system to show that an 'acceptable' operating vacuum has been reached. This makes it very difficult to know reliably when such an apparatus is not performing correctly, and even more difficult to determine the cause of any problem that may have arisen. In general, the only symptom that will be available to indicate a malfunction in such a vacuum system is an increase in pumpdown time. Therefore, it is very important to document the time needed for an instrument of this kind to pump down to the point where the indicator device is activated under standard, easily-reproducible conditions when it can be presumed to be operating properly (Section 3.3.3, p. 112). If the pumpdown time subsequently becomes significantly longer than this standard value it is likely that there is a problem in the vacuum system. It may then sometimes be possible to decide whether the problem is due to a leak or to gas evolution by pumping the vacuum system for a day or two after all specimens and other items that are likely to contribute heavily to internal gas evolution have been removed from it, and then measuring the pumpdown time again. Prolonged pumping in this way will have little effect on the pumpdown time if the problem is a true leak, but might cause it to decrease if the problem is due to internal gas evolution.

10.3 Other problems that can appear to be leaks

Unfortunately, it is also possible for a defective valve, gauge, or pump to cause symptoms that suggest the presence of a vacuum leak. For example, if valve V3 in Fig. 10.1 were to leak it would allow gas from the backing line to return to the vacuum chamber and cause a decrease in the rate of pumpdown and an increase in the ultimate pressure. The performance of this valve can usually be tested by evacuating the system to as low a pressure as possible, closing valves V1, V2, and V3 while the pumps are still running, and then watching the reading of the roughing gauge as air is admitted to the specimen chamber through valve V5. If valve V3 seals properly the gauge reading should remain steady, or possibly decrease slowly, whereas if it leaks the gauge reading will increase.

Symptoms similar to a leak would also occur if either valve V1 or V2 does not open fully, because this will decrease the speed of evacuation by reducing the throughput of the pumping system. Although this problem can be difficult to diagnose, it may be possible to get an indication that it is occurring as follows. First turn off the power to the high vacuum pump and allow it to shut down fully so that it will not be damaged by exposure to air at atmospheric pressure. With valves V1, V2, and V3 all closed, open valve V5 and admit air to the apparatus. Then close valve V5, open valve V3, and note the time required to rough pump the system to a selected pressure low in the rough vacuum range. Now close valve V3, admit air to the apparatus again and note the time needed to rough pump it to the selected pressure when valves V1 and V2 are opened. If it takes significantly longer to pump down through these valves than it did through valve V3, one of these valves is probably not opening properly.

Unfortunately, problems of this general nature actually occur in practice. On one occasion a friend called seeking advice on how to locate a large leak that was thought to be preventing a vacuum evaporator from pumping down much below the rough vacuum range. It was, of course, impossible to diagnose the problem over the telephone, but a trip to the laboratory and a careful examination of the apparatus ultimately revealed the source of the trouble. The apparatus involved was a table-top vacuum evaporator with all the components compactly installed inside a small cabinet. This arrangement made it necessary to locate the rotary-vane roughing pump so close to the oil diffusion pump that only a short piece of rubber tubing could be used to connect them, and this tubing had to make a 120° bend, as shown in Fig. 10.4. Rubber tubing usually will not tolerate such a sharp bend and, as can be seen in the figure, this piece had collapsed, effectively shutting the diffusion pump off from its backing pump. This diagnosis was confirmed by pinching the sides of the tubing together, whereupon the bore opened and the vacuum in the bell jar of the evaporator promptly dropped to a satisfactory level. Thus, the problem was not a bad leak, but a malfunctioning of the pumping system caused by a blockage of the backing line. It was possible to overcome this particular problem without making major changes in the arrangement of the components inside the cabinet by inserting two sturdy springs inside the rubber tube to give it internal support and placing two hose clamps snugly around its midsection to provide external support, as shown in Fig. 10.5, so that it could not easily collapse and close again.

Rubber tubing can also be a site for leaks. In one instance all the symptoms initially indicated that the rotary-vane pump on a vacuum evaporator was malfunctioning. However, as the pump was being removed from the evaporator for servicing it was found that the rubber tubing connected to it had degraded badly and developed cracks that allowed the pressure in the backing line to rise to an unacceptable level. This is quite likely to happen in situations where the tubing is out of sight inside a cabinet and is not inspected for long periods. If a large number of deep, prominent surface cracks become visible when a piece of rubber tubing is bent slightly, it should be replaced.

If a vacuum pump starts to function improperly the first impression is likely to be that the system has developed a leak, because the malfunctioning pump will usually cause the speed of evacuation to decrease, and the initial symptom of this will be a noticeable increase in pumpdown time very similar to that caused by a leak. The performance of the roughing pump in a system such as the one shown in Fig. 10.1 can usually be tested

Fig. 10.4 Collapsed rubber tubing in the backing line to the oil diffusion pump on a table-top vacuum evaporator.

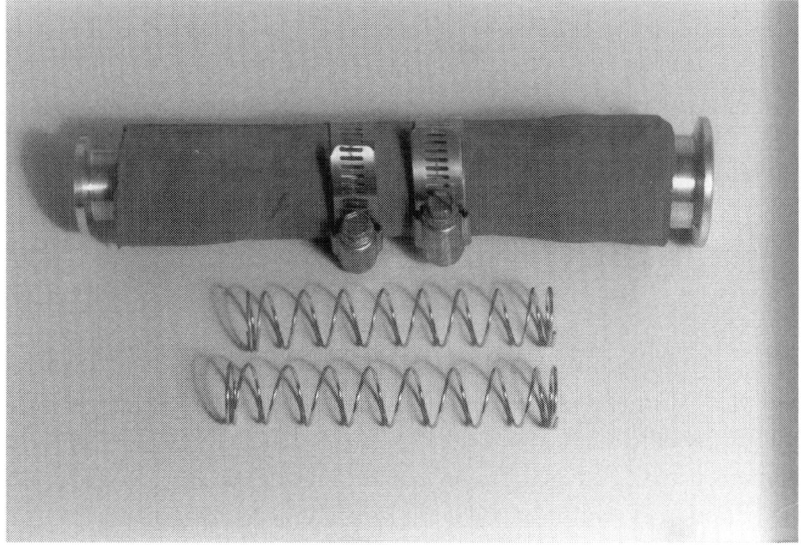

Fig. 10.5 Methods of supporting rubber tubing to prevent it from collapsing.

by evacuating the vacuum system to the lowest possible pressure, closing valves V1, V2, and V3 without admitting air to the apparatus, and noting the ultimate pressure the roughing pump produces as registered by the roughing gauge. If the pump does not produce a steady, unvarying pressure that is near or below 3 Pa (0.05 Torr), it is probably not functioning properly. The level of oil in the pump should then be immediately checked (Section 4.1.1, p. 138), because a deficiency of oil is one of the most common causes for an oil-sealed rotary-vane roughing pump to malfunction. If the oil level is satisfactory and the pump produces a steady unvarying pressure, but does not reach a value as low as 3 Pa (0.05 Torr), it is possible that the oil has become contaminated with water or an organic solvent (Section 4.1.4, p. 141) and needs to be replaced.

Failure to keep rotary-vane pumps properly filled with oil can sometimes lead to rather embarrassing situations. For example, on one occasion the vacuum system on a recently-purchased scanning electron microscope did not function properly. Since the instrument was covered by a service contract and we were involved with other pressing problems at the time, we spent about an hour trying to find the cause, and then called the service engineer. After driving several hundred miles to reach our laboratory and spending half a day trying to find some malfunction in the valve system or the vacuum gauges he finally thought to check the oil level in the rotary-vane pump that backed up the oil diffusion pump. Five minutes later the problem was fixed by adding a cup of oil to this backing pump. This illustrates the fact that it is all too easy to forget that simple problems can occur when dealing with complicated instruments. Of course, this problem could easily have been prevented in the first place by following the proper procedure of routinely checking the level of oil in the backing pump every week or two. However, rotary-vane pumps function so steadily and reliably that it is easy to forget to give them the attention they are due. If a rotary-vane pump periodically makes an unusual barking or gurgling noise, or if the pressure in the line leading to the pump periodically rises and then falls again, the pump will usually be found to be short on oil. A similar periodic fluctuation of the reading of the high vacuum gauge on a unit evacuated by an oil diffusion or a turbomolecular pump is also usually an indication of insufficient oil in the rotary-vane backing pump.

It is somewhat more difficult to test the performance of the high vacuum pump on most vacuum systems. However, some indication of whether or not an oil diffusion pump is performing well in a system such as that shown in Fig. 10.1 can be obtained by watching the reading of the gauge in the backing line immediately after valve V1 is closed while the pressure in the vacuum system is in the upper end of the high vacuum range. At such a high pressure the diffusion pump should be transferring a significant quantity of gas from the high vacuum system into the backing line if it is working properly. Therefore, the pressure in the backing line should decrease rapidly when this supply of gas is reduced by closing valve V1. If the backing pressure does not drop appreciably it is likely that the diffusion pump is not functioning properly. Some of the symptoms that indicate that the various other types of high vacuum pumps are not performing properly are described in the chapters that deal with those pumps. It should be noted, however, that a high level of gas evolution inside a vacuum chamber can sometimes make it appear that the high vacuum pump is not functioning properly. As noted in Section 2.10.4d (p. 74), there are many opportunities for this to occur in scanning electron microscopes and vacuum evaporators, and so it is a good idea to keep a running record of the types of specimens being put into these instruments, particularly if several workers from several different research groups are using them, so that there will be some basis for knowing when a problem might be due to contamination introduced on the specimens.

The sensitivity of most vacuum gauges can decrease over time for various reasons, producing symptoms very similar to those caused by a leak. Contamination of, or damage to, a high vacuum gauge can give rise to similar symptoms. For instance, if air is admitted to a hot cathode ionization gauge while its filament is operating, the filament will usually oxidize before the safety circuit in the gauge controller can turn the filament heating current off. This will coat the interior of the gauge tube with a layer of tungsten oxide which will then adsorb large quantities of gas when the apparatus is brought up to atmospheric pressure. When it is subsequently pumped down, this gas is slowly released inside the gauge tube, causing its reading to be higher than it should be and giving the general impression that the rate of pumpdown is much slower than normal. Organic vapours that find their way into a cold cathode discharge gauge are broken down by the electrical discharge and form carbonaceous deposits which alter the response of the gauge and cause it to change sensitivity and to become unstable. Several incidents in which thermocouple gauges gave misleading readings are described in Section 3.1.2 (p. 87). If there is any likelihood that a vacuum gauge is not performing properly for reasons such as these, it is usually prudent to check it very carefully before testing the entire apparatus for a leak.

It is also necessary to be cognizant of 'natural variations' in pumpdown time for some instruments. For example, it is quite common for the pumpdown time for vacuum evaporators to be longer on days when the humidity is high, than on days when it is low, because their bell jars are usually left openly exposed to the atmosphere while specimens are being loaded into and removed from them. This allows a larger amount of water to adsorb onto the interior surfaces of the bell jar when the humidity is high, and it then takes longer for this water to be pumped away when the evaporator is subsequently evacuated. Of course, a clean bell jar adsorbs less water than one that is heavily covered with evaporated materials. In some locations it is not uncommon to experience high humidity for several consecutive weeks or months. During such times the oil in a rotary-vane pump can become contaminated with water, and this can cause its pumping speed to decrease and its ultimate pressure to increase significantly (Section 4.1.4, p. 141), giving a slower rate of pumpdown and a poorer ultimate vacuum.

10.4 Leak rates

When we are dealing with leaks, it is the rate at which gas molecules enter the vacuum system (i.e. the rate of gas influx q_l, as discussed in Section 2.10, p. 56) that is of concern, and so leak rates are expressed in the same units as gas throughput (see Section 2.1, p. 28). Therefore, leak rates will be given in units of pascal-litres per second (Pa-l/s) in the following discussion. The preferred SI units for leak rates are pascal-cubic metres per second (1 Pa-m^3/s = 10^3 Pa-l/s), but these are seldom encountered in practice. In industrial applications units of millibar-litres per second (1 mbar-l/s = 10^2 Pa-l/s) are widely used in Europe, while units of standard cubic centimetres per second or atmosphere-cubic centimetres per second (1 SCC/s ≈ 1 atm-cm^3/s ≈ 10^2 Pa-l/s) are generally used in the United States, although units of Torr-litres per second (1 Torr-l/s = 133 Pa-l/s) are also sometimes encountered. Factors for converting leak rates given in

Pa-l/s to these other units are: 1 Pa-l/s = 10^{-2} mbar-l/s = 10^{-2} SCC/s = 10^{-2} atm-cm^3/s = 7.5×10^{-3} Torr-l/s = 10^{-3} Pa-m^3/s. As indicated in Table 1.2 (Section 1.3, p. 16), a leak rate of 1 Pa-l/s corresponds to a flow of 2.47×10^{17} gas molecules per second.

It is interesting to make rough estimates of the maximum leak rates $(q_l)_{max}$ that can be tolerated in some common situations. This can be done by making use of eqn. 2.16 (Section 2.6, p. 48 and Section 10.1), $P_u = q/S_e$. Assuming gas influx from gas desorption and gas evolution are small enough to be neglected, and substituting P_o, the lower end of the operating pressure range, for P_u, this equation can be rearranged to give $(q_l)_{max} = S_e{\cdot}P_o$. On this basis, the maximum tolerable leak rate for a high resolution electron microscope evacuated by a sputter-ion pump with $S_e \approx 100$ l/s which must achieve an operating pressure of $P_o \approx 4\times10^{-5}$ Pa (3×10^{-7} Torr), is $(q_l)_{max} \approx 100(4\times10^{-5}) \approx 4\times10^{-3}$ Pa-l/s. To attain a pressure of 10^{-3} Pa (10^{-5} Torr) in the electron gun of the scanning electron microscope discussed in Section 2.5 (p. 42), for which the speed of evacuation is only about 10 l/s, the maximum tolerable leak rate calculated on this basis is $(q_l)_{max} \approx 10\times10^{-3} \approx 10^{-2}$ Pa-l/s. For an ultra-high vacuum system with a 300 l/s ion pump that must reach 10^{-7} Pa (10^{-9} Torr), the maximum leak rate is $(q_l)_{max} \approx 300\times10^{-7} \approx 3\times10^{-5}$ Pa-l/s. These values are undoubtedly larger than could be tolerated in actual practice, however, because the processes of gas desorption and gas evolution also make significant contributions to the total gas load that must be handled by the pumps in actual vacuum systems. Therefore, most systems are designed so that the ultimate pressure is about a factor of 10 less than the anticipated operating pressure (i.e. $P_u \approx 0.1P_o$). Harris (p. 252), for example, suggests that for systems which are continuously evacuated, the total leak rate should not exceed about one-tenth of the throughput capacity Q of the pumping system at the desired operating pressure P_o [that is, $(q_l)_{max} < 0.1Q \approx 0.1S_eP_o$]. Thus, for instruments commonly found in electron microscopy laboratories we are likely to be dealing with leaks in the range from 10^{-5} to 10^{-3} Pa-l/s. Harris (p. 253) calculates that a hole with a diameter of only 6×10^{-3} mm through a vacuum wall 2 mm thick is sufficient to produce a leak rate of 10^{-3} Pa-l/s, which is in the upper end of this range, and so we are not dealing with very large structural defects. Hablanian (Section 11.2) and Harris (Section 14.2) also discuss the magnitude of leaks that are acceptable in sealed devices, such as beverage cans, television tubes, and sealed integrated circuits, where requirements are often somewhat more stringent than for the continuously-pumped vacuum systems found on the electron microscopes and similar laboratory instruments that are of primary concern here.

10.5 The general procedure for locating leaks

When other sources of trouble have been eliminated insofar as possible and it is finally concluded that a leak is probably present, but its location is not obvious, some type of systematic leak detection procedure must be invoked. The most useful of these procedures for apparatus found in electron microscopy laboratories are based on the fact that vacuum gauges have different sensitivities for different gases. Therefore, if some of the air that normally enters a vacuum system through a leak is replaced by another gas that gives a different gauge response, the readings produced by the vacuum gauges on the system will change slightly. This phenomenon can be used to locate a leak by spraying a fine stream of a test gas slowly over the outside of vacuum joints and other likely leak sites while watching the gauge output display carefully for an indication of a change in gauge response. When such a change is noted the rate of gas flow in the test stream is reduced and the region where the response occurred is explored again until the leak is localized as closely as possible.

A suitable test stream can be produced by flowing the test gas through a fine tip cut from a plastic dropper or pipette. A glass tip is not recommended because it can break and leave bits of glass in awkward places. A hypodermic needle can be used if it is blunted so that it does not accidently penetrate the finger of the operator, and if care is exercised to avoid scratching the apparatus with it. The tip should be connected to the tank of test gas by flexible surgical rubber tubing long enough to easily reach all parts of the apparatus being explored. The gas tank must be equipped with a pressure-reducing valve and a needle valve to control the flow of gas in the test stream. All applicable safety regulations must be observed when working with tanks of compressed gas. In particular, *tanks must be secured in such a way that they cannot tip over*, because if this happens the valve on the tank can be broken off, releasing the high pressure gas suddenly and converting the tank into a very destructive and potentially deadly missile.

Start with the valve on the top of the tank closed, the reducing valve set for zero pressure output (this usually involves turning the adjusting control fully counterclockwise so that it no longer exerts pressure against the control diaphragm inside the valve), and the needle valve only snugly shut — needle valves *must not be* tightened fiercely tight, because doing so can damage them. Open the valve on the top of the tank just slightly, and adjust the pressure reducing valve so that the pressure gauge on it barely shows a deflection — the output pressure need not be greater than 10 or 15 kPa (2 or 3 psi). Then carefully open the needle valve just enough to give a gentle stream of gas through the test tip. The rate of flow can be judged by noting the size of the indentation made in the liquid when the tip is held near the surface of a beaker of water. A relatively high flow rate is most useful for initial exploration, while a much lower flow rate is better for finally localizing the site of a leak. Exploration should be done slowly, carefully, and thoroughly starting at the top of the instrument and proceeding systematically downward, being sure that all accessible leak sites are tested.

Some vacuum joints are difficult to test for leaks in the manner just described This would be true, for example, for the vacuum joints shown in Fig. 10.11 (Section 10.11, p. 453). The O-rings in joints such as these are quite inaccessible, and so it is difficult to be certain that the test gas indeed reaches them if it is merely sprayed around the outside of the joints. If, however, a small hole is bored through the clamp ring of the quick-disconnect flange, the test gas can be injected through it into the small cavity that surrounds the O-ring. It will then be retained in close proximity of the O-ring long enough to reveal any leak that might be present. A somewhat similar approach can be used to test the bolted flange joint. A piece of adhesive tape is first run all the way around the outer edges of the flanges so that it covers the joint between them. The groove produced by the bevelled edges of the flanges then forms a small channel all the way around the joint underneath the tape. If two small holes are poked through the tape into this channel on opposite sides of the joint, test gas injected into one hole will flow in this channel around the joint and out the other hole. This will surround the joint with the test gas, and the tape will keep the test gas in the channel long enough for it to have an opportunity to diffuse between the flanges and reach the O-ring. If it is anticipated that a joint of this kind will have to be tested for leaks frequently, it is helpful to file a groove about 2 mm deep running radially outwards from the O-ring groove to the edge of one of the flanges. When the joint is assembled this groove will form a channel through which the test gas can be injected into the O-ring cavity for leak-testing purposes. The O-rings that seal the joints between column elements of most electron microscopes also are usually quite inaccessible because they are covered by the case that surrounds the column and serves as a magnetic shield and decorative cover for it. Such joints can sometimes be tested by injecting the test gas through screw holes in this case. A general survey to determine whether or not a leak is present in a major segment of an apparatus can be carried out by covering that portion of the apparatus with a plastic bag, or by fastening a flexible plastic film around it with adhesive tape, expelling as much air as possible from underneath the plastic, and then injecting the test gas there. It will usually take some time for the test gas to reach the joints of interest, and so such a test must be run for 10–15 min. If a response is noted, an attempt can be made to localize the leak using the test gas probe method described above.

Helium is almost universally used as the test gas in these procedures for several reasons: it is relatively inexpensive; it is commonly found in most laboratories; and it is non-flammable, non-toxic, and non-corrosive. Since helium molecules have a small mass and a small diameter they flow through leaks several times faster than air molecules under most circumstance, effectively increasing the leak rate and enhancing sensitivity. In addition, the atmosphere contains only a trace amount of helium (about 0.0005 percent), and so the background signal from the atmosphere is very low when helium is used as the test gas. Residual gas analysers (RGAs) and mass spectrometer leak detectors (MSLDs) can be adjusted to read only the helium peak at $m/z = 4$, thus providing very high sensitivity and selectivity and quite rapid response. In fact, helium is so widely used as the test gas for the mass spectrometer leak detectors that these devices are commonly called 'helium leak detectors'. As noted in Section 3.1.1 (p. 81), helium has a higher thermal conductivity than most other gases, and so it is a very good test gas when a thermocouple or Pirani gauge is the sensor. Likewise, its ionization cross-section is smaller than those of most other common gases (Section 3.2.1, p. 92), and so helium gives a good response when hot or cold cathode ionization gauges are used. There is one problem that can arise if part of an apparatus is flooded with helium for half an hour or more. Under these conditions the helium can diffuse into rubber O-rings, saturate them, and ultimately penetrate through them into the vacuum system, producing a high background that can severely reduce the sensitivity of a leak testing procedure. If this should occur, the only cure is to blow as much of the helium as possible away from around the apparatus with a jet of air, and then to evacuate the apparatus for several hours until the helium is pumped out of the O-rings.

10.6 Using ionization gauges for leak detection

As discussed in Section 3.2.1 (p. 92), the response of ionization gauges to helium and several other gases is significantly different from their response to air. For this reason these gauges can be quite useful as sensors for leak-detecting procedures. For the same reason, the current in a sputter-ion pump, which is also produced by a gas ionization process, can be used to sense a test gas (Section 7.1.3, p. 285). These ionization devices are capable of detecting leaks smaller than 10^{-4} Pa-l/s, and can therefore serve as leak detecting devices on most instruments in electron microscopy laboratories. In addition, some manufacturers sell amplifiers that are specifically designed to sense and display sudden small changes in the output of ionization gauges and sputter-ion pumps (e.g. the Model VZGLD5 unit from Kurt J. Lesker), thereby facilitating the use of these devices as sensors for leak-testing procedures.

Notwithstanding its many advantages, helium may not always be the best test gas to use with ionization devices. Weston (Chapter 8) points out that for best sensitivity both the response of the gauge and the speed of the high vacuum pump for the test gas must be taken into consideration. Because of its small ionization cross-section, helium causes a decrease in the ion current reading of ionization gauges and sputter-ion pumps when it enters a vacuum system through a leak. If the system has a sputter-ion or cryogenic pump, however, the helium will cause the actual gas pressure in the system to increase, because these pumps have much lower pumping speeds for helium than for air. This will act to increase the ion current reading and oppose the effect of the gauge response, thereby decreasing overall test sensitivity. Argon is a better choice for a test gas under these conditions. Like helium, it is not pumped as fast as air. However, it produces a gauge response which is about 1.5 times greater than air, and this augments, rather than opposes, the effect of the pressure rise caused by the characteristics of these pumps. Roth (p. 457) cites data indicating that propane (C_3H_8) and butane (C_4H_{10}), which have large ionization cross-sections and give responses in ion gauges and ion pumps from five to ten times greater than air, can also be used advantageously as test gases. These gases are conveniently available from hardware and sporting goods stores in small tanks designed for use in torches and camp stoves. They are, however, flammable, somewhat toxic, and a bit noxious, and so must be *used with extreme care*. In fact, their use in this manner may be forbidden under safety and environmental codes in some localities. Cold cathode ionization gauges and sputter-ion pumps are not as well suited for leak testing as hot cathode ionization gauges, because the instabilities that are so often exhibited by cold cathode discharges make it difficult to decide whether a given change in reading is due to a leak or to these random effects.

10.7 Detecting leaks with thermal conductivity gauges

Thermocouple and Pirani gauges can also be used as sensors for leak detection procedures, although they are probably not capable of detecting leaks much smaller than 10^{-2} Pa-l/s. As discussed in Section 3.1.1 (p. 82), the response of these gauges is related to the thermal conductivity of the gas in them, and so a gas that produces a response significantly different from that of air, such as argon, helium, or hydrogen, can be used as a test gas. As shown in Fig. 3.2 (Section 3.1.2, p. 85), most Freons also produce a response considerably different from air. Laboratory dust-off sprayers are commonly filled with a Freon, and so can serve as a convenient source of test gas. Fig. 3.2 also shows that acetone produces a gauge response that is considerably different from that for air. Carbon tetrachloride, chloroform, ether, and several other organic solvents behave in a similar manner. For this reason it is fairly common practice to spray or paint one of these liquids around vacuum joints to test for leaks. This is not as good as using helium, however, because these solvents are toxic and noxious to work with, they can attack O-rings and vacuum grease, and they can be absorbed into and then diffuse through O-rings sufficiently to lead to subsequent problems with gas evolution and system contamination. Interestingly, the cooling of these liquids by evaporation as they flow through a leak and into a vacuum system can cause them to freeze and temporarily plug the leak, making it difficult to locate the site of a leak accurately.

The value of thermal conductivity gauges as leak detecting devices is somewhat limited because they exhibit best sensitivity in only the rough vacuum range, that is, from about 1 to 100 Pa (10^{-2}–1 Torr). Practically, they are most likely to be useful for detecting leaks in rough vacuum systems (e.g. Fig. 4.8, Section 4.1.8, p. 153), which operate only in the rough vacuum range, or in high vacuum systems on which the high vacuum pump is a turbomolecular or oil diffusion pump. Such systems require a roughing pump, and it is common practice to install a thermal conductivity gauge in the pumping line leading into this pump. For example, both the roughing and backing gauges in the system shown in Fig. 10.1 would normally be thermal conductivity gauges. Because the pressure in the backing line is in the range of best gauge sensitivity, these gauges are situated in an optimal location for use in leak detection. It is generally not convenient to use a thermal conductivity gauge for detecting leaks on systems which have cryogenic or ion high vacuum pumps.

Even under favourable circumstances the response of a standard thermal conductivity gauge is small and sluggish when test gas enters the system through a leak. This makes it necessary to work patiently, using a high flow rate of test gas, moving the test probe slowly, and watching the gauge meter carefully. However, a high sensitivity vacuum gauge monitor of the type mentioned in the previous section (e.g. Lesker Model VZGLD5), can make changes in gauge reading much easier to observe and give some improvement in sensitivity. Danielson Associates have achieved very good performance in their Omnibar Leak Sensor by combining features similar to those provided by monitors of this kind with a thermal conductivity tube that is especially designed to give high sensitivity and rapid response to changes in gas composition. This device is capable of detecting leaks over the entire pressure range from atmospheric pressure down to 10^{-5} Pa (10^{-7} Torr). Its sensitivity is advertised to be 5×10^{-1} Pa-l/s near atmospheric pressure and 2×10^{-7} Pa-l/s at 10^{-5} Pa (10^{-7} Torr). The sensor tube is about the size of an ordinary vacuum gauge, and so it can be installed relatively easily on nearly any vacuum system. Because of its wide operating pressure range, it does not require a separate pumping system, but can use the pumps on the system being tested. In addition, its operating pressure range and its sensitivity range are both adequate for detecting leaks on most laboratory vacuum systems.

10.8 Using residual gas analysers for leak detection

A residual gas analyser (RGA) is a very handy device to have on an instrument such as an ultra-high resolution electron microscope. Not only can it be used to evaluate the nature of the gases present in the microscope (Section 3.4.4, p. 123), but it can also serve as a total pressure gauge (Section 3.4.2, p. 119), and as a very effective leak detector. This last capability arises because the control units of most RGAs can be set to monitor the concentration of a single species of gas molecules with high sensitivity, and to produce an audible signal when an increase in the concentration of this gas is detected. Many units also produce a special visual display. If the species monitored in this way is used as the test gas in a leak-detecting procedure, the RGA will show an abrupt increase in reading and will emit an audible signal when the test gas enters the vacuum system through a leak. For the reasons mentioned above helium is the most common test gas, although nearly any convenient gas can be used. As leak detectors, RGAs provide greater sensitivity, faster response, and better selectivity than the total pressure gauges discussed in the previous two sections. They are capable of detecting leaks smaller than 10^{-7} Pa-l/s, and are almost ideally suited for leak testing on most electron microscopes. Perhaps their principal shortcoming is that they usually cannot be operated safely at pressures above about 10^{-2} Pa (10^{-4} Torr), and so cannot be used when a system has a really bad leak. Also, once an RGA is installed on an instrument such as an electron microscope it is inconvenient to remove it for testing on another instrument, even in the same laboratory. These are not serious limitations, however, because electron microscopes seldom develop such large leaks that the pressure in them becomes too high for the operation of an RGA, and other methods are usually available for detecting leaks on other instruments.

10.9 Mass spectrometer leak detectors

Mass spectrometer leak detectors (MSLDs), which are often referred to simply as 'helium leak detectors', were developed on the Manhattan

Project during the latter days of World War II as part of the overall effort to build the atomic bomb. The preparation of the high purity uranium compounds needed for this purpose was carried out by gaseous diffusion. For fairly obvious reasons it was necessary to ensure that the equipment used in this process was as free from leaks as possible, and the method of using helium as a test gas with mass spectrometers as detectors was adapted to test this equipment for leaks, because of the extremely high sensitivity that could be achieved in this way. It was immediately obvious that this was a very valuable new technique, and so it was quickly adapted for general use throughout the vacuum industry. MSLDs are still the most sensitive, the most accurate, the most versatile, and the most widely-used devices available for locating and measuring leaks in vacuum systems. They are capable of detecting leaks smaller than 10^{-8} Pa-l/s.

A helium leak detector consists of a compact, specialized mass spectrometer which is installed in a cabinet with a set of vacuum pumps, traps, gauges, and valves, making it an independent vacuum apparatus which can be moved from one site to another for testing vacuum equipment as desired. In this respect MSLDs are basically different from RGAs, which are normally dedicated to a single instrument. Most current helium leak detectors also differ from RGAs by employing a 90° or 180° magnetic sector mass spectrometer, rather than a quadrupole mass analyser, as the detecting unit, because these units can be made quite compact and provide superior stability, resolution, and sensitivity for helium. The way magnetic spectrometers work is described briefly in Section 3.4.1 (p. 114). Hablanian (Section 10.6 and Chapter 11) and Harris (Section 14.4) describe the design, construction, and use of helium leak detectors in some detail. MSLDs are currently produced in two different designs, direct-flow, and counter-flow.

10.9.1 *Direct-flow mass spectrometer leak detectors*

Leak detectors with the general arrangement of components shown in Fig. 10.6 are called 'direct-flow leak detectors', because gas that enters the test valve can flow directly to the mass spectrometer without passing through either of the vacuum pumps. A distinguishing feature of this design is the liquid nitrogen trap that protects the mass spectrometer from becoming contaminated with oil vapours and other undesirable condensable materials from the vacuum pumps and the system being tested. This

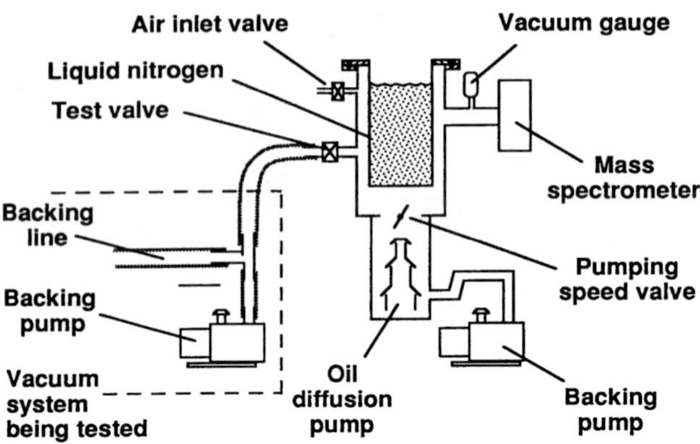

Fig. 10.6 A schematic diagram of a basic direct-flow leak detector, showing the most commonly-used method of attaching a leak detector to a vacuum system that has a turbomolecular or oil diffusion pump.

trap also improves the vacuum in the unit by providing a very high pumping speed for water vapour. The high vacuum required for the operation of the mass spectrometer is produced by an integral pumping system, which in the past usually consisted of a small oil diffusion pump and a small rotary-vane backing pump, as shown. Now, however, units are widely available in which the oil diffusion pump is replaced by a turbomolecular pump. Since the mass spectrometer unit can be made quite compact and the pumps that serve it do not to need to be very large, leak detectors of this simple design can be made light, easily portable, and convenient for use in laboratory applications.

A direct-flow leak detector can be connected to an instrument such as an electron microscope in two different ways. If the instrument has an oil diffusion pump or a turbomolecular pump the usual practice is to connect the leak detector into the line between this pump and its backing pump. For a system such as that shown in Fig. 10.1, for example, valve V4 might be used for this purpose, provided it has enough conductance to permit sufficient gas to flow into the leak detector to give good sensitivity. If the system does not include a suitable, easily accessible valve it is usually relatively easy to insert a Tee joint into the backing line and to connect the leak detector to it with a piece of rubber tubing, as shown in Fig. 10.6, since most backing lines are made of rubber tubing.

An alternative approach is to connect the leak detector into the high vacuum side of the instrument. An existing valve, such as V5 in Fig. 10.1, can be used for this purpose, or a connecting flange can be installed on a port in the instrument. To minimize the possibility of introducing organic contaminants into the high vacuum side of the instrument it is best to use a clean flexible metal bellows tube to make this connection. If rubber tubing is used, a liquid nitrogen trap of the type shown in Fig. 4.6 (Section 4.1.4b, p. 143) should be installed directly on the vacuum flange, and the tubing connected to it. For reasons discussed in Section 4.1.4b, this trap should be kept filled with liquid nitrogen at all times when the tubing is under vacuum. To ensure a good flow of gas into the leak detector, the connecting tube should be at least 25 mm in diameter and as short as possible. This method of connecting directly into the high vacuum side must be used when testing for leaks on instruments that have sputter-ion and cryogenic pumps, because these systems do not have backing pumps. It can, of course, also be used on systems with backing pumps, but the method of connecting into the backing line described above will usually be more convenient and more satisfactory for these systems.

The installation and operation of a direct-flow leak detector on a laboratory instrument is a relatively straightforward procedure. It will usually be necessary to turn off the vacuum pumps on the instrument while the leak detector is being connected to it. After the connection is made the pumps are turned on again and the instrument is pumped down to the lowest pressure possible. The test valve on the leak detector is usually kept closed during this time, and the vacuum pumps of the instrument being tested are used to evacuate the connecting line between it and the leak detector. With a supply of liquid nitrogen conveniently at hand, the leak detector is now started up. The backing pump is turned on first, then the liquid nitrogen trap is immediately filled with liquid nitrogen. It is important to do this before the pressure drops into the range of molecular flow (about 10 Pa or 0.1 Torr) to prevent the mass spectrometer unit from becoming contaminated with oil from the vacuum pumps. The trap should then be faithfully kept supplied with liquid nitrogen at all times when the mass spectrometer is under vacuum, as discussed in Section 4.1.4b (p. 143). The high vacuum pump on the leak detector is now turned on. When the vacuum gauge on the leak detector shows that an acceptable operating vacuum has been reached, usually near the bottom of the 10^{-2} Pa (10^{-4} Torr) range, the mass spectrometer is switched on. After the operation of the mass spectrometer and the pressures in the two systems have stabilized, the test valve is opened and the leak testing operation is carried out, using the general procedures described above in Section 10.5 (p. 434). The test valve is often designed to provide fine control over the rate at which gas is admitted into the leak detector. If so, maximum sensitivity is achieved by opening it as much as possible without causing the pressure in the leak detector to rise above the safe operating range of the mass spectrometer, so that the flow of gas into the leak detector from the system being tested is as large as possible. The sensitivity of many leak detectors can also be varied by manipulation of a valve between the high vacuum pump and the cold trap to vary the rate at which gas is pumped away from the mass spectrometer. Maximum sensitivity is achieved by closing this *pumping-speed valve* as much as possible, without exceeding the operating limit of the mass spectrometer, so that the pressure of the gas admitted from the test system builds up in the mass spectrometer to as high a level as permissible.

To turn the leak detector off, the test valve is closed, the pumping-speed valve is opened fully, and the mass spectrometer and the high vacuum pump are turned off. When the high vacuum pump reaches a condition where it is safe to do so, the air inlet valve is opened and the roughing pump is quickly turned off. The liquid nitrogen reservoir is immediately removed from the liquid nitrogen trap, the liquid nitrogen is poured out, the reservoir is allowed to warm up to room temperature, and all the condensed material that collected on it is thoroughly cleaned off before it is put back into the trap (for reasons discussed in Section 4.1.4b). The opening to the liquid nitrogen trap should be covered with aluminium foil during this time to keep unwanted material out. In the meantime, the leak detector can be disconnected from the instrument being tested.

Many leak detectors have a vacuum system similar to the one shown in Fig. 10.7. This vacuum system differs from the one discussed above principally by including a separate manifold and roughing pump that make it possible to connect it to, and disconnect it from, the apparatus being tested without turning off its high vacuum pump. To start up such a unit, its high vacuum and backing pumps are turned on, the liquid nitrogen trap is filled, and the mass spectrometer is warmed up, in the manner described above, with the test valve, gross leak valve, and counter-flow valve all closed. Meanwhile, the apparatus to be tested is attached to the test port, the roughing pump is turned on, the roughing valve is opened, and the roughing manifold is evacuated to a pressure of about 10 Pa

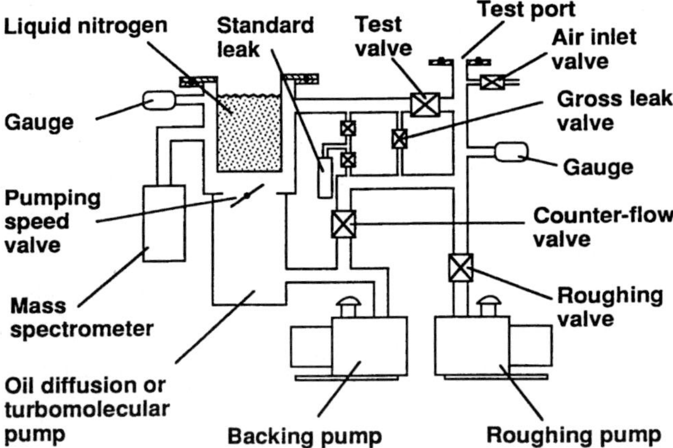

Fig. 10.7 A direct-flow helium leak detector with a separate roughing pump and manifold for the test port, a standard leak, and counter-flow and gross leak valves.

(0.1 Torr). The roughing valve is then closed, the mass spectrometer is turned off, and the test valve is slowly opened. The mass spectrometer is turned on again when the high vacuum pump of the leak detector brings the pressure down into its operating range, whereupon the leak testing operation can be carried out. After leak testing is completed, the test valve is closed, the air inlet is opened, and the test apparatus is detached from the test port. The leak detector can then be moved to another site to test another apparatus without having to shut down its vacuum system.

Leak detectors of this kind are often found in laboratories of large research organizations and in applications laboratories associated with production plants, where they are used for testing a variety of different kinds of apparatus. They are also widely used in manufacturing processes to test items such as aerosol cans, components of refrigeration and air-conditioning systems, food containers, fire extinguishers, batteries, components of vacuum systems, and a host of other products. Such items do not have vacuum systems that can be used to evacuate the connecting line and test port of the leak detector, and so the addition of the roughing pump and test manifold makes these detectors particularly well suited for such applications. In fact, many models of leak detectors have automated operating systems that will carry out the pumpdown process when a 'start' button is pressed, turn the mass spectrometer on and signal when leak detecting can be initiated, and then carry out the venting process when a 'stop' button is pressed. Such uses of leak detectors, and the special methods that are applied in them, are discussed by Hablanian (Chapter 11) and Harris (Chapter 14), and in the Varian book on helium leak detectors.

These more complex leak detectors often have additional features that can be used to modify their operating characteristics for special applications. Gross leak valves, which bypass the regular test valve as shown in Fig 10.7, are found on many models. Whereas the regular test valve is usually designed to control the flow of gas into the leak detector under conditions of molecular flow, the gross leak valve contains a tube or orifice small enough to admit gas to the mass spectrometer from the roughing manifold while the pressure in it is in the range of viscous flow. Using this valve instead of the regular test valve makes it possible to begin leak testing while the pressure in the roughing manifold is still in the rough vacuum range, or to test apparatus that cannot or need not be pumped down into the high vacuum range. Some models of leak detectors now have a counter-flow bypass valve that can be used to convert to the counter-flow mode of operation, as shown in Fig. 10.7.

In many applications it is necessary to determine the leak rate numerically. To facilitate this a standard helium leak is often incorporated into the leak detector for calibrating the response of the mass spectrometer. Standard leaks are usually constructed by sealing a slightly porous glass tube into a metal cylinder that contains helium under pressure. The valves associated with this device are usually arranged somewhat as indicated in Fig. 10.7, so that the helium which diffuses through the leak while it is not in use can be pumped away by the roughing pump before the leak is opened to the spectrometer to begin the calibration process. Naturally, with the current popularity and versatility of microprocessors, many models of leak detectors now have a variety of computer-automated features.

10.9.2 *Counter-flow mass spectrometer leak detectors*

Counter-flow leak detectors take advantage of the fact that oil diffusion pumps and turbomolecular pumps do not pump helium very efficiently. As discussed in Section 5.6 (p. 200) and Section 6.1.2 (p. 232), these pumps usually develop compression ratios of the order of 10^9 or greater for oxygen and nitrogen, the major atmospheric gases, whereas the compression ratio for helium is usually only of the order of 10^4, and may be as low as 10^2 in some pump designs. The practical consequence of this is that if helium is introduced into the backing line of these pumps it can readily diffuse through them in the direction opposite to that of the primary gas flow (hence, the 'counter-flow' designation). In this way a small concentration of helium in the backing line can produce a concentration at the inlet port of the pump that is large enough to be detected by a mass spectrometer.

Fig. 10.8 shows the arrangement of components for a basic counter-flow leak detector. To start up a unit of this kind the roughing valve is closed, the backing valve is opened, and the rotary-vane and high vacuum pumps are turned on. When the high vacuum pump establishes an operating vacuum, usually near the bottom of the 10^{-2} Pa (10^{-4} Torr) range, the mass spectrometer is turned on. The apparatus to be tested is then attached to the test port, the backing valve is closed, the roughing valve is opened, and the rotary-vane pump is used to bring the pressure in the test port down into the backing pressure range of the high vacuum pump of the leak detector, usually below 50 Pa (0.5 Torr). At this point the backing valve is opened again and the leak-detecting procedure is carried out using helium as the test gas. If helium enters the apparatus being tested through a leak, some of it will find its way through the test port and into the backing line of the leak detector, whereupon it will diffuse backward through the high vacuum pump and be detected by the mass spectrometer. As is true for the direct-flow leak detectors discussed above, counter-flow detectors are available with an added roughing pump and manifold, and some models also have a standard leak for calibration purposes and a gross leak bypass line for testing at high pressures, as shown schematically in Fig. 10.9. Units

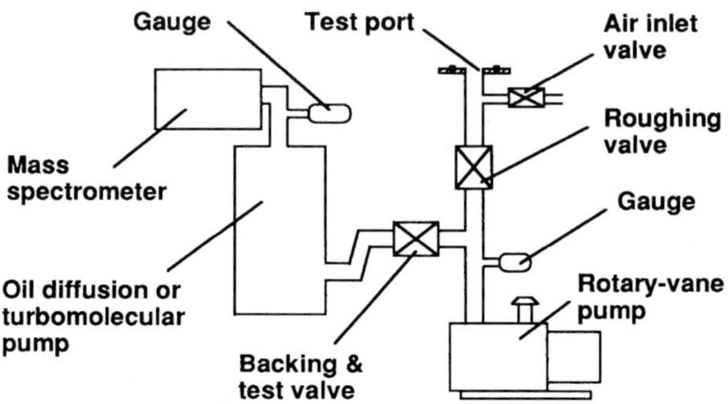

Fig. 10.8 A basic counter-flow helium leak detector.

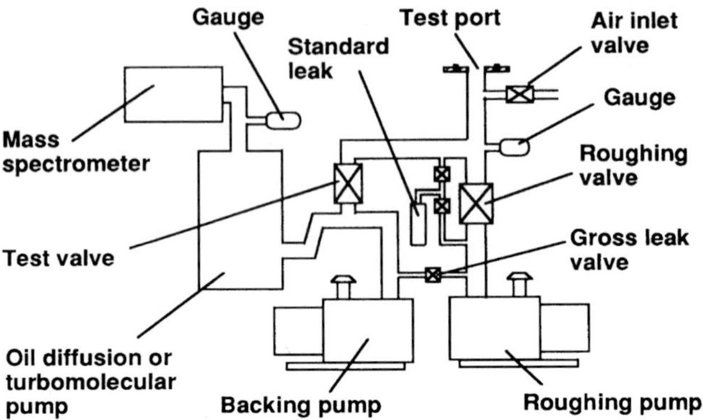

Fig. 10.9 A counter-flow helium leak detector with a separate roughing pump and manifold for the test port, a standard leak, and a gross leak valve.

that are automated to facilitate running tests in quick succession are available for production applications.

The counter-flow design affords several advantages over the traditional direct-flow design, the principal one being that it does not require a liquid nitrogen trap. In effect, the high vacuum pump substitutes for the liquid nitrogen trap in preventing contamination from the system being tested from reaching the mass spectrometer. This simplifies design, reduces construction cost, and generally makes the units significantly smaller, lighter, and more easily portable. In addition, it eliminates the bother and expense of keeping a trap filled with liquid nitrogen while the leak detector is in use. Interposing the high vacuum pump between the test port and the mass spectrometer also allows the pump to buffer the mass spectrometer from bursts of gas that may be accidentally admitted from the test apparatus. This can be a significant advantage, because it helps prevent damage to the mass spectrometer and reduces the frequency with which the detector is accidentally shut down because of pressure fluctuations.

One important advantage of counter-flow leak detectors is that the gas sampled from the test system is introduced into the rough vacuum side of the leak detector, and so the pressure in the test port does not need to be brought down into the high vacuum range. This considerably reduces the time needed to pump down before leak testing can be started, and makes it possible to test vacuum systems that have large leaks or are otherwise difficult to pump down into the high vacuum range. Normally, leak testing can be started at a pressure of about 40 Pa (0.3 Torr), the maximum backing pressure acceptable for operating most oil diffusion and turbomolecular pumps, but can be carried out at pressures as high as 100 Pa (1 Torr) by using a gross leak bypass line. Interestingly enough, counter-flow leak detectors usually have better sensitivity than direct-flow detectors when used at test port pressures above about 1 Pa (10^{-2} Torr).

The major disadvantage of conventional counter-flow leak detectors is the fact that the test port is connected more or less directly to an oil-sealed, rotary-vane pump, and so any vacuum system that is connected to the test port is in danger of becoming contaminated with oil from this pump (see Section 4.1.5, p. 144). This means that counter-flow leak detectors are not well suited for applications in which they must be connected to the high vacuum side of instruments or devices that must be kept free of oil contamination, such as high resolution electron microscopes that are evacuated by ion or cryogenic pumps. They can, however, be used on high vacuum instruments that are evacuated with oil diffusion or turbomolecular pumps by connecting them into the backing line of these pumps in the general manner shown in Fig. 10.6. There are several ways to overcome this oil contamination problem. It is, of course, possible to install a liquid nitrogen trap into the line between the leak detector and the apparatus, but this approach brings with it the cost and inconvenience of the liquid nitrogen trap in the direct-flow detectors. In a more direct approach, Danielson Associates have developed a leak detector, sold under the Helibar trade name, with a totally oil-free pumping system that is based on their Tribodyn compound pump (Section 6.2.3b, p. 269). Balzers have also eliminated this problem in their

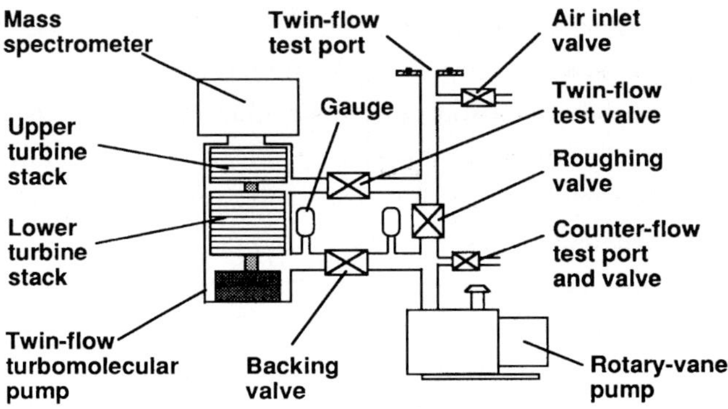

Fig. 10.10 The arrangement of components in the Balzers HLT 160 twin-flow leak detector.

Model HLT 160 leak detector by utilizing a patented design that allows the test gas to enter into the middle of the turbine stack of a turbomolecular pump, as indicated schematically in Fig. 10.10. With this arrangement the several stages of turbine blades above this 'twin-flow inlet' protect the mass spectrometer against contamination by material from the test apparatus, while the stages below it bring the pressure at the test port into the high vacuum range, permitting easy connection into high vacuum systems. It is, of course, alternatively possible to introduce the test gas into the backing line of the turbomolecular pump through the counter-flow port and valve, as is done in conventional counter-flow leak detectors. The roughing valve is kept closed in this mode of operation to prevent contamination of the twin-flow port with oil from the roughing pump.

10.10 Repairing leaks

If a leak is found in a vacuum instrument the question then arises as to the best means to use to repair it. About the best general advice that can be given on this matter is to proceed very carefully, cautiously, and thoughtfully. On complicated instruments such as electron microscopes, many different considerations are involved in the design of each part, and so making a change that superficially might appear acceptable in repairing a leak can very well impair some other function. In general, the best procedure to follow if a leak is found in such an instrument is to contact the manufacturer's service organization for advice and assistance. In the past it was common practice to use various types of lacquers and leak sealants to stop leaks in vacuum systems. This approach should not be used on modern vacuum instruments, because: these materials often do not seal permanently; they can become a source of gas evolution and contamination if they reach the inside of a vacuum system; and, since they are liquids they can accidentally flow into locations where they can adversely affect some function not related to making a vacuum seal.

Actually, the vacuum systems on most electron microscopes are now so well designed and constructed that leaks are unlikely to develop in any place except in mechanisms that transmit items or motion through the vacuum wall, such as doors to photographic film chambers, specimen exchange mechanisms, and aperture centring devices. On many recent models the vacuum seals for aperture centring mechanisms are made with flexible metal bellows, and should give long, trouble-free service if used with due care and respect. If a metal bellows does fail, however, the resulting leak is usually large enough so that there is no question about its existence or location. The only way to repair such a leak satisfactorily is to replace the metal bellows, and this usually requires delicate welding or brazing operations which can best be done by the manufacturer. When O-rings are used to make the vacuum seals in these mechanisms, they become the most likely source of leaks. Because it is relatively simple to take care of problems involving O-rings, the use and care of them is discussed briefly in the following section.

10.11 O-ring seals

O-rings came into prominence during World War II in response to a need for gaskets of standard sizes and shapes for use in military machinery. Shortly thereafter they were widely adopted by manufacturers of vacuum equipment for use in demountable vacuum seals on equipment designed for operation in the rough and high vacuum ranges. Prior to that time, gaskets with square or rectangular cross-section were used. Usually, these were cut from rubber sheet stock, and replacement was expensive and difficult because sizes and shapes were not standardized.

The size of an O-ring is specified by giving its cross-sectional diameter d (i.e. the diameter measured through the actual rubber, also sometimes called the width w of the O-ring), and its circumferential internal diameter ID. When selecting an O-ring, care must be exercised to be sure that it is of the proper size for the joint in which it is to be used. Attempting to use an O-ring with a smaller cross-sectional diameter than the joint was designed for will result in insufficient compression of the O-ring and a high probability that the joint will leak. An O-ring with a cross-sectional diameter larger than the joint was designed for will usually prevent the joint from being tightened properly, and may cause the O-ring to be sheared or otherwise physically damaged. If the internal diameter of an O-ring is too small, it will have to be stretched to fit into the joint, and this will decrease its cross-sectional diameter, make surface irregularities more prominent, and keep it constantly under tension, all of which make it likely that a poor seal and a short service life will result. If an attempt is made to use an O-ring with an internal diameter that is too large for the groove in which it must fit, the O-ring will wrinkle, twist, and pop out of the groove, giving at best a poor sealing action and often making it impossible to assemble the joint. Some problems may be encountered in obtaining O-rings of the proper size in the United States, because here most suppliers stock O-rings to meet industrial needs, and these are invariably sized on the basis of nominal fractions of an inch, whereas most electron microscopes and many other vacuum instruments are manufactured in Europe and Japan where O-rings with metric dimensions are used. It is unusual to find an O-ring sized in inches that will properly fit a groove designed for one with metric dimensions. Fortunately, most manufacturers will supply proper replacement O-rings for their instruments. O-rings should not be washed with an organic solvent such as acetone or alcohol, because the solvent will penetrate into them and then will later be released into the vacuum system, giving undesirably high levels of outgassing and contributing to contaminating the system with organic materials. O-rings are usually cleaned by the manufacturer and shipped in sealed envelopes, and so new O-rings should not need cleaning before being put into use. If, however, an O-ring does need to be cleaned, the best procedure is to wipe it with a clean, lint-free cloth pad moistened with a little high vacuum grease.

Although O-rings are made from a variety of different elastomers to meet the requirements of various industrial applications, most of those used in electron microscopes and other laboratory instruments are now made of the synthetic fluoroelastomer Viton that was introduced in the 1950s by the Du Pont Company. This material is a linear polymer of vinylidene fluoride and hexafluoropropylene with a molecular mass of about 60 000 g/mol. It is tough, long-wearing, and highly resistant to atmospheric oxidation and to most solvents, lubricants, and common chemicals. Its service temperature range extends from below –30°C to above 200°C. Because of these unusual properties it is widely used for seals in industrial lubricating, hydraulic, and fuel systems. It is particularly useful in vacuum applications because it has a very low permeability to most gases, and a very low outgassing rate after being baked out in a vacuum (see Section 2.10.4a, p. 66). O-rings made of butyl rubber are sometimes used because they are less expensive and have a lower gas permeability than those made of Viton. O-rings made of polyimide rubber and of Teflon are also sometimes used for special purposes.

A large number of different designs have been developed for making static and dynamic O-ring seals. Some of the more frequently used of these are described and illustrated by Harris (Chapter 12) and in the Varian booklet on Basic Vacuum Practice (Chapter 6). Anyone involved in the design of vacuum apparatus will also find the detailed discussion of the theory and practice of designing vacuum seals given by Roth (Chapter 7) interesting and useful. A large fraction of the static O-ring seals on vacuum apparatus currently found in electron microscopy laboratories use flat flanges to compress an O-ring in a rectangular groove, as shown schematically in Fig. 10.11a. In this design the O-ring groove usually has a depth D that is about seven-tenths the cross-sectional diameter of the O-ring (i.e. $D = 0.7d$) and a width W that is about 25 percent greater than the diameter of the O-ring (i.e. $W = 1.25d$). With these dimensional relationships proper sealing is achieved when the two metal flanges are brought into contact. This gives the joint good mechanical strength and stability and deforms the O-ring enough so that it fills all of the groove except for small regions at the corners. Under these conditions the O-ring makes a wide band of contact with all four surfaces of the groove, thereby effecting a good vacuum seal. Dovetail grooves are sometimes used in applications where it is necessary to prevent the O-ring from being dislodged from the groove accidentally.

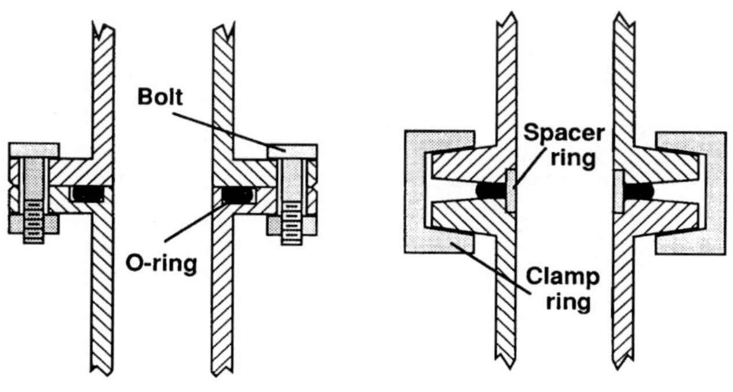

Fig. 10.11 Two common types of static O-ring seals.

Various methods other than the bolted flanges shown in Fig. 10.11a are used to hold joints with O-ring seals together. For example, for joints between the lenses of a transmission electron microscope, the weight of the lenses is usually sufficient to effect the seal, and for the doors of the photographic film chambers of transmission electron microscopes and the specimen chambers of scanning electron microscopes, atmospheric pressure is more than sufficient to produce a good seal (Section 1.2, p. 13). Fig. 10.11b shows a type of seal very frequently used for vacuum joints in tubes up to about 60 mm in diameter. In this design the O-ring is held in place by a metal spacer ring while it is compressed between two slightly tapered flanges. The flanges, in turn, are held in place by an internally tapered C-shaped clamping ring which is tightened around the tapered outer surfaces of the flanges. This clamping ring is usually split diametrically, hinged on one side, and then held together and wedged against the flanges by a bolt or some other device designed to make it possible to take the joint apart quickly and conveniently. Fig. 10.12 shows three common ways in which O-rings are used to effect a vacuum seal around sliding and rotating shafts. As was true for the static seals just described, the dimensions of the grooves in these seals are chosen so that the O-rings are

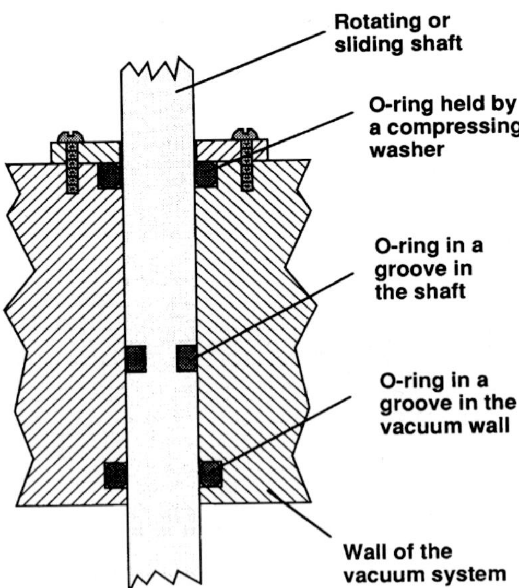

Fig. 10.12 Three types of O-ring seals commonly used on rotating and sliding shafts.

elastically compressed between the shaft and the vacuum wall sufficiently to prevent the movement of air molecules past them.

O-rings must be given periodic service and maintenance if they are to perform properly. O-rings used in static seals of the general type shown in Fig. 10.11a, which are subjected primarily to compressive loading and which are well protected from the degenerative effects of the atmosphere and ultraviolet light, tend to last for several years. As a general rule, such O-rings only need to be inspected when these vacuum joints are opened for servicing, whereupon it is usually prudent to replace the O-rings. An O-ring should certainly be replaced if it shows any signs of permanent deformation or of surface crazing or cracking. O-rings used in some quick-disconnect joints of the type shown in Fig. 10.11b are particularly susceptible to surface deterioration, because they are constrained on only three sides. In such joints the outer surface of the O-ring is fully exposed to the atmosphere, and is constantly stretched to a considerable degree, particularly if the design of the joint allows the flanges to be clamped together too tightly. Under these conditions the outer surface of the O-ring becomes very susceptible to oxidative degradation, which leads to premature embrittlement and cracking. Fig. 10.13 shows four O-rings that had been been misused and neglected in over-tightened joints of this kind.

Never use a metallic object, such as a needle or screwdriver, to dig an O-ring out of its groove! Instead, use a tool that cannot possibly scratch

Fig. 10.13 A set of badly deteriorated O-rings

the bottom of the groove, such as your fingernail, a small diameter stick sharpened to a flat point (bamboo is quite strong, and works well), or the corner of a plastic card. Alternatively, pushing on an O-ring circumferentially at two points about a centimetre apart will sometimes cause a small segment of it to rise up out of the groove enough so that a toothpick can be slipped under it. The tiniest scratch across the inner surface of an O-ring groove can produce a surprisingly big leak. Such a scratch cannot be reliably filled with vacuum grease, but must be carefully polished away with a fine abrasive. If such a polishing operation has to be carried out, the final stages of it should be done with a grade of alumina or diamond polishing compound that produces scratches that are at least as fine as those existing in other parts of the groove. The polishing motion should be circumferential so that any scratches produced, even very fine ones, run parallel to the axis of the groove, and the polishing action should be tapered away for several millimetres on each side of the site of the original scratch. The groove should, of course, be cleaned thoroughly after polishing is completed. A number of important comments and suggestions on cleaning procedures are contained in Section 2.10.4c (p. 69).

Most people are under the impression that it is necessary to 'grease' O-rings, and this brings us to the question of vacuum greases and the lubrication of O-rings. The important principle to remember here is that the function of vacuum grease in this application is only to lubricate the O-ring so that it seals properly in the joint, and that the grease cannot and should not be expected to contribute to plugging small leaks or otherwise enhancing the sealing action. It is advisable not to lubricate O-rings in the vacuum seals on electron guns to minimize the possibility of introducing organic materials into the gun chamber where they can contaminate the ceramic insulator and cause high voltage instabilities. In fact, O-rings in static seals of the general type shown in Fig. 10.11a ideally should not need to be lubricated or 'greased' to effect a good vacuum seal, if the O-ring groove is properly designed and if an O-ring of the proper size is used in it. The same is true for the easily demountable joint of Fig. 10.11b. However, it is common practice, and somewhat beneficial, to grease O-rings used in both of these types of static seals very lightly, because doing so helps ensure that they can slide enough to give a uniform sealing action around the entire diameter of the joint. For this purpose only a few monolayers of grease is sufficient. Place a small dab of vacuum grease on a small piece of lint-free cloth or tissue, spread it uniformly around the O-ring, and then wipe off all visible excess so that only enough remains to give the O-ring a slightly shiny or lustrous appearance. O-rings that are used to seal sliding and rotating shafts must be lubricated to prevent sticking, erratic motion, backlash, and wear. Again, however, a large amount of vacuum grease is not needed, nor is it beneficial. Perhaps the best practice is to lubricate the O-ring lightly, as just described, before it is placed in the groove. Then, after the O-ring has been properly seated in its groove, spread a barely-visible ring of grease uniformly into the cavities between the O-ring and the wall of the groove, on each side of the groove, all the way around the joint. A thin, barely-visible film of grease is also spread uniformly around the part of the shaft that must slide through the O-ring as the shaft is inserted. After the shaft is in place, residual grease should be wiped off exposed parts of the shaft so that it does not get on the hands of people using the apparatus, and so that it does not collect dust and dirt which can then be carried into the seal and cause a leak. Because vacuum greases are so widely used, some of the types that are most commonly available, together with a recently introduced fluid lubricant for use on sliding and rotating shafts, are discussed briefly in the following sections.

10.12 Vacuum greases

As the molecular mass of the molecules of most classes of organic compound increases, the compounds change successively from being gases, to thin liquids, to oils, to greases, to waxes, and finally to solid polymers, at room temperature. Thus, it is possible to prepare lubricating oils and greases from a variety of different types of chemical compound by selecting molecules with an appropriate range of molecular masses. For the ordinary hydrocarbons, compounds having 30–40 carbon atoms per molecule and molecular masses of about 350–500 g/mol are oils (see Section 5.4.1, p. 183), while those with 50–100 carbon atoms per molecule and molecular masses of about 1000–1500 g/mol have greasy characteristics. It is, of course, common practice to incorporate various additives into greases to enhance their properties. One very important characteristic of greases is their tendency to 'bleed'. This involves the separation of oily components (i.e. those of lower molecular mass) from the grease after a period of time, and particularly when put into service under a vacuum or at a high temperature. Although there may be exceptions, greases with high softening and melting points or high service temperatures usually contain smaller amounts of oily low molecular mass components, and are less likely to bleed than greases with low service temperatures and low melting points. Vacuum greases based on hydrocarbon, synthetic ester, silicone, and perfluorinated polyether compounds are currently available from a number of suppliers. Some of the more widely-used commercial brands of each type are listed in Table 10.1.

Table 10.1 Characteristics of some vacuum greases

Grease	Vapour pressure at 20°C (Pa)	Service temperature range (°C)
Hydrocarbon-based		
Apiezon AP-100	below 10^{-8}	10 to 30
Apiezon AP-101	below 10^{-3}	−40 to 180
Apiezon H	3×10^{-7}	up to 250
Apiezon L	below 10^{-8}	melts at 47
Apiezon M	3×10^{-7}	below 30
Apiezon N	1×10^{-7}	below 30
Apiezon T	7×10^{-7}	0 to 120
Synthetic ester-based		
Celvacene-Light	below 10^{-4}	melts at 90
Celvacene-Medium	below 10^{-4}	melts at 120
Celvacene-Heavy	below 10^{-4}	melts at 130
Silicone-based		
Dow Corning High Vacuum	6×10^{-5}	−40 to 250
Perfluorinated polyether-based		
Krytox LVP	below 10^{-11}	20 to 300
Braycote 803	below 10^{-9}	−54 to 260

10.12.1 Hydrocarbon-based greases

The series of Apiezon hydrocarbon-based greases (GEC Alsthom) listed in Table 10.1 are widely available from companies that handle supplies for electron microscopy in the United States and Europe. Of these, greases L and M are reported to be pure hydrocarbon materials with no additives. Grease M, which is the least expensive, is recommended as a general purpose laboratory lubricant. The other greases, and particularly greases H and T, contain various types of additives to increase stiffness and to extend the service temperature range. Greases AP-100, L, M, and N all melt in the temperature range from 40 to 50°C, and thus cannot be used in applications where temperatures much above 30°C are likely to be encountered. Because they consist mainly of hydrocarbon compounds, these greases are all very good lubricants, and they can all be easily cleaned up with common solvents. The Apiezon product line also includes Sealing Compound Q which is a low cost, putty-like material with a vapour pressure in the 10^{-2} Pa (10^{-4} Torr) range that is useful for temporarily sealing holes and joints in rough vacuum lines.

10.12.2 Synthetic ester-based greases

The Celvacene synthetic ester-based greases are produced by CVC Vacuum Products. These are general purpose vacuum greases with good lubricating properties that are suitable for use from room temperature to moderately high temperatures. They are soluble in acetone, methyl-ethyl ketone, chloroform, and other common organic solvents.

10.12.3 Silicone-based greases

The Silicone High Vacuum grease manufactured by the Dow Corning Corporation is the most widely available silicone grease in the United States. Its base is a poly-dimethylsiloxane polymer, and it contains a small amount of colloidal silica as a thickening agent. It has a wide service temperature range, a very low vapour pressure, virtually no tendency to bleed, and excellent resistance to oxidation and to most common chemicals and solvents. While this grease is a good lubricant in applications where metal surfaces rub against rubber, like most silicone compounds it is not recommended as a lubricant in metal bearings and in many applications involving metal rubbing on metal (particularly steel on steel). Like the silicone diffusion pump fluids (Section 5.4.3, p. 184) it is not soluble in most common solvents, but can be dispersed in toluene, carbon tetrachloride, perchloroethylene, and kerosene. It is not widely used on electron microscopes because, like the silicone oils, it tends to break down into highly insoluble, non-conducting, siliceous products under electron bombardment.

10.12.4 *Perfluorinated polyphenyl ether greases*

Two greases compounded by combining a perfluoroalkylpolyether oil of high molecular mass with a fluorocarbon resin thickener are widely available. These are the Braycote 803 high vacuum grease (Bray Products Div., Castrol, Inc.) and the Krytox LVP (low vapour pressure) grease (Du Pont). These greases are excellent vacuum lubricants, even for applications involving metal-to-metal contact, and have virtually no tendency to bleed at ambient temperatures. They have very low vapour pressures and wide service temperature ranges, and are very resistant to oxidation and to attack by most common chemicals and solvents. They are not readily soluble in any common solvent, but can be dispersed in highly fluorinated solvents such as trichloro-trifluoro-ethane (Freon-113). These greases are perhaps the best currently available for use on electron microscopes, but probably should not be used on any vacuum seal in or near the electron gun because of the possibility that some of the perfluorinated base compound might get onto the high-voltage insulator and cause microdischarges, as described for the perfluorinated oils in Section 5.4.5 (p. 186). It is best to assemble O-ring seals in the electron gun without lubrication.

10.12.5 *A fluid high vacuum lubricant*

Sputtered Films Inc. have recently introduced a fluid lubricant (TorrLube) that is especially formulated for lubricating O-rings on sliding and rotating shafts. This material is advertised as having a vapour pressure well below 10^{-6} Pa (10^{-8} Torr), an outgassing rate less than that of baked Viton, and excellent lubricating properties for all high vacuum applications (including metal sliding on metal, and ball-bearings). For O-ring seals on rotating shafts, a few drops of this lubricant is advertised to have a service life more than 30 times that of most vacuum greases.

10.13 Concluding remarks

In a way, confronting a problem involving a leak in a vacuum system is a lot like making a trip to the dentist — the anticipation and expectation are usually far worse than the event itself. A thoughtfully planned, systematic approach will usually make it possible to locate a leak rather quickly, and the repair of leaks on modern vacuum equipment is usually a relatively straightforward matter of replacing either an O-ring or a defective part. It is very important to service joints involving O-ring seals properly, and to use them carefully. Above all, with modern vacuum instruments, avoid applying excessive amounts of vacuum grease to these joints. If vacuum grease is indeed needed to make a joint seal properly, then the joint itself is defective and should be repaired. Vacuum grease should be looked on as merely a lubricant, and should not be used in the hope that it will serve as a primary sealant. In many instances, problems that initially appear to be due to leaks will turn out to be due to some other cause. Several examples of this have been related in this chapter. The best way to prepare for these events is to develop a thorough familiarity with the characteristics of a vacuum system when it is known to be functioning properly. Regularly record such items as pumpdown time, ultimate pressure, time of recovery after specimen exchange, and rate of pressure rise, so that it will be evident when a malfunction is occurring. Also keep records of the kinds of specimens that are put into an apparatus, and of any unusual operations that are performed on or with it, because this information will often point to the cause of a problem and prevent a lengthy search for a non-existent leak. Finally, if you have to have a leak, may it be a moderately big one, because these are usually the easiest ones to find and fix.

A Appendix 1: References

Reference books

Agar, A.W., Alderson, R.H. and Chescoe, D. (1974) Principles and practice of electron microscope operation. In Practical methods in electron microscopy, Vol. 2 (Glauert, A.M. ed.). Elsevier, Amsterdam

Berman, A. (1992) Vacuum engineering: calculations, formulas, and solved problems. Academic Press, Inc., San Diego

Bozzola, J.J. and Russell, L.D. (1992) Electron microscopy: principles and techniques for biologists. Jones and Bartlett, Boston

Dushman, S. and Lafferty, J.M. (1962, 2nd edn.) Scientific foundations of vacuum technique. John Wiley, New York

Guthrie, A. (1963) Vacuum technology. John Wiley, New York

Hablanian, M.H. (1990) High vacuum technology: a practical guide. Marcel Dekker, New York

Hall, C.E. (1966, 2nd edn.) Introduction to electron microscopy. McGraw-Hill, New York

Harris, N.S. (1989) Modern vacuum practice. McGraw-Hill, London

Lichte, H. (1991) Electron image plane off-axis holography of atomic structures. In Advances in optical and electron microscopy, Vol. 12 (Mulvey, T. and Sheppard, C.J.R., eds.), pp. 25–91. Academic Press, London

O'Hanlon, J. F. (1980) A user's guide to vacuum technology. John Wiley, New York

Reimer, L. (1984) Transmission electron microscopy, Springer-Verlag, Berlin

Robards, A.W. and Sleytr, U.B. (1985) Low temperature methods in biological electron microscopy. In Practical methods in electron microscopy, Vol. 10 (Glauert, A.M., ed.). Elsevier, Amsterdam

Roth, A. (1982, 2nd edn.) Vacuum technology. North-Holland, Amsterdam

Spence, J.C.H. (1988, 2nd edn.) Experimental high-resolution electron microscopy. Oxford University Press, Oxford

Stuart, R.V. (1983) Vacuum technology: thin films and sputtering. Academic Press, New York

Van Atta, C.M. (1965) Vacuum science and engineering. McGraw-Hill, New York

Varian Associates (1980) Introduction to helium mass spectrometer leak detection. Varian Associates, Inc., Palo Alto

Varian Associates (1989, 2nd edn.) Basic vacuum practice. Varian Associates, Inc., Palo Alto

Weston, G. F. (1985) Ultra-high vacuum practice. Butterworths, London

Further reading

Carpenter, L.G. (1983, 2nd edn.) Vacuum technology: an introduction. Adam Hilger, Bristol

Chambers, A., Fitch, R.K. and Halliday, B.S. (1989) Basic vacuum technology. Adam Hilger, Bristol

Lewin, G. (1965) Fundamentals of vacuum science and technology. McGraw-Hill, New York

Turnbull, A.H., Barton, R.S. and Riviere, J.C. (1962) An introduction to vacuum technique. John Wiley, New York

Training booklets published by:

Varian Associates, Inc: Diffusion pumps; Turbomolecular pumps

Leybold Vacuum Products Inc: Vacuum technology — its foundations, formulae, and tables; Vacuum Vademecum

The American Vacuum Society: Handbook of electron tube and vacuum techniques; The physical basis of ultrahigh vacuum; Vacuum technology and space simulation

Additional titles in this series

Practical Methods in Electron Microscopy, edited by Audrey M. Glauert and published by Elsevier, Amsterdam:

Alderson, R.H. (1975) Design of the electron microscope laboratory, Vol. 4

Butler, E.P. and Hale, K.F. (1981) Dynamic experiments in the electron microscope, Vol. 9

Chandler, J.A. (1977) X-ray microanalysis in the electron microscope, Vol. 5, Part II

Goodhew, P.J. (1985) Thin foil preparation for electron microscopy, Vol. 11

Glauert, A.M. (1975) Fixation, dehydration and embedding of biological specimens, Vol. 3, Part I

Lewis, P.R. and Knight, D.P. (1992) Cytochemical staining methods for electron microscopy, Vol. 14

Misell, D.L. (1978) Image analysis, enhancement and interpretation, Vol. 7

Misell, D.L. and Brown, E.B. (1987) Electron diffraction: an introduction for biologists, Vol. 12

Reid, N. and Beesley, J.E. (1991) Sectioning and cryosectioning for electron microscopy, Vol. 13

Williams, M.A. (1977) Autoradiography and immunocytochemistry, Vol. 6, Part I

Williams, M.A. (1977) Quantitative methods in biology, Vol. 6, Part II

Willison, J.H.M. and Rowe, A.J. (1980) Replica, shadowing, and freeze-etching techniques, Vol. 8

Appendix 2: List of suppliers

The following list of suppliers includes those that are directly referred to in the text and those that otherwise seemed to me to be most directly related to the subject matter covered in this book. It would have been virtually impossible, and probably of little use, to have attempted to include the names of all of the numerous companies that deal in one way or another with vacuum equipment and accessories for electron microscopy, so some choices had to be made. This was often a difficult process. Compilation of this list was also complicated by the fact that companies merge, change names, and disappear so frequently in today's highly competitive economic environment. I apologize in advance for any serious errors or omissions I may have made, and I welcome suggestions in this regard so that appropriate changes can be made if a revision of this book is produced. I might also comment that in the United States the Thomas Publishing Company annually produces the *Thomas Register of American Manufacturers*, which includes an extensive listing of manufacturers of a wide range of different types of products. A similar listing can be found in the telephone directories and buyers guides that are published annually by such periodicals as *R&D Magazine*, *Physics Today,* and *American Laboratory*. Similar publications are undoubtedly produced in other countries. Such resources can be found by inquiring at most major libraries.

General electron microscopy supplies

Agar
Electron Microscopy Sciences
EMCorp
Energy Beam Sciences
Fisher
Fisons
Fullam
Ladd
M.E. Taylor
Polysciences
SPI/Structure Probe
Taab
Technical Products
Ted Pella
Tousimis

Carbon coaters and evaporators

Agar
Bal-Tec
Cressington
Denton
Edwards
Electron Microscopy Sciences
EMCorp
Emitech
Energy Beam Sciences
Fisons
Fullam
M.E. Taylor
Microfield
SPI/Structure Probe
Taab
Technical Products
Ted Pella
Tousimis
VG Microtech/Fisons

Cleaning supplies

Agar
Alconox
Branson Ultrasonics
Copper Clad Products
Electron Microscopy Sciences
EMCorp
Emitech
Energy Beam Sciences
Fisher
Fullam
International Products
Ladd
Pierce Chemical
Procter & Gamble
SPI/Structure Probe
Taab
Technical Products
Ted Pella
Tousimis
Westley Products

Cryogenic specimen preparation apparatus for SEM

Bal-Tec
Cressington
Electron Microscopy Sciences
Emitech
Energy Beam Sciences
Fisons
Oxford Instruments
RMC
SPI/Structure Probe
Ted Pella
VG Microtech/Fisons

Electron microprobe analysers

Cameca
JEOL
R.J. Lee
Spectra-Source

Electron microscopes, scanning

Amray
Cameca
CamScan
Hitachi
JEOL
Leica
Noran
Philips
R.J. Lee
Secondary Images
Spectra-Source
Topcon
Wintron
Zeiss

Electron microscopes, scanning transmission UHV

Hitachi
JEOL
VG Instruments

Electron microscopes, scanning, variable pressure

ElectroScan
Hitachi

Electron microscopes, scanning probe

Burleigh
Digital Instruments
DME
Hitachi
LK Technologies
McAllister
Park Scientific
Perkin-Elmer
RHK Technology
Technical Instruments
TopoMetrix
VG Instruments/Fisons
WYKO/Leica

Electron microscopes, transmission

Hitachi
JEOL
Philips
Spectra-Source
Topcon
Wintron
Zeiss

Freeze drying units

Agar
Bal-Tec
Denton
Edwards
Electron Microscopy Sciences
Emitech
Energy Beam Sciences
Fisons
GeneVac
Microfield
VG Microtech/Fisons

Freeze fracture and etching units

Agar
Bal-Tec
Cressington
Fisons
JEOL
Microfield
RMC
VG Microtech/Fisons

Ion beam mills

Anatech
Bal-Tec
EMCorp
Gatan
VCR Group

Leak detectors

Advanced Vacuum
Alcatel
Ametek
Balzers
Chell
Cooke
Danielson
Duniway
Edwards
EPM
Evey
Lesker
Leybold
McGrath
MSVS
MVAK
UTI
Vacuum Instrument
Vacuum Technology
Varian
Veeco

Plasma asher and etcher units

Denton
Electron Microscopy Sciences
Emitech
Energy Beam Sciences
Fisons
Fullam
Microfield
Oxford Instruments
Plasma Sciences
SPI/Structure Probe
Taab
VG Microtech/Fisons

Residual gas analysers

Advanced Vacuum
Ametek
Balzers
Chell
HTE
LDS
Lesker
Leybold
MSVS
NGS/MKS
UTI
Vacuum Technology
Varian
VG Instruments/Fisons

Sputter coaters

Agar
Anatech
Bal-Tec
Cooke
Cressington
Denton
Edwards
Electron Microscopy Sciences
EMCorp
Emitech
Energy Beam Sciences
Fisons
Fullam
Ladd
M.E. Taylor
Microfield
Plasma Sciences
SPI/Structure Probe
Taab
Ted Pella
Tousimis
VG Microtech/Fisons

Traps and filters for backing pumps

Advanced Vacuum
Balston
Balzers
Cooke
Duniway
Edwards
Evey
Fisher
Key
Kinney
Lesker
Leybold
MSVS
MVAK
Stokes
M.E. Taylor
Ted Pella
Thermionics
Welch

Vacuum evaporators and coating units

Agar
Bal-Tec
Cooke
Danielson
Denton
Edwards
Electron Microscopy Sciences
EMCorp
Emitech
Energy Beam Sciences
Evey
Fisons
Fullam
GeneVac
Ladd
McGrath
Microfield
SPI/Structure Probe
Taab
Ted Pella
Thermionics
Tousimis
Wintron
VG Microtech/Fisons

Vacuum gauges

Agar
Alcatel
Balzers
Cooke
CVC
Duniway
Edwards
Evey
Fil-Tech
Granville-Phillips
GeneVac
HTE
HPS/MKS
Key
Kinney
LDS
Lesker
Leybold
McGrath
MKS Instruments
MSVS
Perkin-Elmer
Taab
M.E. Taylor
Teledyne
Televac
Terranova
Varian
Veeco
Welch

Vacuum oils, greases, and sealants

Advanced Vacuum
Agar
Alcatel
Ausimont
Avo Biddle
Balzers
Bray
Chell
Cooke
CVC
Danielson
Dow Corning
Duniway
Du Pont
Edwards
Electron Microscopy Sciences
Emitech
Evey
Fisher
Fisons
Fullam
GEC Alsthon
Key
Kinney
Ladd
LDS
Lesker
Leybold
McGrath
Monsanto
MSVS
MVAK
SPI/Structure Probe
Sputtered Films
Taab
Ted Pella
Tousimis
Varian
Welch

Vacuum pumps, bulk getter

Barry Scientific
Danielson Associates
SAES Getters
UltraPure Systems

Vacuum pumps, claw and Roots blowers

Alcatel
Balzers
Edwards
Kinney
Lesker
Leybold

Vacuum pumps, cryogenic

APD Cryogenics
Balzers
CTI
CVI
EBARA
Edwards
Leybold

Vacuum pumps, diaphragm

Air Dimensions
Balzers
Barnant
Danielson
Elnik
Gast
GeneVac
KNF Neuberger
Leybold

Vacuum pumps, diaphragm-molecular drag

Alcatel
Balzers
Danielson

Vacuum pumps, dry roughing systems

Alcatel
Balzers
Danielson
Edwards
Leybold
Varian

Vacuum pumps, ion

Duniway
Lesker
Leybold
Perkin-Elmer
Varian

Vacuum pumps, molecular drag

Alcatel
Balzers
Danielson
Leybold
Osaka Vacuum

Vacuum pumps, oil diffusion

Advanced Vacuum
Alcatel
Balzers
Cooke
CVC
Duniway
Edwards
Evey
Key
Kinney
LDS
Lesker
McGrath
MSVS
Varian

Vacuum pumps, piston

Gast
MEDO
Varian

Vacuum pumps, rotary-vane

Advanced Vacuum
Alcatel
Balzers
Cooke
CVC
Duniway
Edwards
Evey
Fisher
Fullam
GeneVac
Key
Kinney
Ladd
LDS
Lesker
Leybold
McGrath
MSVS
MVAK
SPI/Structure Probe
Stokes
Taab
Ted Pella
Varian
Welch

Vacuum pumps, sorption

Duniway
Evey
McGrath
MDC
Lesker
Perkin-Elmer
Varian
VG Instruments

Vacuum pumps, titanium sublimation

Balzers
Duniway
Lesker
Leybold
Perkin-Elmer
Varian

Vacuum pumps, turbomolecular

Alcatel
Balzers
Danielson
Duniway
Edwards
Evey
Key
Lesker
Leybold
McGrath
MSVS
Osaka
Perkin-Elmer
Seiko Seiki
Varian
Welch

Vacuum pumps, venturi

Gast

Vacuum valves, fittings, and accessories

Alcatel
Balzers
Cooke
CVC
Duniway
Edwards
GeneVac
Granville-Phillips
HPS
Huntington
Key
Kinney
LDS
Lesker
Leybold
McGrath
MDC
MKS Instruments
Perkin-Elmer
Thermionics
Varian
Veeco
VG Instruments
Welch

Appendix 3: Addresses of manufacturers and suppliers

The following is a list of the addresses of the suppliers and manufacturers of electron microscopy and vacuum supplies and equipment contained in Appendix 2. I apologize for any errors that may be in this list; however, compiling it has been a rather challenging task, because over the several years this book was being written a number of companies have changed names and addresses, and it has been very difficult to keep up-to-date on all such events. I will be pleased to know of any corrections or changes to this list, and also of the names and addresses of other suppliers, for inclusion in later editions of this book.

Advanced Vacuum Company, Inc.
1215 Business Parkway
N. Westminster, MD 21157, U.S.A.
Tel: 410-876-8200; Fax: 410-876-5820

Agar Scientific, Ltd.
66A Cambridge Road
Stansted, Essex CM24 8DA, U.K.
Tel: 0279-813519; Fax: 0279-815106

Air Dimensions, Inc.
1015 West Newport Center, Suite 101
Deerfield Beach, FL 33441, U.S.A.
Tel: 305-428-7333; Fax: 305-360-0987

Alcatel Vacuum Products
67 Sharp Street
Hingham, MA 02043-4348, U.S.A.
Tel: 617-331-4200; Fax: 617-331-4230

Alcatel Vacuum Technology (U.K.), Ltd.
The Business Centre
The Vansittart Estate
Windsor, Berkshire SL4 1SE, U.K.
Tel: 0753-830422; Fax: 0753-851202

Alcatel CIT Vacuum Technology
55 rue Edgar Quinet
92240 Malakoff, France
Tel: 1-40-92-3000; Fax: 1-40-92-0450

Alconox, Inc.
215 Park Avenue South
New York, NY 10003-9990, U.S.A.
Tel: 212-473-1300; Fax: 212-353-1342

Ametek, Inc.
Analytical Instruments Division
150 Freeport Road
Pittsburgh, PA 15238, U.S.A.
Tel: 412-828-9040; Fax: 412-826-0399

Amray, Inc.
160 Middlesex Turnpike
Bedford, MA 01730, U.S.A.
Tel: 617-275-1400; Fax: 617-275-0740

Anatech, Ltd.
5510 Vine Street
Alexandria, VA 22310, U.S.A.
Tel: 703-971-9200; Fax: 703-971-4818

APD Cryogenics, Inc.
1919 Vultee Street
Allentown, PA 18103, U.S.A.
Tel: 215-791-6700; Fax: 215-791-0440

APD Cryogenics
IGC (Europe), Ltd.
5 Jupiter House
Calleva Industrial Park
Aldermaston, Berkshire RG7 4QW, U.K.
Tel: 0734-819373; Fax: 0734-817601

Ausimont U.S.A., Inc.
44 Whippany Road
Morristown, NJ 07962-6250, U.S.A.
Tel: 201-292-6250; Fax: 201-292-0886

Avo Biddle Instruments
510 Township Line Road
Blue Bell, PA 19422, U.S.A.
Tel: 215-646-9200; Fax: 215-643-2670

Balston, Inc.
260 Neck Road
P.O. Box 8223
Haverhill, MA 01835-0723, U.S.A.
Tel: 508-374-7400; Fax: 508-374-7070

Bal-Tec Products, Inc.
984 Southford Road
P.O. Box 1221
Middlebury, CT 06762, U.S.A.
Tel: 203-598-3660; Fax: 203-598-3658

Bal-Tec, Inc.
FL-9496 Balzers
Furstentum, Liechtenstein
Tel: 075-388-5611; Fax: 075-388-5660

Balzers Aktiengesellschaft
FL-9496 Balzers
Furstentum, Liechtenstein
Tel: 075-388-4411; Fax: 075-388-42761

Balzers
8 Sagamore Park Road
Hudson, NH 03051-4914, U.S.A.
Tel: 603-889-6888; Fax: 603-889-8573

Balzers High Vacuum, Ltd.
Bradbourne Drive
Tilbrook
Milton Keynes MK7 8AZ, U.K.
Tel: 0908-373333; Fax: 0908-377776

Barnant Company
28W092 Commercial Avenue
Barrington, IL 60010, U.S.A.
Tel: 800-637-3739; Fax: 702-381-7053

Barry Scientific, Inc.
P.O. Box 173
58 Streeter Road
Fiskdale, MA 01518, U.S.A.
Tel: 508-347-9855; Fax: 508-347-8280

Branson Ultrasonics Corporation
41 Eagle Road
Danbury, CT 06813-1961, U.S.A.
Tel: 203-796-0400; Fax: 203-796-0381

Bray Products Division
Castrol, Inc.
16715 Von Karmen Avenue, Suite 230
Irvine, CA 92714, U.S.A.
Tel: 714-660-9414; Fax: 714-660-9374

Burleigh Instruments
Burleigh Park
Fishers, NY 14455, U.S.A.
Tel: 716-924-9355: Fax: 716-924-9072

Burleigh Instruments, Ltd.
9 Allied Business Centre
Coldharbour Lane
Harpenden, Hertfordshire AL5 4UT U.K.
Tel: 0582-766888; Fax: 0582-767888

Burleigh Instruments, GmbH.
Bergstrasse 104-106
D-6102 Pfungstadt, Germany
Tel: 06157-3047; Fax: 06157-7530

Cameca
103 Blvd. Saint-Denise, B.P. 6
92403 Courbevoie Cedex, France
Tel: 1-334-3060

Cameca Instruments, Inc.
2001 West Main Street
Stamford, CT 06902, U.S.A.
Tel: 203-348-5252; Fax: 203-348-5516

CamScan
Saxon Way, Bar Hill
Cambridge CB3 8SL, U.K.
Tel: 0954-780926; Fax: 0954-789829

CamScan USA, Inc.
500 Thomson Park Drive, Suite 508
Mars, PA 16046, U.S.A.
Tel: 412-772-7433; Fax: 412-772-7434

Chell Instruments, Ltd.
Tudor House, Grammar School Road
North Walsham, Norfolk NR28 9JH, U.K.
Tel: 0692-402488; Fax: 0692-402488

Cooke Vacuum Products
13 Merritt Street
S Norwalk, CT 06854, U.S.A.
Tel: 203-853-9500; Fax: 203-838-9553

Copper Clad Products, Inc.
600 South 9th Street
P.O. Box 312
Reading, PA 19603, U.S.A.
Tel: 215-374-6572; Fax: 215-375-3557

Cresssington Scientific Instruments, Ltd.
34 Chalk Hill
Watford WD1 4BX, U.K.
Tel: 0923-220499; Fax: 0923-816646

Cressington Scientific Instruments, Inc.
508 Thomson Park Drive
Mars, PA 16046, U.S.A.
Tel: 412-772-0220; Fax: 412-772-0219.

CTI-Cryogenics
9 Hampshire Street
Mansfield Corporate Center
Mansfield, MA 02048, U.S.A.
Tel: 508-337-5000; Fax: 508-337-5169

CTI-Cryogenics, SA
4, rue Rene Razel
F-91892 Orsay Cedex, France
Tel: 1-6985-3900; Fax: 1-6985-3725

CTI-Cryogenics, Ltd.
Avenue Two, Unit 4
Station Lane Industrial Estate
Witney, Oxford OX8 6YD, U.K.
Tel: 0993-776436; Fax: 0933-772067

CVC Products
525 Lee Road, Box 1886
Rochester, NY 14603-1886, U.S.A.
Tel: 800-448-5900; Fax: 716-458-0424

CVI, Inc.
P.O. Box 2138
Columbus, OH 43216, U.S.A.
Tel: 614-876-7381; Fax: 614-876-5648

Danielson Associates, Inc.
1989A University Lane
Lisle, IL 60532, U.S.A.
Tel: 708-960-0086; Fax: 708-960-0546

Denton Vacuum, Inc.
2 Pin Oak Lane
Cherry Hill, NJ 08003, U.S.A.
Tel: 609-424-1012; Fax: 609-424-0395

Digital Instruments, Inc.
520 East Montecito Street
Santa Barbara, CA 93103, U.S.A.
Tel: 805-899-3380; Fax: 805-899-3392

DME A/S
Transformervej 12
DK-2730 Herlev, Denmark
Tel: 42-84-9211; Fax: 42-84-9197

Dow Corning Corporation
P.O. Box. 0994
Midland, MI 48686-0994, U.S.A.
Tel: 517-496-4000

Duniway Stockroom Corporation
1600 N. Shoreline Drive
Mountain View, CA 94043, U.S.A.
Tel: 415-969-8811; Fax: 415-965-0764

Du Pont Company
Chemicals and Pigments Department
Wilmington, DE 19898, U.S.A.
Tel: 800-441-9442

Du Pont de Nemours (U.K.), Ltd.
Wedgwood Way, Stevenage
Hertfordshire SG1 4QN, U.K.
Tel: 0438-734716; Fax: 0438-734376

EBARA Technologies, Inc.
Cryopump Operation
3560 Bassett Street
Santa Clara, CA 95054-2704, U.S.A.
Tel: 408-496-2888; Fax: 408-496-2801

Edwards High Vacuum International
Manor Royal, Crawley
West Sussex RH10 2LW, U.K.
Tel: 0293-528844; Fax: 0293-533453

Edwards High Vacuum International
One Edwards Park
301 Ballardvale Street
Wilmington, MA 01887, U.S.A.
Tel: 508-658-5410; Fax: 505-658-7969

Edwards/Temescal
16 rue Champ Lagarde
78000 Versailles, France
Tel: 39-53-4823; Fax: 39-50-5416

Electron Microscopy Sciences
321 Morris Road, Box 251
Fort Washington, PA 19034, U.S.A.
Tel: 215-646-1566; Fax: 215-646-8931

ElectroScan Corporation
66 Concord Street
Wilmington, MA 01887-9857, U.S.A.
Tel: 508-988-0055; Fax: 508-988-0062

Elnik Systems
3 Edison Place
Fairfield, NJ 07004-3511, U.S.A.
Tel: 201-882-8033; Fax: 201-882-8037

EMCorp
P.O. Box 67285
Chestnut Hill, MA 02167, U.S.A.
Tel: 617-965-6340; Fax: 617-964-1052

Emitech, Inc.
3845 FM, 1960-West, Suite 345
Houston, TX 77068, U.S.A.
Tel: 713-893-2067; Fax: 713-893-8443

Emitech, Ltd.
11 Enterprise Centre
Newtown Road, Ashford
Kent TN24 0PD, U.K.
Tel: 0233-646332; Fax: 0233-640744

Energy Beam Sciences, Inc.
11 Bowles Road, Box 468
Agawam, MA 01001, U.S.A.
Tel: 413-786-9322; Fax: 413-789-2786

EPM, Environmental, Inc.
834 E. Rand Road, Suite 6
Mt. Prospect, IL 60056, U.S.A.
Tel: 708-255-4494; Fax: 708-255-1959

Ernest F. Fullam, Inc.
900 Albany Shaker Road
Latham, NY 12210-1491, U.S.A.
Tel: 518-785-5533; Fax: 518-785-8647

Evey Engineering, Inc.
714 Southwest Boulevard
Vineland, NJ 08360, U.S.A.
Tel: 508-372-5511; Fax: 508-521-1631

FEI Company
19500 N.W. Gibbs Drive, Suite 100
Beaverton, OR 97006, U.S.A.
Tel: 503-690-1500; Fax: 503-690-1509

FEI Europe, Ltd.
Brookfield Business Park
Twentypence Road, Cottenham
Cambridge CB4 4PS, U.K.
Tel: 0954-50526; Fax: 0954-51856

Fil-Tech, Inc.
6 Pinckney Street
Boston, MA 02114, U.S.A.
Tel: 617-227-1133; Fax: 617-742-0686

Fisher Scientific
711 Forbes Avenue
Pittsburgh, PA 15219-4785, U.S.A.
Tel: 412-562-8300; Fax: 421-562-5344

Fisher Scientific
New Enterprise House
St. Helen's Street
Derby DE1 3GY, U.K.
Tel: 0332-200755; Fax: 0332-2965546

Fisons Scientific Equipment
Bishop Meadow Road
Loughborough, Leicestershire LE11 0RG, U.K.
Tel: 0509-231166; Fax: 0509-231893

Fisons Instruments, Inc.
450 National Avenue
Mountain View, CA 94034, U.S.A.
Tel: 415-926-8780; Fax: 415-961-8656

Gast Manufacturing Corporation
2300 M-139 Highway, Box 97
Benton Harbor, MI 49022, U.S.A
Tel: 616-926-6171; Fax: 616-927-0808

Gast Manufacturing Corpororation
Halifax Road, Cressex Estate
High Wycombe, Bucks HP12 3SN, U.K.
Tel: 0494-023571; Fax: 0494-436588

Gatan, Inc.
6678 Owens Drive
Pleasanton, CA 98588-3334, U.S.A.
Tel: 510-463-0200; Fax: 510-463-0204

Gatan, Ltd.
3 Saxon Way E, Oakley Way
Corby, Northamptonshire NN18 9EY, U.K.
Tel: 0536-743150; Fax: 0536-743154

Gatan, GmbH.
Ingolstaedter Strasse 40
D-80807 Munich, Germany
Tel: 089-352374; Fax: 089-3591642

GEC Alsthom (M&I), Ltd.
P.O. Box 136
Manchester M60 1AN, U.K.
Tel: 0618-722431; Fax: 0618-752695

GeneVac Sales Development, Ltd.
6 The Sovereign Centre, Farthing Road
Sproughton, Ipswich IP1 5AP, U.K.
Tel: 0473-240000; Fax: 0473-461176

Granville-Phillips Company
5675 Arapahoe Avenue
Boulder, CO 80303-1398, U.S.A.
Tel: 303-443-7660; Fax: 303-443-2546

Hitachi Naka Works
882 Ichige, Katsuta-shi
Ibaraki-ken, Japan 312
Tel: 292-73-2111; Fax: 292-74-6771

Hitachi/Nissei Sangyo America, Ltd.
755 Ravendale Drive
Mountain View, CA 94043, U.S.A.
Tel: 415-969-1100; Fax: 415-961-7259

Hitachi/Nissei Sangyo Co., Ltd.
Ivanhoe Road
Hogwood Industrial Estate
Finchampstead, Wokingham
Berkshire RG11 4QQ, U.K.
Tel: 0734-328632; Fax: 0734-328779

Hitachi/Nissei Sangyo, GmbH.
Berliner Strasse 91
40880 Ratingen 1
Dusseldorf, Germany
Tel: 2102-4530; Fax: 2102-473168

HPS Vacuum Products Division
MKS Instruments, Inc.
5330 Sterling Drive
Boulder, CO 80301-9989, U.S.A.
Tel: 303-449-9861; Fax: 303-442-6880

HTE, Inc.
Vacuum Systems & Components
1001 R. Shary Circle
Concord, CA 94518, U.S.A.
Tel: 510-687-1410; Fax: 510-687-1414

Huntington Laboratories
1040 L'Avenida
Mountain View, CA 94043-9763, U.S.A.
Tel: 415-964-3323; Fax: 415-964-6153

International Products Corporation
201 Connecticut Drive
P.O. Box 70
Burlington, NJ 08016-0070, U.S.A.
Tel: 609-386-8770; Fax: 609-386-8438

JEOL, Ltd.
1-2 Mushashino 3-chome
Akishima, Tokyo, Japan
Tel: 0425-43-1111; Fax: 0425-46-3353

JEOL (Europe) S.A.
16 Avenue de Colmar
92500 Reuil-Malmaison, France
Tel: 1-4749-6700; Fax: 1-4752-1041

JEOL (U.K.), Ltd.
JEOL House, Silver Court
Watchmead, Welwyn Garden City
Hertfordshire AL7 1LT, U.K.
Tel: 0707-377117; Fax: 0707-373254

JEOL U.S.A., Inc.
11 Dearborn Road
Peabody, MA 01960, U.S.A.
Tel: 508-535-5900; Fax: 508-536-2205

Key High Vacuum Products, Inc.
36 Southern Boulevard
Nesconset, NY 11767, U.S.A.
Tel: 516-360-3970; Fax: 516-360-3973

Kinney Vacuum Company
495 Turnpike Street
Canton, MA 02021, U.S.A.
Tel: 617-828-9500; Fax: 617-828-5612

KNF Neuberger, Inc.
P.O. Box 4060
Princeton, NJ 08543, U.S.A.
Tel: 609-799-4350; Fax: 609-799-3117

LAB-6, Inc.
4890 Opeongo Road
Box 259, RR #3
Woodlawn, Ontario K0A 3M0, Canada
Tel: 613-832-4050

Ladd Research Industries, Inc.
P.O. Box 1005
Burlington, VT 05402, U.S.A.
Tel: 802-658-4961; Fax: 802-863-6478

LDS Vacuum Products, Inc.
980R Sunshine Lane
Altamonte Springs, FL 32714, U.S.A.
Tel: 407-862-4643; Fax: 407-862-8723

Leica/Cambridge, Ltd.
Clifton Road
Cambridge CB1 3QH, U.K.
Tel: 0223-411411; Fax: 0223-412776

Leica, Inc.
111 Deerlake Road
Deerfield, IL 60015, U.S.A.
Tel: 708-405-0123; Fax: 708-405-0147

Leica Vertrieb, GmbH.
Lilienthalstrasse 39/45
Postfach 1651
D6140 Bensheim 1, Germany
Tel: 66251-1360; Fax: 6251-136155

LK Technologies, Inc.
3910 Roll Avenue
Bloomington, IN 47403, U.S.A.
Tel: 812-332-4449; Fax: 812-332-4493

Kurt J. Lesker Company
1515 Worthington Avenue
Clairton, PA 15025, U.S.A.
Tel: 412-233-4200; Fax: 412-233-4275

Kurt J. Lesker Company
444 Old London Road
Hastings, East Sussex TN3 55BB, U.K.
Tel: 424-719101; Fax: 424-421160

Leybold Vacuum Products, Inc.
5700 Mellon Road
Export, PA 15632-8900, U.S.A.
Tel: 412-327-5700; Fax: 412-733-5960

Leybold, Ltd.
Waterside Way, Plough Lane
London SW17 7AB, U.K.
Tel: 081-947-9744; Fax: 081-947-0210

Life Cell, Inc.
3606-A Research Forest Drive
The Woodlands, TX 77381, U.S.A.
Tel: 713-367-5368; Fax: 713-363-3360

McAllister Technical Services
West 280 Prarie Avenue
Coeur d' Alene, ID 83814, U.S.A.
Tel: 208-772-9527; Fax: 208-772-3384

E. McGrath, Inc.
35 Osborne Street
Salem, MA 01970-2599, U.S.A.
Tel: 508-744-3546; Fax: 508-741-4020

MDC Vacuum Products Corporation
23842 Cabot Boulevard
Hayward, CA 94545, U.S.A.
Tel: 510-887-6100; Fax: 510-887-0626

MEDO U.S.A., Inc.
808-A North Central Avenue
Wood Dale, IL 60191, U.S.A.
Tel: 708-860-0500; Fax: 708-860-9473

M. E. Taylor Engineering, Inc.
21604 Gentry Lane
Brookeville, MD 20833, U.S.A.
Tel: 301-774-6246; Fax: 301-774-6711

Microfield Scientific, Ltd.
Crowsnest, Rectory Lane
Kingston Bagpuize
Oxford OX13 5AT, U.K.
Tel: 0865-821348; Fax: 0865-821459

MKS Instruments, Inc.
6 Shattuck Road
Andover, MA 01810, U.S.A.
Tel: 508-975-2350; Fax: 508-975-0093

Monsanto Company
800 North Lindbergh Boulevard
St. Louis, MO 63167, U.S.A.
Tel: 314-694-1000

MSVS, Ltd.
6 Old Station Industrial Park
Lower Farmington
Alton, Hampshire GU34 3DP, U.K.
Tel: 0420-587244; Fax: 0420-587241

MVAK Technologies, Inc.
299 Ridgedale Avenue
East Hanover, NJ 07936, U.S.A.
Tel: 201-887-5551; Fax: 201-887-1146

NGS Associates, Inc.
24 Walpole Park South
Walpole, MA 02081, U.S.A.
Tel: 508-660-1770; Fax: 508-668-5822

Noran Instruments, Inc.
2551 Beltline Road
Middleton, WI 53562, U.S.A.
Tel: 608-831-6511; Fax: 608-836-7224

Osaka Vacuum, Ltd.
911 Bern Court, Suite 140
San Jose, CA 95112, U.S.A.
Tel: 408-441-7658; Fax: 408-441-7660

Osaka Vacuum, Ltd.
6th Fl. Keihan-Yodoyabashi Bldg.
2-25 Kitahama 3-chome
Chuo-ku, Osaka 541 Japan
Tel: 203-3981; Fax: 222-3645

Oxford Instruments, Ltd.
Old Station Way
Enysham, Oxford OX8 1TL, U.K.
Tel: 0865-882855; Fax: 0865-881567

Oxford Instruments, Inc.
130A Baker Avenue
Concord, MA 01742, U.S.A.
Tel: 508-369-9933; Fax: 508-369-6616

Park Scientific Instruments
1171 Borregas Avenue
Sunnyvale, CA 94089-1304, U.S.A.
Tel: 408-747-1600; Fax: 408-747-1601

Perkin-Elmer Corporation
Physical Electronics Division
6509 Flying Cloud Drive
Eden Prarie, MN 55344-9983, U.S.A.
Tel: 612-828-6100; Fax: 612-828-6322

Perkin-Elmer, Ltd.
Post Office Lane
Beaconsfield
Buckinghamshire HP9 1QA, U.K.
Tel: 0494-676161; Fax: 0494-679331

Perkin-Elmer, GmbH.
Physical Electronics Europe
Brukmannering 40
D-8042 Oberschleissheim, Munich, Germany
Tel: 089-315-7170; Fax: 089-315-3117

Perkin-Elmer/ULVAC-PHI, Inc.
2500 Hagisono, Chigasaki-Shi
Kanagawa-ken 253, Japan
Tel: 0467-856522; Fax: 0467-854411

Philips Electron Optics
ATI Unicam
York Street
Cambridge CB1 2QU, U.K.
Tel: 0223-374422; Fax: 0223-374266

Philips Electronic Instruments, Corporation
85 McKee Drive
Mahwah, NJ 07430, U.S.A.
Tel: 201-529-3800; Fax: 201-529-2252

Philips Electron Optics.
Building AAE, P. O. Box 218
5600 MD Eindhoven, The Netherlands
Tel: 40-766234; Fax: 40-766164

Philips Nihon Corporation
Shuwa Shinagawa Bldg.
26-33 Takanawa 3-chome
Minato-ku, Tokyo 108, Japan
Tel: 5611

Pierce Chemical Company
P.O. Box 117
Rockford, IL 61105, U.S.A.
Tel: 815-968-0747; Fax: 815-968-7316

Plasma Sciences, Inc.
7200H Telegraph Square Drive
Lorton, VA 22079, U.S.A.
Tel: 703-550-7888; Fax: 703-339-9860

Polysciences, Inc.
400 Valley Road
Warrington, PA 18976-2590, U.S.A.
Tel: 215-343-6484; Fax: 215-343-0214

Polysciences, Ltd.
Handelstrasse 3, Postfach 1130
69208 Eppelheim, Germany
Tel: 6221-765767; Fax: 6221-764620

Procter & Gamble Company
1 Procter & Gamble Plaza
Cincinnati, OH 45202, U.S.A.
Tel: 513-983-1100; Fax: 513-983-4500

RHK Technology, Inc.
1750 W. Hamlin Road
Rochester Hills, MI 48309, U.S.A.
Tel: 313-656-3116; Fax: 313-656-8347

R. J. Lee Group, Inc.
350 Hochberg Road
Monroeville, PA 15146, U.S.A.
Tel: 412-325-1776; Fax: 412-733-1799

RMC, Inc.
4400 S. Santa Rita Avenue
Tucson, AZ 85714, U.S.A.
Tel: 602-889-7900; Fax: 602-741-2200

RMC (Europe), Ltd.
P.O. Box 65, Abingdon
Oxford OX13 5RF, U.K.
Tel: 0865-821442; Fax: 0865-821459

SAES Getters/USA, Inc.
1122 E. Cheyenne Mtn. Blvd.
Colorado Springs, CO 80906, U.S.A.
Tel: 719-576-3200; Fax: 719-576-5025

SAES Getters, S.p.A.
Via Gallarate 215
I-20151 Milano, Italy
Tel: 2-30201; Fax: 2-3340-3636

SAES Getters G.B., Ltd.
5 Southern Court, South Street
Reading, Berkshire RG1 4QS, U.K.
Tel: 734-566856; Fax: 734-504074

Secondary Images, Inc.
2371 Emery Lane
Winchester, OH 45697, U.S.A.
Tel: 513-927-5373; Fax: 513-927-5373

Seiko Seiki Instruments U.S.A., Inc.
1150 Ringwood Court
San Jose, CA 95131, U.S.A.
Tel: 408-922-5932; Fax: 408-922-1959

Seiko Seiki Co., Ltd.
3-1, Yashiki 4-chome
Narashino-shi, Chiba 275, Japan
Tel: 0474-79-2655; Fax: 0474-72-8936

Spectra-Source, Inc.
2284-C Ringwood Avenue
San Jose, CA 95131, U.S.A.
Tel: 408-955-9040; Fax: 408-955-9041

SPI Supplies Division
Structure Probe, Inc.
569 E. Gay Street, Box 656
West Chester, PA 19381-0656, U.S.A.
Tel: 215-436-5400; Fax: 215-436-5755

Sputtered Films, Inc.
314 Edison Avenue
Santa Barbara, CA 93103, U.S.A.
Tel: 805-963-9651; Fax: 805-963-2959

Stokes Vacuum
5500 Tabor Road
Philadelphia, PA 19120, U.S.A.
Tel: 215-831-5400; Fax: 215-831-5420

Taab Laboratories Equipment, Ltd.
3 Minerva House,
Calleva Industrial Park
Aldermaston, Berkshire RG7 4QW, U.K.
Tel: 0734-817775; Fax: 0734-817881

Technical Products, Inc.
154 Haddon Avenue
Westmont, NJ 08108, U.S.A.
Tel: 609-858-0451; Fax: 609-858-1251

Technical Instrument Company
348 Sixth Street
San Francisco, CA 94103-4788, U.S.A.
Tel: 415-431-8231; Fax: 415-431-6491

Ted Pella, Inc.
P.O. Box 492477
Redding, CA 96049-2477, U.S.A.
Tel: 916-243-2200; Fax: 916-243-3761

Teledyne Hastings-Raydist
P.O. Box 1436
Hampton, VA 23661, U.S.A.
Tel: 804-723-6531; Fax: 804-723-3925

Televac
2337 Philmont Avenue
P.O. Box 662
Huntington Valley, PA 19006-0662, U.S.A.
Tel: 215-938-4444; Fax: 215-947-7464

Terranova Scientific, Inc.
5841 Bell Road
Auburn, CA 95603, U.S.A
Tel: 961-889-1100; Fax: 916-269-2877

Thermionics Laboratory, Inc.
22815 Sutro Street, Box 3711
Hayward, CA 94540, U.S.A.
Tel: 510-538-3304; Fax: 510-538-2889

Topcon Corporation
75-1 Hasunuma-cho
Itabashi-ku, Tokyo 174, Japan
Tel: 3966-3141; Fax: 3966-5106

Topcon Europe, B.V.
Esse Bahn 11
2908 LJ Capelle a/d IJssel, The Netherlands
Tel: 458-5077; Fax: 458-5045

Topcon Technologies, Inc.
6940 Koll Center Parkway
Pleasanton, CA 94566-3100, U.S.A.
Tel: 510-462-2212; Fax: 510-846-2803

TopoMetrix Corporation
5403 Betsy Ross Drive
Santa Clara, CA 95054, U.S.A.
Tel: 408-982-9700; Fax: 408-982-9751

Tousimis Research Corporation
2211 Lewis Avenue
Rockville, MD 20851, U.S.A.
Tel: 301-881-2450; Fax: 301-881-5374

UltraPure Systems, Inc.
P.O. Box 60160
Colorado Springs, CO 80960, U.S.A.
Tel: 719-576-3355; Fax: 719-540-0734

UTI Instruments
2030 Fortune Drive, Suite C
San Jose, CA 95131, U.S.A.
Tel: 408-428-9400; Fax: 408-428-0823

Vacuum Instrument Corporation
2099 Ninth Avenue
Ronkonkoma, NY 11779, U.S.A.
Tel: 516-737-0900; Fax: 516-737-0949

Vacuum Technology, Inc.
1003 Alvin Weinberg Drive
Oak Ridge, TN 37830, U.S.A.
Tel: 615-481-3342; Fax: 615-481-3788

Varian Vacuum Products Division
121 Hartwell Avenue
Lexington, MA 02173, U.S.A.
Tel: 617-861-7200; Fax: 617-860-5437

Varian, Ltd.
128 Manor Road
Walton-on-Thames
Surrey KT12 2QF, U.K.
Tel: 0932-243741; Fax: 0932-228769

Varian GmbH, Vacuum Products
Kuehnstrasse 71 D
22045 Hamburg 70, Germany
Tel: 40-6696031; Fax: 40-6682282

VCR Group
250 East Grand Avenue, Suite 31
South San Francisco, CA 94080, U.S.A.
Tel: 415-875-1000; Fax: 415-875-7111

Veeco Instruments, Inc.
Terminal Drive
Plainview, NY 11803, U.S.A.
Tel: 516-349-8300; Fax: 516-349-8321

Veeco, GmbH.
Wissenschaftliche Gerate
Siemensstrasse I
D-8044 Unterschleibheim, Munich, Germany
Tel: 089-310-1035; Fax: 089-317-3440

Veeco Instruments, Ltd.
Colne Way Court, Colne Way
Watford, Hertfordshire WD2 4NE, U.K.
Tel: 0923-210044; Fax: 0923-35130

VG Instruments/Fisons
32 Commerce Center
Danvers, MA 07923, U.S.A.
Tel: 508-524-1000; Fax: 508-777-0678

VG Instruments/Fisons
Sidney Little Road
Churchfields Industrial Estate
Hastings, East Sussex TN38 9PX, U.K.
Tel: 0424-851291; Fax: 0424-851489

VG Microtech/Fisons
Bellbrook Business Park
Bolton Close, Uckfield
East Sussex, TN22 1QZ, U.K.
Tel: 0825-761077; Fax: 0825-768343

Welch Vacuum Technology, Inc.
7300 N. Linder Avenue, Box 183
Skokie, IL 60076-0183, U.S.A.
Tel: 708-676-8800; Fax: 708-677-8606

Westley Products Division
Blue Coral Inc.
1215 Valley Belt Parkway
Cleveland, OH 44131, U.S.A.
Tel: 216-641-5490; Fax: 216-351-0270

Wintron, Inc.
620 Alden Road, Suite 102
Markham, Ontario, Canada L3R 9R7
Tel: 416-479-9016; Fax: 416-479-9081

WYKO/Leica, Inc.
111 Deerlake Road
Deerfield, IL 60015, U.S.A.
Tel: 800-248-0123; Fax: 708-405-0147

XEI Scientific
3124 Wessex Way
Redwood City, CA 94061, U.S.A.
Tel: 415-369-0133; Fax: 415-363-1659

Carl Zeiss (Oberkochen), Ltd.
P.O. Box 78, Woodfield Road
Welwyn Garden City
Hertfordshire AL7 1LU, U.K.
Tel: 0707-331144; Fax: 0707-373210

Carl Zeiss, Inc.
One Zeiss Drive
Thornwood, NJ 10594, U.S.A.
Tel: 914-681-7745; Fax: 914-681-7446

Carl Zeiss
P.O. Box 1380
D-73446 Oberkochen, Germany
Tel: 7364-20-2700; Fax: 7364-20-4670

Subject index